Engineering Materials

This series provides topical information on innovative, structural and functional materials and composites with applications in optical, electrical, mechanical, civil, aeronautical, medical, bio- and nano-engineering. The individual volumes are complete, comprehensive monographs covering the structure, properties, manufacturing process and applications of these materials. This multidisciplinary series is devoted to professionals, students and all those interested in the latest developments in the Materials Science field, that look for a carefully selected collection of high quality review articles on their respective field of expertise.

Indexed at Compendex (2021) and Scopus (2022)

Jiří Militký · Mohanapriya Venkataraman
Editors

Fibrous Structures for Sustainable Future

Springer

Editors
Jiří Militký
Department of Material Engineering
Faculty of Textile Engineering
Technical University of Liberec
Liberec, Czech Republic

Mohanapriya Venkataraman
Department of Material Engineering
Faculty of Textile Engineering
Technical University of Liberec
Liberec, Czech Republic

ISSN 1612-1317 ISSN 1868-1212 (electronic)
Engineering Materials
ISBN 978-981-95-1381-9 ISBN 978-981-95-1382-6 (eBook)
https://doi.org/10.1007/978-981-95-1382-6

This Springer imprint is published by the registered company Springer Nature Singapore Pte Ltd.
The registered company address is: 152 Beach Road, #21-01/04 Gateway East, Singapore 189721, Singapore

Contents

Bio-based Fibers and Fibrous Wastes

Jiří Militký, Mohanapriya Venkataraman, and Dana Křemenáková

1 Introduction

Production of textiles is amongst the oldest industrial sector of world economy and it is instantaneously associated with the development of human society [70]. It impacts developments in many areas, in particular clothing, protection and fashion, but also for technical applications. An important aspect of textiles is the fact that consumption of apparel is directly related to the total population size. In addition, technical textile and fibrous structures are used in composites and in knowledge intensive applications. These include special materials for construction in automobiles and the transportation industry, civil engineering, architecture and building industry, special agro-textiles, geo-textiles and a wide variety of other applications [67].

Clothing production follows of population size and is supported by economically oriented activities as fast fashion (design and production of inexpensive clothes that are quickly available to consumers who follow the latest fashion trends, often with little or no consideration for environmental and social impacts), which is supporting the purchase-discard strategy i.e. waste product generation. The overproduction/overconsumption of garments is source of many environmental problems such as global emissions of CO_2 and other greenhouse gases, huge water and energy consumption, use non-sustainable raw materials (fibers, dyestuffs and other chemicals) and microplastics generated pollution of environment.

J. Militký (✉) · M. Venkataraman · D. Křemenáková
Department of Material Engineering, Faculty of Textile Engineering, Technical University of Liberec, Liberec, Czech Republic
e-mail: jiri.militky@tul.cz

M. Venkataraman
e-mail: mohanapriya.venkataraman@tul.cz

D. Křemenáková
e-mail: dana.kremenakova@tul.cz

J. Militký and M. Venkataraman (eds.), *Fibrous Structures for Sustainable Future*, Engineering Materials, https://doi.org/10.1007/978-981-95-1382-6_1

The specificity of the textile field is that practically the same materials, structures, and technologies are used for very different applications [75]. Textile production need to be changed globally to create a sustainable system that will be more environmentally friendly, based on sustainable oriented design, sustainable raw materials, use less energy, and generate less pollution and waste. For textiles, the extended producer responsibility (EPR) is the basis on which the manufacturer bears responsibility for the product until the end of its life has long been discussed and partially developed. Therefore, it is necessary to focus on materials and processes that will be oriented toward the main needs of a future society.

The development in the textile branch is and will be critically dependent on the following factors [7]:

1. Serious financial imbalances.
2. Recession in US and Europe.
3. Oil price volatility.
4. Energy and water costs on the increase.
5. Sustainability and Environmental Awareness.
6. China and India becoming dominant economic players, Africa and certain Asian and Subcontinent countries will gain importance.
7. Slower textile consumption internationally, less than 2% growth forecasted thru 2020.

The demand for textile products that have the epithet "bio" is growing despite their significant economic disadvantage. These "bio" products are not in themselves safer from the point of view of human health (as some "pseudo-expert" sources try to inform), but during their production or the preparation of fiber raw materials considers the requirements of preservation and protection of the environment. It is therefore not about protecting the health of an individual, but about promoting a good state of the environment, which is important for the whole society.

In this chapter, the main ways of selection of fibers following requirements of sustainable future development including their suitability for clothing/textile applications, availability for whole society, ecology of their production and waste treatment are presented. The benefits and weak points of different strategies are critically discussed. It is shown that the selection of fibers from natural sources is not automatically equivalent to use of fibers sustainable for future society.

2 Sustainability and Circular Economy

In the Brundtland report is sustainability defined as "sustainable development meets the needs of the present without compromising the abilities of future generations to meet their own needs" [1]. "Sustainable development" require also to secure environmental and social needs of society besides maintaining economic growth. Sustainability has been critically influenced by the decreasing of nonrenewable resources, deterioration of ecological environment, needs for better living environments, and use of more environmentally friendly products. Sustainable manufacturing comprising both sustainable products and sustainable production processes is comprehensively discussed in encyclopedia of sustainable technologies [2]. Utilization of modern digital technologies as artificial intelligence, internet of things and cybersecurity for solving tasks connected with sustainability concept is described in the book [13]. In textile field are often sustainable fibers equivalent to fibers of natural origin or fibers from waste without detailed investigations of their real sustainability [11, 80, 82]. The challenge is to use principles of green chemistry for preparation of fibers from bio-based polymers [44].

The impacts of products or processes at all stages of life cycle, from design to material selection, manufacturing and final disposal is evaluated by life cycle assessment (LCA) [81]. LCA evaluate the interaction between all the input/output flows in the system. The flows are defined as the input of materials (materials resources, process media and energies) and outputs (e.g. waste air emissions, microplastics release, waste heat, by products etc.). LCA is a mathematical model that relates of input data (process data, characterization factors etc.) with an output result at a desired level (inventory analysis, characterization, normalization, weighting) [41]. Modern LCA is a scientific tool implementing rigorous statistical treatment of data, uncertainty and sensitivity analysis [41]. Based on detailed analysis of LCA studies about environmental impact of clothing production was revealed that the raw material extraction stage generally has the highest environmental impact [78], either just one or a few stages of the product life cycle, and/or on a specific type of product. The linear economy model of design product, fabrication of products, using and maintaining of products, and disposing of products are not considering the environmental, societal and economic impacts and consequences [45]. Nowadays the replacement of classical linear economy to circular economy is supported by EU initiatives [110]. Circular economy is mainly oriented on resources, especially secondary ones (wastes) via 6R principles (**reduce, recover, reuse, recycle, redesign, remanufacture**—see Fig. 1) [45].

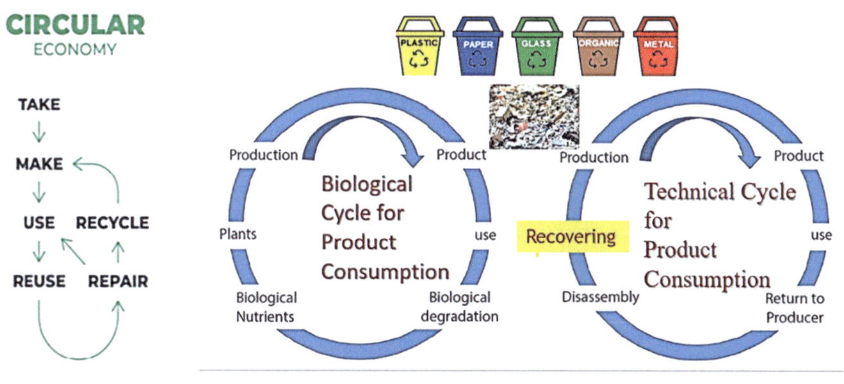

Fig. 1 Principle of circular economy

The principles of circular economy (CE) are tools for realization of the European Green Deal (see Regulation (EU) [94]). Circular economy is realized as proper combination of two major cycles.

Technical cycle: recovers and restores products, components, and materials through strategies like reuse, repair, remanufacture, or recycling.
Biological cycle: the consumption of resources is high and the products are designed to back-feed into the system by different processes like composting and anaerobic digestion. This cycle regenerates living systems, such as soil, which provide renewable resources for the economy.

Circular economy should be used from these reasons:

- the scarcity of resources obliges the search for new ways to optimize the use of available ones;
- the price of resources puts a noticeable pressure on the economy;
- there is a need for sustainable materials, to alleviate the pressure on current resources particularly from petroleum-based products.

There were published some vision of a future economy that transforms from being circular to sustainable [110] where battle for resources will be replaced by resources sufficiency. Today the battle for resources related to critical raw materials, carbon-based energy production and non-polluted water is prevailing. In future will be resources produced by environmentally friendly ways, resources will be circulated in controlled manner and non-renewable materials will be replaced by renewable ones. Most of energy sources will be sustainable from natural based resources (sun, wind, waves, geothermal sources etc.). The world will be managed by data enabling resource sufficiency. Sustainable development is long-term strategy including economic, human (social) and environmental (material) resources [56]. Major parts of this strategy are:

(a) society becomes aware of a common responsibility,
(b) environmental protection becomes an integrated search for solutions,
(c) industry prevents pollution by means of proactive actions.

Sustainability is therefore oriented as well on needs of future society (as are: scarcity of fresh water and agri-lands, environmental friendliness, use of renewable energies and viable technologies). Changing the current fashion industry to sustainable fashion was the main aim of comprehensive research program Mistra Future Fashion in years 2011–2019 realized in Sweden [76]. This program used principles of circular economy and life cycle assessment mainly for proposal of new methods for design, raw materials selection, production, maintenance during use and recycling related to sustainable fashion. In frame of this research program there were published Mistra future fashion reports covering critical assessment of current knowledge in many fields related to sustainability as e.g. "Environmental impact of textile fibers" [97].

3 Fibers and Sustainability

A typical feature of the fibers is that their thickness is several orders of magnitude smaller than the length. The thickness of common fibers is usually in the range $d = 10^{-6}$–10^{-4} m and the length in the range $l = 10^{-2}$–10^{-1} m. The ratio $l/d \sim 10^3$ shows that the predominant dimension is length [75]. In the case of natural fibers, the length and thickness are determined by the conditions of fiber growth and are only indirectly influenced by humans. With chemical and synthetic fibers, it is possible to change not only the length and thickness, but also the shape of the cross section. This is particularly important for the category of microfibers (thickness on the order of 2 μm). Thanks to this small thickness, microfibers have some unique properties. E.g. the transport of moisture takes place by a capillary mechanism, so the fiber bundles of hydrophobic fibers are able to easily transport moisture without swelling of the fibers and diffusion through their mass. Fibers have a number of specific features that distinguish them from other materials [114].

Depending on how it was obtained, the starting polymer is fibers grouped into five groups [75]:

(a) Natural—created in nature from natural polymers (practically cellulose or peptides only),
(b) Chemical from natural polymers (chemical fibers)—created artificially from natural polymers,
(c) Chemical from synthetic polymers (synthetic fibers)—created artificially from synthetic polymers,
(d) Metallurgical—from inorganic metallic and non-metallic materials by metallurgical technology,
(e) Ceramic fibers based on single crystals or whiskers, etc.

Fiber characteristics depend on their origin, natural (cotton, bast fibers, and wool) or man-made and molecular arrangements (see Fig. 2).

Natural fibers are difficult and time-consuming to change their geometry and properties. It can be achieved through selective breeding, gene manipulation, etc., to enable good processability (such as spinning and processing of fibrous mixtures). On the other hand, the properties of man-made chemical and synthetic fibers (viscose, polyamides, polyesters, acrylic fibers, polyolefins, etc.,) can be modified with relative ease by changing the conditions of production (spinning, drawing, heat setting, etc.,); geometry (fineness, cross-section form, texturing, etc.,), spinning conditions (spinning speed, drawing conditions, conditions of heat treatment, etc.,), or through chemical modification [73].

Around 60% of global fiber production is used for the clothing/fashion industry. The rest being used for technical textiles, medical textiles and hygienic textiles mainly.

Polyester consumption was 51% (54 million tons) in 2018, followed by cotton at 25% (26 million tons) Polyester dominates production (see Fig. 3) due to its performance characteristics and cost-efficiency but not due to sustainability.

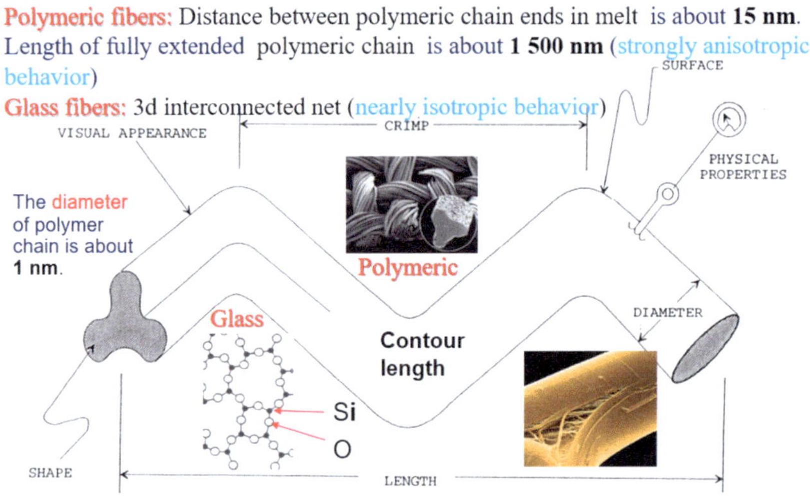

Fig. 2 Fibers characterization

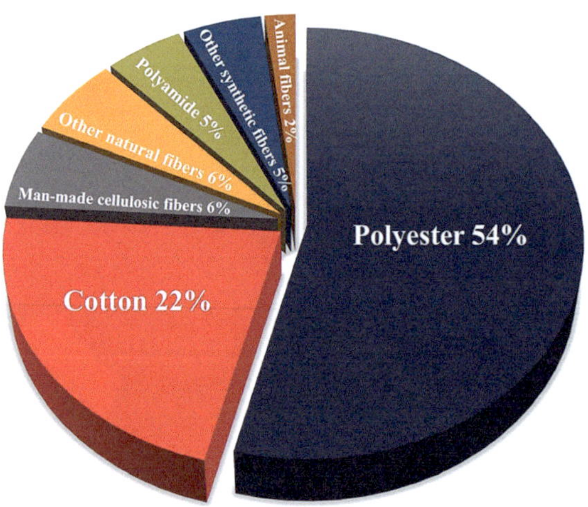

Fig. 3 Textile production by fiber type (reused from Seifali Abbas-Abadi et al. [101] under CC-BY Permission)

The selection of sustainable fibers is not so simple. Usually the natural fibers and fibers based on the renewable resources are nominated as more sustainable.

In fact, the majority of lifetime energy use of a garment is in usage-phase (washing, drying, ironing). According to gross energy required for production of 1 kg of manmade fibers is the one of the best choice the polyester fiber. Major limitation of polyester as sustainable fiber is requirement to use of non-renewable sources. But in reality, the situation is not so critical. Approximate fiber production, energy consumption, freshwater consumption and CO_2 emissions for cotton, polyester, non-cotton cellulosic, polyamide, wool and hemp are discussed in Niinimäki et al. [85].

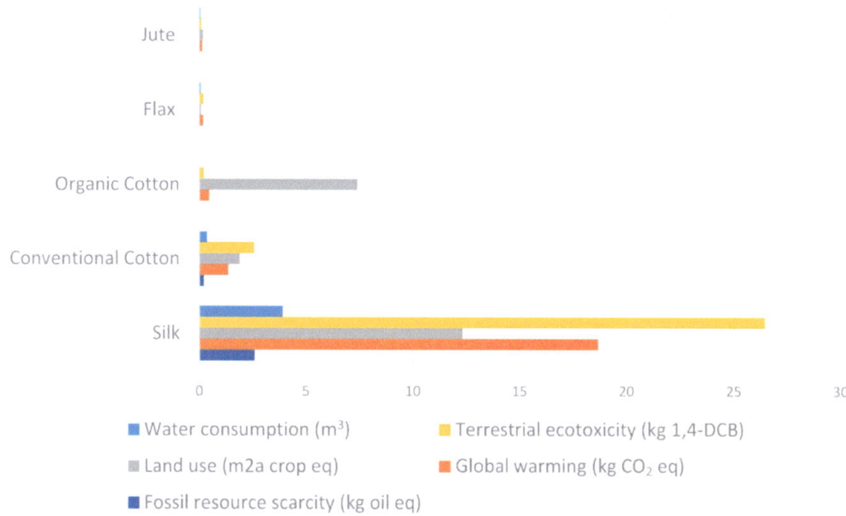

Fig. 4 Environmental impacts of common types of natural fibers (reused from Gonzalez et al. [35] under CC-BY Permission)

The environmental impact of production of some natural fibers are shown in Fig. 4 [35]. Natural fibers require less energy but more water during production than synthetics (polyester and polyamide). Extraordinary water consumption is for cotton. CO_2-production values are higher for polyamide and wool only [97].

It is visible that cotton use is potentially limited by surplus of water necessary for growing. The waste water from cotton plants is very highly polluted. Limitation of wool is the land use mainly. Usually the fibers from renewable resources (cellulosic, see Fig. 5) are automatically classified as sustainable fibers without investigation of all aspect of sustainability (time of production energy consumption, land and water use, maintenance cost, recycling.).

Plant based fibers can be obtained from different parts of plants (see Fig. 5):

1. From **bast–phloem–bark** surrounding the stem of certain plants, e.g. flax, hemp.
2. From **leaf–hard**. These are not traditional leaves, but more fleshy leaf plants such as pineapple and agave used for sisal.
3. From **seeds and fruits of plants** often attached to hairs or fibers or encased in a husk such as cotton, coir and kapok.
4. From **wood** sourced from trees, grasses, and bamboo. These fibers are stiffer due to their increased lignin and xylan content and usually need an extra process to soften them. They are often used as raw material for viscose fibers preparation.

Plant based fibers can be obtained from different parts of plants (see Fig. 5).

Production of plant-based fibers, excluding cotton, is slowly changing. Cotton alone was in year 2022 about 22% of all plant-based fibers [65].

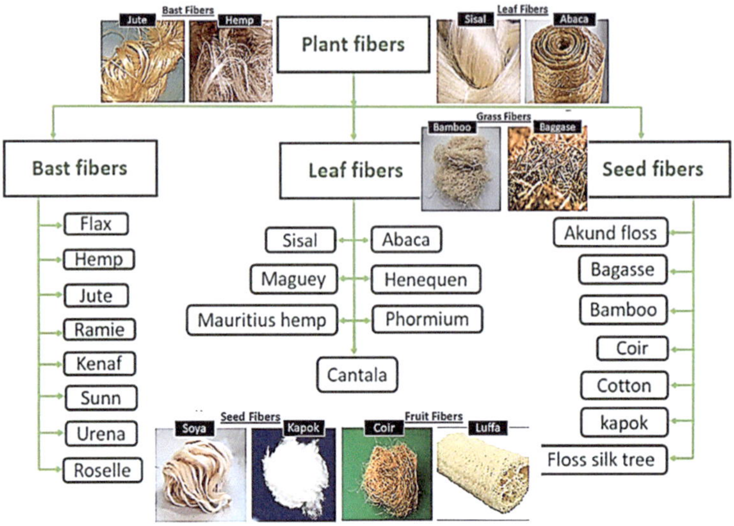

Fig. 5 Classification of plant fibers

Most of these fibers are composed of cellulose, hemicelluloses, lignin, and other compounds (see Table 1 and Fig. 6). Non-cellulosic parts are removed during transformation to textile products or during separation from seeds, bark, stems, or leaves. Especially, pectin is serving as a natural glue joining elementary fibers to technical fibers, which are spinnable [50].

Hemicelluloses can be divided into four groups of polysaccharides [27]:

1. xyloglucans (xylans);
2. manno glycans (mannans);

Table 1 Plant fibers composition

Fiber	Cellulose	Hemi-celluloses	Lignin	Pectin
Cotton	92–93	5–6	N	< 1
Jute	65–70	12–13	12	0.2
Flax	62–70	15–17	2–3	1–4
Sisal	65–73	12–13	10–11	0.8–2
Hemp	67–74	16–18	3–4	1–2
Coir	43	< 1	45	4
Ramie	70–76	13–15	0.6–1	2
Nettle	40	28	17	8
Kapok	53	29	20	N
Bamboo	35–45	20–25	30	N
Abaca	64	20	4–6	0.5

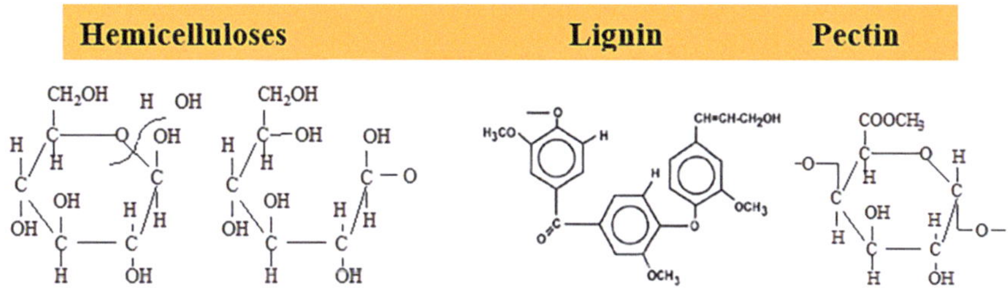

Fig. 6 Non-cellulose components of plant fibers

3. xyloglucans (XG);
4. mixed-linkage β-glucans.

All these hemicelluloses are amorphous and are mechanically week. Common features of bast fibers are:

(a) technical fibers are multicellular, consist of a number of elementary fibers glued together by pectin or lignin.
(b) fibers have a similar chemical composition: cellulose 65–85%; ash 1–5%; fats, waxes 2–4%; pectin, hemicelluloses 2–10%; lignin 1–20%; moisture 10%.
(c) fibers have a similar microscopic appearance.
(d) fibers have similar properties: are quite strong (strength increases when wet); have little tensile strength; resistant to the effects of fresh and salt water; difficult to bleach and dye.

The crystallinity of the selected bast fibers are shown in Table 2.

Major limitation of bast fibers is their multicellular structure where are strong cellulosic units (ultimate fibers) glued by weak, often easily degradable compounds (pectin, lignin) and there are present number of voids and pores. This structure is supported removal of fibrous elements during their use. Well known is so called dustiness of jute where having ultimate fibers very short. Small length or elementary (ultimate) fibers (for bamboo 4–8 mm only), see Table 3, is frequently critical factor for textile processing and for durability in practical use.

Only few of plant fibers are suitable for textile processing. Some of the plants providing fibers suitable for textile processing, such as flax and hemp, can also be used for removing heavy metals (cadmium, lead, copper, etc.) from contaminated soil. An overview of exotic plants and trees from which fibers suitable for textile composites (here glue is replaced by matrix material) can be obtained is given in the work [54]. Some physical properties of selected plant fibers are presented in Table 4.

Table 2 Crystallinity of typical bast fibers

Fiber	Hemp	Sisal	Jute	Flax	Bamboo
Crystallinity (%)	88	71	72	91	70–75

Table 3 Geometry of typical plant fibers

Fiber	Length tech./length ult. (mm)/(mm)	Width tech./width ult. (mm)/(mm)	Fineness tech. (tex)
Flax	(200–1400)/(4–77)	(0.04–0.62)/(5–76)	0.18 – 2
Jute	(1500–3600)/(1–6)	(0.03–0.14)/(5–25)	1.4 – 3
Hemp	(1000–3000)/(5–55)	(0.16)/(10–51)	0.34 – 2.2
Sisal	(600–1000)/(1–8)	(0.1–0.46)/(7–47)	2 – 40
Abaca	(1000–2000)/(3–12)	(0.01–0.28)/(6–46)	4.2 – 30
Coir	(150–350)/(0.3–1)	(0.1–0.45)/(15–24)	–

Table 4 Some physical properties of selected plant fibers

Fiber	Density (kg/m^3)	Humidity at 65% RH (%)	Porosity (%)
Flax	1540	12	10
Jute	1500	13.8	14–15
Hemp	1480	12	11
Sisal	1200–1450	14	17
Abaca	1480	14	17–21
Coir	1150–1330	10.5	High
Bamboo	850–1100	6–8	High

The low fiber densities compared to the cellulose density of 1560 kg/m^3 indicate porosity. It is evident that coir and bamboo fibers have the highest porosities (see Fig. 7).

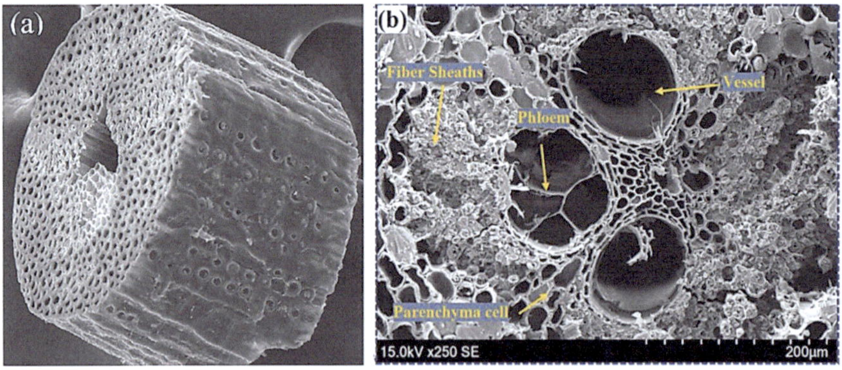

Fig. 7 Structure of **a** coir fiber (resued from Jawahir and Bradley [45] under CC-BY Permission) and technical bamboo fiber (reused from Cui et al. [26] under CC-BY Permission)

Table 5 Mechanical properties of typical plant fibers

Fiber	Strength (MPa)	Breaking strain (%)	Modulus (GPa)
Flax	700–1250	1.6–3	45–70
Jute	200–860	1.7	3–60
Hemp	630–780	2–2.6	19–44
Sisal	800–0.855	2–5.4	2–14
Abaca	400–980	2.5–3	6–45
Coir	106–770	16–20	2–55
Bamboo	140–800	2.8–4	10–40
Ramie	220–938	2–4	9–128
Banana	52–910	1–4	3–35

Plant based fibers are stiff (relatively high initial modulus) strong (high break stress) with low elongation to break (see Table 5). They are usually spun by cotton type spinning lines (rotor technology) and used for technical applications.

For clothing purposes are suitable cotton fibers mainly. Other plant fibers are suitable as reinforcements for composites and concrete [40]. One potentially beneficial application is utilization of these fibers as source for preparation of nanocellulose particles and regenerated cellulosic fibers. Main limitation of regenerated fibers fabrication is complicated preparation of cellulose solution. One promising solvent are ionic liquids [77, 106, 115]. There are plenty of papers dealing with use of plant fibers in technical sectors as reinforcement of composites for biomedical applications [3, 100], automotive industry [99], aerospace [63], and textile resources [49]. Different plants can be used as source of biomass or for preparation of regenerated cellulose fibers [118].

Based on these facts it is evident that natural fibers alone cannot be used for replacement fossil based synthetic fibers and they are not be able to solve requirement for global sustainably. They will be henceforward in future used as material for specific application in some countries where plants are growing or as source of cellulose/nanocellulose. Plant fibers are prepared by nature for other purposes and therefore they have often not suitable geometry (length and thickness) and morphology required for textile applications. Therefore, it is more convenient to prepare textile fibers industrially from biopolymers or bio monomers.

4 Bio Based Fibers and Biopolymer-Based Fibers

Products from bio-based materials are attractive alternative to common products from natural fibers and fossil-based materials due to their natural sustainability. These products are basic element of circular bio-based economy.

Circular, bio-based economy is using of renewable and waste resources as raw materials (see Fig. 8). Fossil based carbon is substituted by carbon from biomass and recycled sources.

According to ASTM D6866-06 are bio-polymers a special polymer using during synthesis products of natural origin. Polymers are generally called to be "biodegradable" if they have the ability to break down, safely and relatively quickly, by biological means, into raw materials of nature. These polymers can be also used as sustainable replacement of dyestuffs and finishing agents by sustainable colorants [37] and biomacromolecules [62].

Biodegradation is generally part of a cascade of physical and chemical phenomena by which a material is broken down into small parts and eventually converted to substances that either are reused in biological cycles or accumulate. Not all biopolymers are easily biodegradable and vice versa not all biodegradable polymers are bio-based. There are defined three basic categories of biomaterials:

Bio-based—these are materials that are based on carbon originating from biological (non-fossil) sources. These materials may or may not be biodegradable. Bio-based materials are identifiable by C 14 carbon marking.

Biodegradable—these are materials that decompose into carbon dioxide, water and biomass as a result of microbial attack.

Biopolymers—these are materials that are bio-based and biodegradable. These are both plastics and fibers as well as non-plastic materials (proteins, lipids, DNA, etc.). These materials are used by living organisms during their synthesis (ASTM D6866-06).

In general, three basic factors influence the development of biopolymers. They are starting raw materials, durability and processing after the period of use.

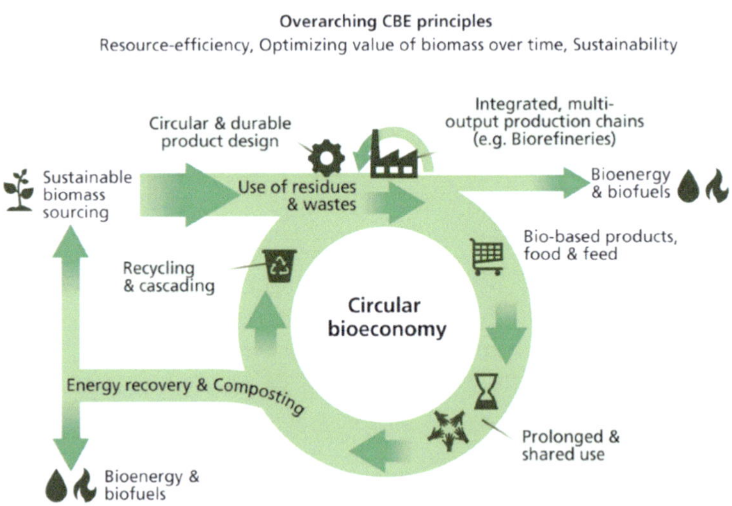

Fig. 8 System of circular bioeconomy (reused from Mehta et al. [66] under CC-BY Permission)

A. Initial raw materials

Most bio-based polymers and bio-monomers are produced from corn, although cellulose from cotton and trees or from some biomass. E.g. approx. 2.5 kg of corn is needed to produce 1 kg of polylactic acid (PLA). Covering the global needs of the PLA requires roughly 230,000 km^2 of agricultural land. In addition, bioethanol is mainly produced from corn. Agricultural policy will therefore affect the availability of this raw material for biopolymers.

B. Durability

Mostly, durability is narrowly assessed with respect to biopolymers only. These "bio" products are not safer for human health but during their preparation the protection of environment is considered.

Here, the service life is limited in many cases by the relative ease of degradation under conditions of use and maintenance. However, it is also about issues related to for example the cultivation of corn, where fossil fuels, fertilizers and pesticides are used during cultivation.

The term "biodegradable" was introduced by ASTM for cases of degradation of a material caused by biological activity, especially the action of enzymes, which significantly change its chemical structure. Generally:

> "Biodegradable" means the degradation of macromolecules when dispersed in vivo, without significant release from the body. Biological action results in degradation and the formation of by-products.
> "Bioerodible" means the gradual degradation of the surface and subsequent in vivo absorption. This can lead to the complete elimination of both the original material and the low molecular weight products formed during bioerosion.
> "Bioabsorbable" means the degradation of the volume and subsequent in vivo absorption. "Bioresorbable" means the dissolution of the material at varying rates in body fluids without chain breakage and weight loss.
> "Biocompatibility" is generally the ability of a material to function in the environment of a biomedical host system without interference.

In general, materials must be biocompatible and non-toxic. Biodegradability is advantageous especially in solving waste problems, but it limits the use of products in practical applications, where, for example, during treatment (washing), conditions for degradation processes are actually set.

Biodegradability is characterized by reduction of molecular weight of polymers (fastest process), decrease of strength (slower process) and loss of weight (slowest process). These changes also result in a deterioration of other mechanical properties, which can be characterized by conventional methods. For characterization of durability was introduced Deborah number D defined by relation

$$D = \frac{\text{time to degradation}}{\text{service life}}$$

For degradable polymers is *D* approaching to zero and for non-degradable polymers is *D* approaching to infinity. Biodegradable polymers are commonly degraded by microorganisms such as bacteria, fungi and algae.

Enzymatic degradation normally occurs in natural polymers (e.g. collagen enzymatically degraded by collagenases, pectin by pectinases, proteins by proteases, cellulose by cellulases etc.). Enzymatic degradation also occurs in synthetic polymers (polyesters degraded by esterases, etc.). Biodegradable polymers can be hydrolyzed passively or actively by catalytic action of enzymes. Synthetic polymers with functional groups are more likely to undergo non-specific enzymatic degradation. Some types of bacteria promote enzymatic degradation of polymers.

In many cases, oxidative degradation, hydrolytic degradation or photodegradation also occur. Soil erosion, water contamination and CO_2 overproduction can therefore occur. In some cases is use of biodegradable resources complicated by other factors. For example, the effective genetically modified corn varieties will likely continue to encounter environmental groups and legal barriers, even if they are not intended for food use (similar to genetically modified cotton).

C. Processing after the period of use

As a number of materials are gradually replaced by polymers, landfills become clogged with materials that degrade extremely slowly. The general opinion is that degradability is a positive factor. However, landfills are built to limit degradation. These are mostly anaerobic storage conditions without access to air. If the material degrades under these conditions, methane production usually occurs. In addition, the degradation of biopolymers usually requires quite specific conditions such as higher temperature (over 60 °C) and the presence of moisture. This can be ensured in special landfills and composts, but it will have little effect on ordinary solid waste outside of landfills, which is increasingly dangerous for the environment. In general, waste from biopolymers is more voluminous and lighter, which can be used in the construction of equipment for their collection and processing. One of way how to promote degradation of common fibers (PET) is use of degradation enhancing additives. PET has been modified by incorporating additives, including calcium carbonate (~ 1%) acting as a nucleation site for degradation and a combination of different biodegradable polymers like polycaprolactone (PCL), polyhydroxybutyrate (PHB), and PLA (~ 60%) to the PET (~ 40%) melt-spinning dope.

By processing of biomass, it is possible to obtain some important monomer and intermediates for synthesis of advanced biopolymers [23] (see Fig. 9).

Sustainable polymers are then created by using of monomers from renewable resources, with low energy costs, minimum use of harmful reagents, and controlled release of harmful by-products. The furan platform has become one of the most relevant the renewable resources of polymer chemistry. The 2,5 acid, used as monomer for preparation of poly (ethylene furanoate) (PEF) is more favorable than the bio-based production of terephthalic acid (TPA) [109].

The comparison of structure of PET and PEF is shown in Fig. 10.

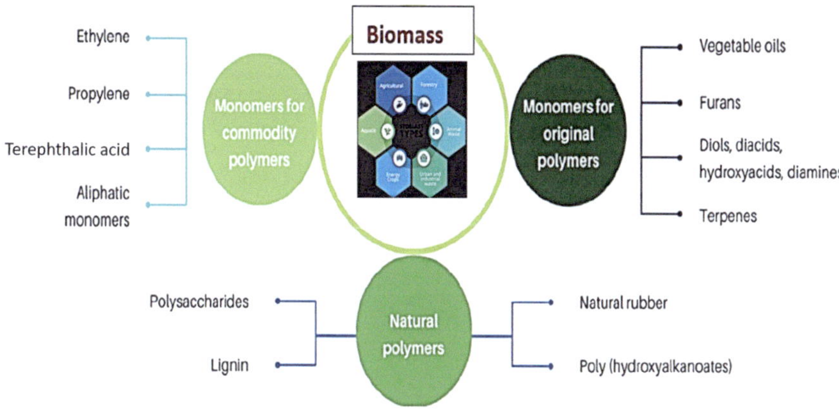

Fig. 9 Monomers and intermediates from biomass (reused from Burgess et al. [20] under CC-BY Permission)

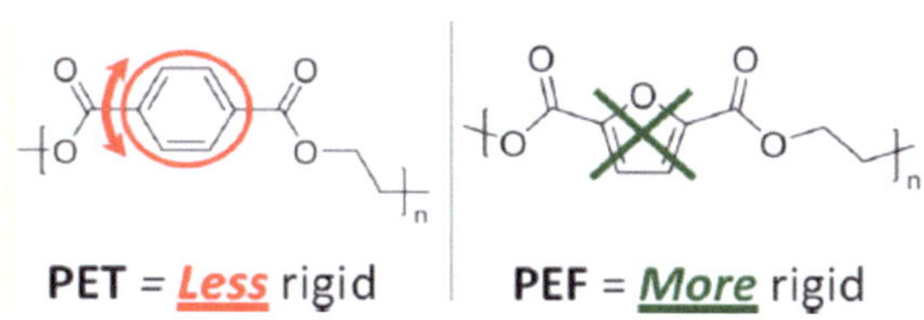

Fig. 10 Polymeric chain of PET (less rigid due to flipping of benzene ring) and PEF (more rigid due to presence of furan ring)

PEF is desirable, due to renewable monomers (from renewable plant sugars as fructose and glucose) and has the improved barrier, mechanical, and thermal properties compared to PET. Polarity of the furan ring in PEF imparts the increased equilibrium water solubility of ~ 1.8 averaged over the entire concentration range compared to PET. Permeability reduction is ~ 2.8 for water in PEF compared to PET at 35 °C. PEF exhibits ~ five-fold reduced water diffusion coefficient [20, 109]. The gas barrier of PEF to O_2, CO_2, and H_2O are 6–11, 8–19, and 2–3 times higher than of PET [18–20, 105].

Economical, large-scale production of 2,5-furandicarboxylic acid (FDCA), as monomer for preparation poly (ethylene furanoate) (PEF) is used e.g. by company Avantium (The Netherlands). PEF can be recycled by typical PET recycling methods, which is an important advantage. PEF has less released energy from the combustion process than PET, lesser CO_2 release, and soot yields compared to PET. PEF as a viable alternative to PET for the packing foils and beverage containers. The density and thermal properties of PET and PEF are shown in the Table 6.

There are some limiting factors for the preparation of PEF fibers, such as large brittleness during spinning. PEF monofilament is more difficult to draft, and it shows poor mechanical strength. The mass spinning temperature of PEF is higher than for PES (for molecular mass 84,000, is PEF melt spinning temperature is 320 °C) [109]. Natural draw ratio is lower for PET than for PEF. This could be related to

Table 6 Some properties of amorphous PET and PEF

Polymer	Density (kg/m^3)	T_g (°C)	T_m (°C)	Decomposition (°C)
PEF	1430	85	211	389
PET	1335	76	247	413

the easier change in local conformations of the benzenic ring compared to the furan one. High-speed spinning process, which was originally designed for PET, was also successfully applied to PEF [109].

Bioplastics consumption is rapidly growing [15]. In the year 2021, 2.42 million tons of bioplastics were used (see Fig. 11) [64].

Strictly speaking all fibers from natural resources are bio based but their geometry and structure are not controlled. In some cases, the monomers prepared from natural sources for creation of synthetic polymers and then fibers with controlled geometry are advantageous. Standard approach is to prepare monomers replacing monomers from non-renewable resources for creation of common synthetic fibers as polyesters [5]. More challenging is to prepare new types of polymers from bio-intermediaries or polymers created by microorganisms [28, 57, 83]. Basic ways to prepare bio-based polymers.

Bio based polymers can be classified according to their origin into three main classes:

Class 1: Extracted directly from biomasses with or without modification Example: starch modified polymers and polymers derived from cellulose.

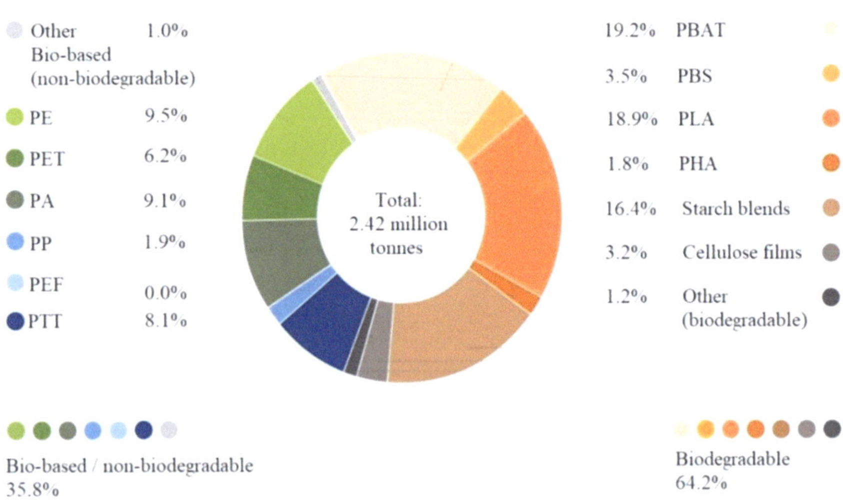

Fig. 11 Global production capacities of bioplastics in 2021 (reused from Markevičiūtė and Varžinskas [64] under CC-BY Permission)

Class 2: Produced directly from microorganisms (e.g. bacteria, fungi etc.) in their natural or genetically modified state. Example: Polyhydroxyalcanoates (PHAs).
Class 3: Obtained with the participation of bio-intermediaries produced with renewable raw materials. Examples: PLA; bio-polyethylene (BPE) by polymerization of ethylene produced from bio-ethanol; bio nylon via diacids from biomasses; and bio-polyurethanes from polyols of vegetable origin.

Some biopolymers are suitable for creation of textile fibers [98]. During the textile processing of fibers from biopolymers, problems can generally arise both with the construction of textile structures (many of these fibers are fragile, not very extensible, etc.) and with possible finishing and dyeing. This "together with often the relatively high price of biopolymer fibers" leads to the fact that their industrial use is currently limited to special applications.

One of promising kind of biopolymers used also for fiber production and special applications are chitin and chitosan. Chitin is natural material widely available in nature (second after cellulose). Chitosan is product of chemical or enzymatic deacetylation (removal of acetyl groups) of chitin (see Fig. 12). These bio-based polymers are linear polyamines characterized by biocompatibility, biodegradability, low allergency, hemostatic action and other medical activities as antitumor and spermicidal [96]. Chitin is produced by aquatic organism (in crustacean shells), terrestrial organism (some insects), microorganism (fungal and yeast) and mushrooms.

The shells of crayfish and shrimp contain 20–50% chitin. The shells are usually firstly decalcified in hydrochloric acid and then heated in a 1–2 mol/l NaOH solution to remove proteins and pigments [4, 96]. The chitin remains as a white powder.

Chitin occurs in two modifications. α-chitin has an antiparallel arrangement and β-chitin has a parallel arrangement. Chitin can be dissolved in concentrated NaOH

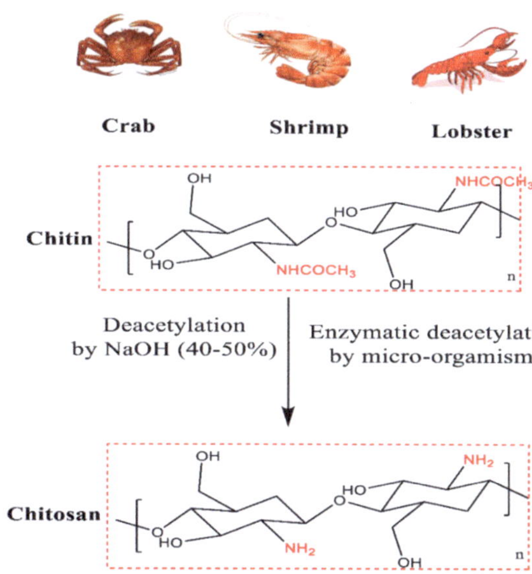

Fig. 12 Chitin and chitosan (reused from Ibrahim et al. [43] under CC-BY Permission)

Table 7 Mechanical properties of chitin and chitosan fibers

Fiber	Strength (cN/dtex)	Break deformation (%)
Chitosan	0.78–1.24	13–28
Chitin	0.79	27

under intense cooling. Due to strong intermolecular hydrogen bonds, α-chitin is poorly soluble and does not swell in common solvents. It is soluble in special solvents such as dimethylacetamide containing 5–10% LiCl and hexafluoroacetone. Its direct use is therefore complicated and expensive. The β-chitin modification is easily soluble, e.g. in formic acid. It swells very strongly in water [4]. Butyryl-chitin with a high degree of substitution is well soluble in organic solvents. It was first prepared in Japan [48]. A simple method for the preparation of dibutyrylchitin (DBC) fibers, with a degree of substitution of about 2, was developed at the Technical University of Lodz [107, 108, 116].

Unlike chitin, chitosan has excellent fiber-forming properties and can be easily spun. It is also easy to produce chitosan films and foam materials. A solution containing 5% chitosan and 2% acetic acid is used for spinning chitosan fibers. 10% NaOH is used as a precipitation bath in combination with 30% sodium acetate solution at 30–40 °C. Coagulation is followed by washing to remove excess alkali and drawing to a drawing ratio of around 3. Drawing is preferably carried out in hot water. The resulting fineness ranges from about 3 dtex and above. The mechanical properties of chitosan fibers compared to chitin fibers are given in Table 7.

The main disadvantage of chitosan fibers is their low wet strength [116]. To improve it, crosslinking (with epichlorohydrin) or preparation of mixed fibers can be chosen. With regard to similar preparation conditions, it is most advantageous to use a mixture of chitin with polyvinyl alcohol. The resulting fibers already have improved mechanical properties and increased water absorption capacity. Another possibility of increasing biological activity can be achieved by modifying the NH_2 group. Acylation using acid anhydrides and grafting using linear unsaturated hydrocarbons are commonly used [16].

A special area of use of biopolymers is medical textiles [3, 75]. In this area, it is necessary to distinguish between medical textiles approved for use in medicine and textiles that only have a certain medical effect (e.g. antibacterial) or they contribute in a certain way to the health of the wearers. For medical purposes, it is of course possible to use these fibers in a number of standard textiles and textile structures, especially for common clothing purposes, bed linen, and common home textiles [75]. However, applications where biocompatibility and in vivo biodegradability are used are advantageous. Here, emphasis is placed on special properties (possibility of influencing healing processes, etc.). Examples of biopolymer-based fibers produced industrially are:

Biodegradable and bio-based

- Polylactide acid—PLA, fiber **Ingeo** (*NatureWorks LLC*)
- Starch based materials, bioplastic **Mater-bi** (*Novamont*)

- Polyhydroxyalkanoates—PHA/PHB, bioplastics (*Mirel, Tianan*)
- Cellulosed based materials, foils **NatureFlex** (*Innovia*)
- PLA/Starch blends, compostable biopolymer **Ecoflex** (*BASF*)
- Chitosan fiber produced in Japan (*Mitsubishi Rayon, Fuji spinning*)

Biodegradable but not bio-based

- Special polyesters (*BASF*)
- Polycaprolactone (*Perstorp*)

Bio-based but non-biodegradable

- PTT—polyester, fíbre **Sorona** (*DuPont*)
- PFA (polyester from furandicarboxylic acid [29]), Fibers (*Avantium*)

In the group of synthetic fibers, some biopolymers, like poly (lactic acid) (PLA), poly (butylene succinate) (PBS), bio-PET, poly (ethylene furanoate) (PEF), will also be included in mass production of fibers. Replacing fossil-based fibres with biobased fibers is really advantageous because fossil-based fibers at their processing after end of service will generate greenhouse gases emissions. Biobased materials of natural origin are carbon-neutral, because the CO_2 that they can emit at their end of life (because of incineration or degradation due to microorganisms) is balanced by the CO_2 consumed during growing of the plants.

PLA [36, 61, 84], being both renewable and biodegradable in industrial composting plants, can represent an interesting perspective for replacing fossil-based polymers such as PET and PP [87, 104]. Raw material, i.e. lactic acid, exists in two basic optically active stereoisomers, *L* and *D*. These isomers affect both the degree of crystallinity and the melting point. Lactic acid obtained by fermentation is 99.5% only the *L*-form. It is a colorless and strongly hygroscopic with melting point 18 °C for the mixed *L*, *D* form it is 53 °C. Homopolymer of L-lactic acid (PLLA) is semi-crystalline, with a relatively high strength or modulus and low elongation. Statistical copolymer of *D*, *L* forms (PLDLA) is an amorphous polymer composed of randomly arranged both isomeric forms. This material has low strength and higher elongation. It undergoes degradation much faster, so it is suitable for special applications (controlled drug delivery). Poly-(L)lactic acid has about 37% of the crystalline phase, melting point 175–180 °C. In the presence of the *D*-form the melting point drops to 130 °C. The glass transition temperature of PLLA is Tg = 60 °C. The average molecular weight of PLA is around 300,000. The methyl group in PLA increases its hydrophobicity (low water sorption). The rate of hydrolysis of PLA is slow. Complete degradation in vivo occurs after more than two years. Copolymers of PLA with Polyglycolic acid are also used, which are significantly more hydrophilic (VicrylTM and Polyglactin 910TM) [61].

PLA does not require enzymes for degradation by hydrolysis of ester bonds. Degradation strongly depends on the size and shape of the material, the proportion of *D* and *L* forms and temperature. Thermal degradation of PLA is manifested at temperatures above 200 °C by oxidative chain cracking and trans esterification reactions. Also, the conditions of composting on an industrial scale, i.e. relatively high temperatures (over 50 °C) and high relative humidity, promote the degradation of PLA [111].

High molecular weight poly (L-lactic) acid is used to produce PLA fibres. The fibres are often also referred to as PLLA. Both melt spinning (at 190–240 °C) and solution spinning (in chloroform/toluene mixture) can be used. This is followed by drawing at higher temperatures (160 °C). The mechanical properties of PLA fibres are strongly dependent on the preparation conditions. The strength of conventional fibres ranges from 0.5 to 0.8 GPa and the modulus is 7–10 GPa. Strengths of up to 2.3 GPa can be achieved. The elongation of the fibres is 18–25%. Using dry-jet-wet spinning, where chloroform and methanol are used as the solvent and precipitant, PLA fibres with a strength of up to 0.60 GPa and a modulus of 8.2 GPa can be prepared [36]. PLLA fibers are semicrystalline (degree of crystallinity up to 60%) and their biological resistance depends on the level of mechanical properties. During spinning and drawing, the molecular weight of the fibers decreases by up to 1/3. Compared to classic polyesters, PLA-based fibers are more hydrophilic (wettability 0.6%) and have a lower density (1250 kgm^{-3}). They are resistant to UV radiation and their melting point can be varied within wide limits. They are non-flammable and have good recovery capabilities. The disadvantage of PLA fibers is mainly their low abrasion resistance. The hydrolysis of PLA fibers occurs at 98% relative humidity and temperatures above 60 °C [47]. At ambient temperatures is hydrolysis very slow [47].

The hydrolyzed form is very quickly microbiologically degradable. In addition to applications in medicine, PLA fibers are beginning to be used mainly for packaging but also in clothing textiles [61]. Kanebo Japan supplies PLA fibers to the market under the name "Corn Fiber" with comparable mechanical properties as classical PET fibers.

Biobased polyamides were developed to replace standard fossil-based polyamides [117]. Polyhydroxyalcanoates (PHA), created by microorganisms, can be used for producing natural bioplastic fibres [52], but their high cost and difficult processability is currently directing them mainly towards biomedical applications.

Fibres based on biopolymers such as proteins [25] and polysaccharides [24, 30], obtained by waste from the agro-food chains, are also used to promote the application of circular economy principles. Casein from waste milk has high barrier properties to non-polar molecules and can be used for fabricating paper coatings, adhesives, food packaging, textiles, and fibres [31]. Cellulose recovered from biomass [51] can be simply converted into regenerated cellulose fibres for textiles. Major advantages of biofibers are:

- Biodegradability
- Preparation from renewable resources
- Low cost and energy use to produce
- Low specific weight, high specific strength related to common plastics
- Safer for handling and manufacturing compared to common synthetic fibers
- Cost savings due to reduced machine wear (non-abrasive compared to synthetic fibers)

- Eco friendliness—carbon neutral (Plants take out as much carbon as they will produce when burned or composted).

On the other hand, biofibers overall quality (durability, comfort) required for clothing purposes mainly is often lower than quality of common, fossil based, fibers.

5 Fibers and Wastes

Growing consumer awareness of the importance of environmental care leads to increased demand for recyclable fibers with minimal impact on polluting our planet. Waste from these fibers will serve as raw materials for next loop of production. Generally, wastes treatment is focused to two main strategies:

1. product specific: product reuse and life extension,
2. material specific: post used waste and resources.

They are realized in so called spiral loops

(i) reuse loop: extending the life of the product by multiple uses,
(ii) repair loop: after completion of a preceding product cycle, damage control or repair is required to make the product ready for successive cycles,
(iii) reconditioning: use of the used product as a raw material for the manufacturing of a new product,
(iv) recycling loop: waste generated is collected and reprocessed to prepare start material for the new products. Especially recycling of textiles to lower quality products (down cycling) is not very challenging, as the products require less stringent compositional and mechanical properties. In order to recycle textiles into higher quality products (up cycling), fibers are needed.

The demand for textiles has increased significantly in the last decade due to rise in living standards of people. However, increased demands of textiles also brought the challenges to dispose significant amount of wastes generated during the processing and end of life textile materials [42, 119] (see Fig. 13).

Currently about 2.8 million tons of textiles are already separately collected in the EU and this amount is expected to be doubled by 2025. The percentage of recycled fibers in global fiber production is still less than 10% and major part are comprised recycled PET fibers [65]. When analyzing textile waste, it is important to divide it into returnable and non-returnable waste with regard to the possibility of their use as secondary textile raw materials [9, 10]. Returnable waste is used again in the same

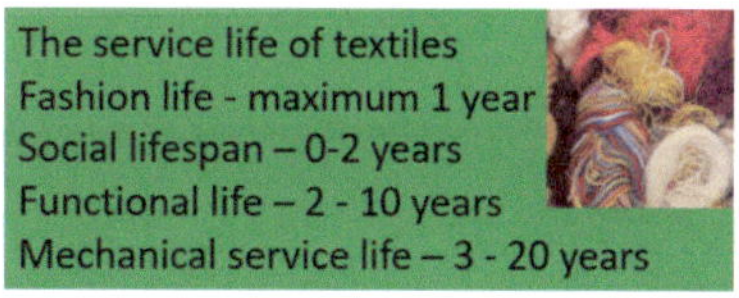

Fig. 13 The service life of clothing textiles

products, during the production of which it was created. Non-returnable waste can be used in products other than the products from which it was produced. In general, textile waste is classified as pre-consumer (more precisely textile industrial waste) or post-consumer (more precisely collection textile waste). Pre-consumer textile wastes are fibrous residues from the production of textiles and textile products. These are textile composite waste, textile scraps, fiber waste and thread waste [39].

Fibrous textile waste is essentially all textile fibers that are discarded during their production process (these are chemical fibers unfinished with insufficient length or with varying degrees of damage or contamination) or fibers discarded during the processing process [39].

Spinning waste is in the form of yarn scraps of different lengths, skeins, **remains of slivers, rovings** etc. **Spinning** waste is produced during yarn processing, e.g. by spinning, drafting and carding. They are also created during processing into flat textiles **(thread waste)** [29].

Textile scraps are pieces of textiles of various shapes and sizes. This is textile waste produced during the production of flat textiles. Textile scraps are divided according to the type of fibers, the type of fabric, the size of the scraps and the color of the scraps [39].

On the other hand, post-consumer textile waste (usually the remains of used clothing) is normally disposed of by landfilling or incineration. Incineration of textile waste is energy-friendly. Textile waste has a calorific value of 14,500 kJ/kg, which is comparable to, for example, brown coal of 15,000–20,000 kJ/kg. Both landfilling and incineration have a negative impact on the environment. During the biological decomposition of textiles in landfills, methane is released into the air, and CO_2 is produced during combustion, and toxic gases can also be released [42, 119]. Separate problem is contamination of land and water by release of leachable components as well as heavy metals such as Pd, Ni, Sn, Zn, and Sb trapped in the plastic wastes and washed of by rainwater.

The general opinion is that degradability is a positive factor for waste treatment. In general, biopolymer waste are bulkier and lighter, Landfills are built to limit degradation. Most often, there are anaerobic storage conditions without access to air. If the material degrades under these conditions, methane is usually produced. In addition, the degradation of biopolymers usually requires very specific conditions, such as higher temperatures (above 60 °C) and the presence of moisture. This requires special landfills and composts, but this will have little effect on regular solid waste outside landfills, which is increasingly dangerous for the environment.

Textile waste from textile products can be divided into categories according complexity of their processing:

1. Used clothing that can be pre-sorted or as part of municipal waste. For pre-sorted waste, the main problem is both the mixture of different fibers and the presence of non-fibrous materials (buttons, zippers, membranes, plastic and metal parts, etc.). Municipal clothing waste is normally heavily polluted and can be attacked by various microorganisms. These wastes are normally created only

after the physical or moral wear of clothing textiles or termination of their real applicability.
2. Used home textiles, such as carpets, covers, curtains, etc., which are usually multi-layered and contain a predominant proportion of non-fibrous materials in some layers. In some cases (e.g. replacement of floor coverings in US department stores) there will be a local need to process a large volume of practically the same waste, which will facilitate the search for an economically advantageous solution. These wastes are created over a longer time horizon.
3. Waste from industrial textiles (e.g. construction textiles) and composites containing a fiber component again contain a significant proportion of other polymeric or of non-polymeric components that are normally combined in one layer from different parts (matrix and fillers). They can be also textiles with sewn-in yarns that are fixed on the reverse side with a coating (latex, Hot Melt, etc.), such as bathroom mats. These wastes can often be easily used for the creation of other structures (non-woven fabrics, filling materials, insulation, etc.).

In a number of cases (e.g. the automotive industry) it is a requirement that the problem of waste processing is already solved at the product design stage, which leads to the selection of materials in order to facilitate the process of their recycling or further processing. These wastes are created in the medium to long term period time.

Traditionally, some **pre-consumer textile wastes** are used in the production of yarns, non-woven fabrics and industrial cotton wool [39]:

- During yarn production, textile secondary raw materials are processed on shredding machines into fibers that are used as an admixture to primary fibers. When processing fiber waste, the length of the fibers must be checked in particular with regard to spinability [55]. The adding of secondary raw material is often main source of release of fibrous fragments during washing.
- The production of non-woven textiles currently uses the largest share of textile secondary raw materials. Non-woven textiles for the technical industry do not place such an emphasis on color or surface appearance as is the case, for example, in home textiles. Non-woven textiles made from secondary raw materials are mainly used as heat and sound insulation material in the construction industry, sound insulation and filling material in the construction of vehicles and aircraft. These materials are suitable as insulating material in the electrical industry. They are also used in the clothing, footwear and housing industries [39].
- Industrial cotton wool (waffle) is formed by densifying the fibers into a fleece, in which the fibers are held together by natural cohesion. In order to preserve the characteristic features, industrial cotton wool is sometimes strengthened by applying a binder to the surface of the fleece. Bandage cotton wool is made from cotton raw material, a fiber mixture is used for the production of industrial cotton wool. Textile waste is also used as a secondary raw material for non-textile processing, e.g. for the production of paper, special boards, flat textile boards, as fillers and reinforcing materials [39].
- Various textile waste is used as the basic material for the production of flat textile boards, especially waste of hard felt, textile floor coverings, non-woven, laminated

and layered textiles or cord fabric from tires. Textile waste is crushed or chopped. Waste is combined with other substances (foils, plastics) where their fusion is used. Surface boards are produced from these secondary raw materials, the properties of which are determined by the basic material, the chemical binder, as well as the production method or surface treatment [29].

Conventional practice is to discard the **post-consumer textile waste** (garments) after use, which causes waste and pollute environment. There are professional collectors, charity organizations, and municipalities that collect used garments and either resell them or donate to needy ones. Non-reusable garments are recycled. Discarded garments are usually either incinerated or send to landfill as solid waste (see Fig. 14).

The Eco Circle program [120], developed by Teijin, distributes PET and cotton textile products to participating organizations. The restricted distribution, high PET content requirement and waste collection network circumvents the difficulty of dealing with textiles of blended fibers. In so doing it limits its commercial impact considerably. The aim of EcoMeTex (Ecodesign methodology for recyclable textile coverings used in the European construction and transport industry) was to develop a tailored Ecodesign methodology [8]. It only focuses on carpets. In Textiles for Textiles (T4T) (ECOINNOVATION project) an automatic sorting installation for post-consumer textiles and clothing based on fiber composition (cotton, wool, polyester, blend, etc.) and color was developed. The sorting was based on NIR-spectroscopy and was able to obtain separated textiles with a consistent quality in terms of composition and color.

There exist a serious barrier limiting the recycling of textile waste:

Economic disadvantages: Many recycled textile wastes are unsuitable for use due to the widespread production of lower-grade products from textile recycling. Prices of recycled textile fibres are influenced by the high cost of the recycling processes.

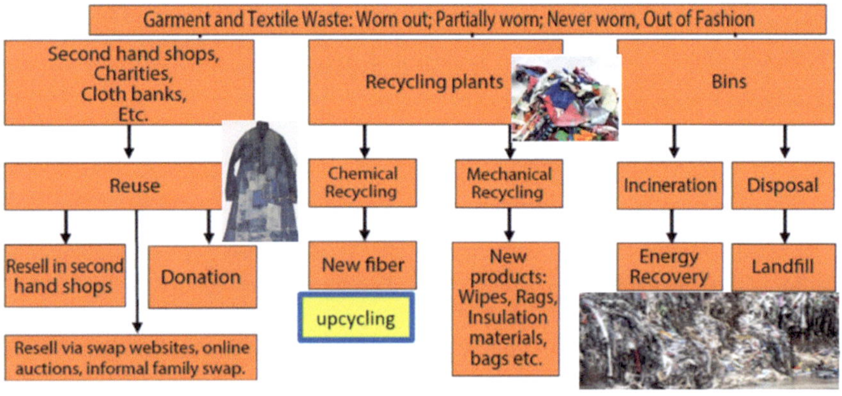

Fig. 14 Basic ways of post-consumer fibrous wastes treatment (reused from Militký et al. [72] under CC-BY Permission)

Composition of textile products: The base components of many textile products make them unsuitable for recycling. Difficulties in separating the mix of various polymer with different properties. Increased textile fiber strength complicates the shredding of the fibres.
Non-availability of recyclable textile materials: The quantity of textile waste that is suitable and accessible for recycling is insufficient. Limited quantity of textile waste is collected and sorted for recycling.
Technological limitations: The lack of technologies for sorting textile waste in preparation for recycling. Most methods cannot separate dyes and other contaminants from the original fibers.
Lack of information and limited public participation: Inadequate knowledge about textiles. The knowledge (ignorance of what to recycle) and the attitude (non-commitment to the ideals of recycling) barriers.

Waste management is strictly regulated by EU directives. Used and waste textiles should be collected separately from other wastes starting from 1 January 2025 to maintain their reusability and for high-quality recycling. This should cover the whole territory of EU States and not be used only in areas where the waste collection is profitable. The digital product passport (DPP), a digital identity card for products, components and materials including technical performance; materials and their origins; repair activities; recycling capabilities and life cycle environmental impacts is introduced. General routes of polymer waste recycling are shown in Fig. 15.

Mechanical recycling of textiles into lower quality products (down-cycling) is not very demanding, as the products require less stringent composition and mechanical

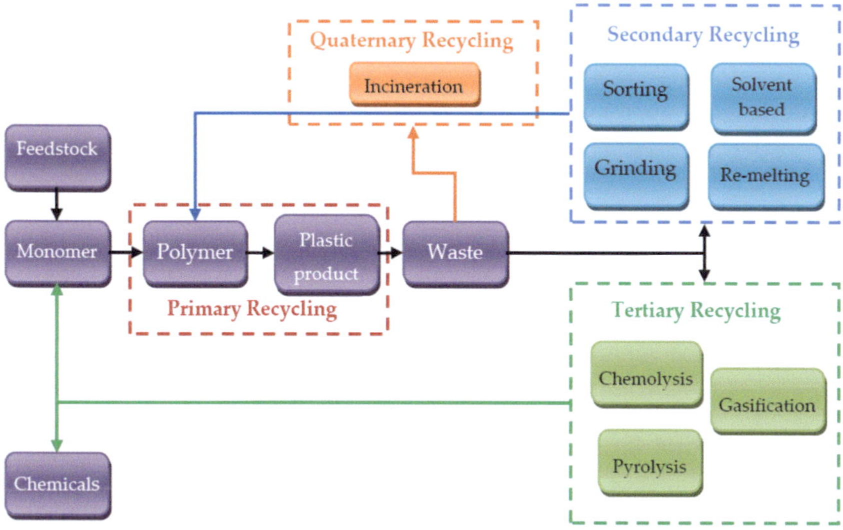

Fig. 15 Routes of polymer waste treatment (reused from Siddiqui et al. [102] under CC-BY Permission)

properties. Technology for the effective mechanical recycling of certain types of polymeric waste is cryogenic crushing. The essence is the cooling of the material to a temperature at which the waste material becomes brittle to be easily crushed into dust of small (μm) dimensions by mechanical action. The cryogenic medium is usually solid carbon dioxide CO_2 (− 80 °C), liquid nitrogen N_2 (− 195.8 °C) or liquid helium He (− 268.93 °C).

Biodegradable fiber can be simply degraded by microorganisms. If not properly managed (e.g. be compostable fibers) they could become sources of environmental pollution, resulting in sicknesses and diseases and possibly death to humans and other living organisms as well [93].

To recycle textiles into higher quality products (up cycling), the synthesis of recycled fibers is obviously needed.

Most advanced chemical recycling processes are applicable mainly to fibers such as polyesters, polyamides, or polypropylene. In these processes, fiber molecules are broken down, and the feedstock is polymerized.

By far, the main fibers consumed in textile and garment production are PET, cotton, and their blends. Consequently, they are most predominant in textile multicomponent waste. There are still difficulties in selecting an industrial process for recycling of PET in the presence of other fibers, for example, cotton and wool. The use of blends in the textile sector is preferred due to superior tailored properties of finished products (garments). The Achilles' heel for recycling is efficient recovery of pure components from intimate blends of various fiber combinations. This remains a major stumbling block towards commercialization.

Cellulose can be chemically recycled or removed from mixtures by using concentrated inorganic or organic acid at room temperature. In industry the process to form regenerated cellulosic fibers from wood by using organic solvents is well known and widely used.

It is reported that the reduced sugar yields of 64.3–73.9% was obtained by cotton cellulose hydrolysis using concentrated sulphuric acid at room temperature and fermentative hydrogen production [33]. Also, the potential of textile cotton waste for energy production has been suggested. However, the main disadvantage of these concepts is ecological. This is primarily due to use of very high concentrated acids which decompose the cellulose.

Further, cellulose can be hydrolyzed/decomposed at low acid concentrations (0.07–0.1 mass.%) of sulphuric acid at above 200 °C with yield of up to 90% [22]. This ecologically acceptable method is energetically unacceptable due to high temperature reaction requirements.

Researchers are still trying to find green solvents for the dissolution and regeneration of cellulose such as ionic liquids [113].

Kosan et al. [53] investigated the properties of cellulose solutions in different dissolving liquids such as N-methyl morpholine-N-oxide and ionic liquids with varied cations and anions.

Originally driven by environmental reasons, recycling of postconsumer PET bottles into textile fibers is now becoming commercially attractive using extrusion polymer melt technology. On other hand, the recycling/recovery of textile fibers from textile waste is not commercially attractive thus far due to the availability of waste primarily in an intimate blend form. In 1989 it was reported that Eastman Chemicals had used solvolysis to recover DMT (dimethyl terephthalate) from Kodak films in a plant with 5000 tons/year capacity; a few years later, Coca Cola (Hoechst Celanese) and PepsiCo Inc. (Goodyear) also announced the development of plants for recycling of PET bottles by chemical recycling. In 2011 Toray announced that it has formulated a strategy to expand the fibers and textiles recycling business [86].

Poly (ethylene terephthalate) (PET) is one of most frequent polymers appeared in waste (packing material, plastic bottles, foils, textiles etc.) and it can be subjected to chemical degradation [88]. PET is not suitable for pyrolysis because liberates gas (20–40 mass.%).

PET Mechanical Recycling

Principle of mechanical recycling of thermoplastic is re-extrusion of their melts. Main limitation is here photodegradation, hydrolysis, and thermo-oxidation which cause the breakage of polymer chains into low molecular weight ones. Extrusion process, which is main part in plastic mechanical recycling, induce also severe shear deformation from screw. Therefore, mechanical properties of recycled products were degraded due to changes of morphology and conformation. The simple new twin screw extruder unit with a blank space ("Molten Resin Reservoir") can ensure improvement of mechanical properties of recycled thermoplastics [90].

For textile purposes are strategical polyester (mainly polyethylene terephthalate—PET) fibers due to their versatility, durability and stability [73, 114]. It is therefore natural to produce also recycled polyester fibres from various types of suitable wastes [79]. Mechanical recycling can be used effectively for as non-dyed pre-consumer textile waste or used PET bottles. Used PET bottles create the major part of polyester waste available for recycling today and are also a major environmental problem and challenge. General mechanical recycling steps for polyester waste by re-melting are shown in Fig. 16.

The design of PET bottles should be chosen to best suit subsequent recycling (see Fig. 17). Therefore, there is an effort to cooperate between bottle producers and companies involved in recycling. It is important that all materials used for bottle construction can be suitably separated by flotation. The production of recycled PET fibers does not differ significantly from the conventional production of virgin PET fibers from granulate polymer.

Spinning of mechanically recycled PET fibers from waste bottles has some specific features:

- Use of a special screw design for melting flakes with low bulk density.
- Create a degassing zone on the extruder to remove gaseous components of partial degradation of PET.
- Allows partial degradation of PET during re-melting as PET for bottles has a high molecular weight (solid-state post-polycondensation),
- Include a post-polycondensation reactor to equalize the output molecular weight of PET,

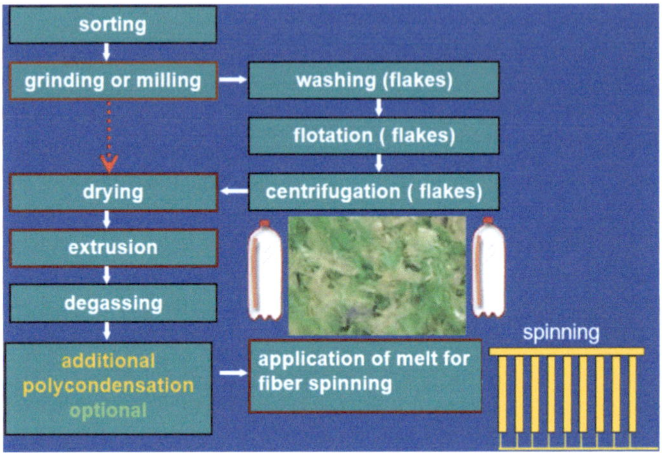

Fig. 16 Mechanical recycling of PET waste

Fig. 17 Materials for PET bottles

- Addition a device for adding color concentrates for mass dyeing the fiber.
- Use of compact lines for continuous spinning and drawing because the recycled PET fibers are mainly designated for the production of less demanding products, as are nonwovens.

The main limiting factor influencing quality of recycled fibers is composition of the PET flakes. Their degradation products (and the rest of other polymers) discolor the melt and the fibers are not pure white. Frequent presence of PVC is dangerous because it splits off hydrogen chloride, which causes degradation and creates charred residues at the spinning temperature. Discoloration can also be caused by poor sorting of transparent and colored bottles. During direct melting of PET flakes, the quality of the melt (molecular weight of PET) fluctuates, and thus the final physical and mechanical properties of the fiber also fluctuate. When mixing flakes of different origin, the polymers are modified in various ways (due to presence of isophthalic acid, diglycols, etc. and crystallization retarders), and thus the quality of the fiber fluctuates. Mechanically recycled PET fibers can be used in virtually all known applications of polyester fibers in the textile industry, because the current level of recycling processes allows for the production of fibers of comparable quality. The most common application of mechanically recycled PET staple fibers till yet are nonwovens for the furniture industry, automotive industry, construction industry, etc.

PET Chemical Recycling

This type of recycling leads to creation of monomers and intermediates realized by glycolysis, metanalysis, hydrolysis (alkali, acid, neutral), ammonolysis, and aminolysis mainly (see Fig. 18 and Table 8) by mechanism of ester bonds cleavage (nucleophilic attack—see Fig. 19) [43, 58, 60, 88]. Monomers are used for polycondensation of PET by standard conventional methods and therefore their molecular weight variation is low and properties are influences by conditions of spinning drawing and heat treatment mainly [68].

The depolymerization catalysts as ester-forming sulfonate or phosphate compounds, polyhydric alcohols, or tributylhexadecylphosphonium bromide (especially for NaOH hydrolysis) are frequently used.

Hydrolysis needs special conditions as pressure, temperature, and a long-time process. The cost of product purification is high.

Methanolysis needs low pressure but purity of products is not high.

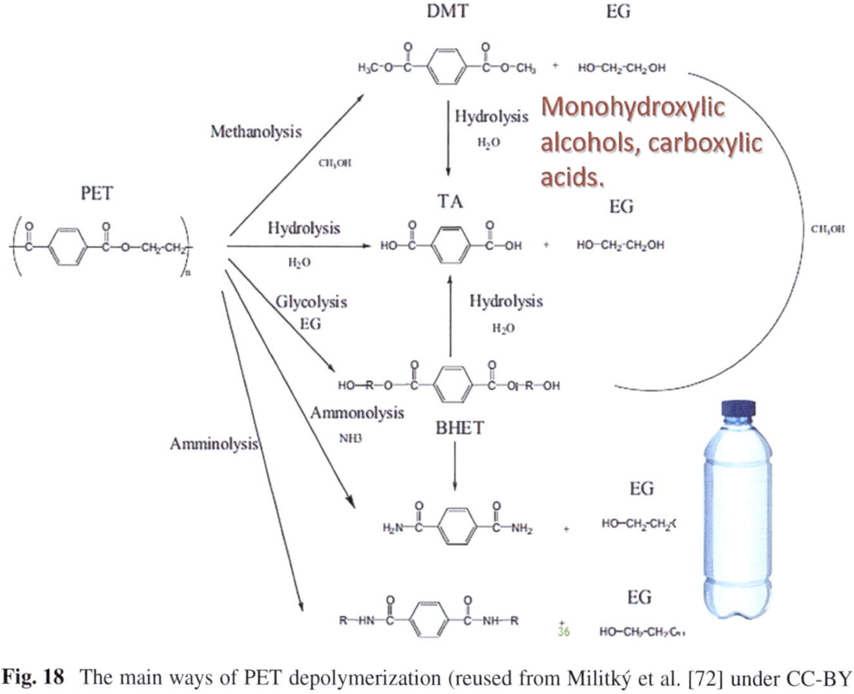

Fig. 18 The main ways of PET depolymerization (reused from Militký et al. [72] under CC-BY Permission)

Table 8 Pros/Cons of the chemical recycling methods of poly (ethylene terephthalate)

Aspects of evaluation	Chemical recycling methods					
	Methanolysis	Glycolysis	Hydrolysis	Aminolysis	Ammonolysis	Other methods
1. Flexibility in utilizing a variety of wastes	Low: well-defined industrial wastes (−)	Medium (±)	Medium (±)	Medium (±)?	Medium (±)	Low or medium (±)
2. Degradation conditions	High parameters; $t \leq 300$ °C; $p \leq 4$ MPa; separation; purification of the products (−)	$t \leq 280$ °C (+)	Acidic, $t \leq 100$ °C; alkaline $t \leq 250$ °C; neutral, $t \leq 300$ °C; $p \leq 4$ MPa (±)	$t \leq 100$ °C (+)	$t \leq 200$ °C; $p \leq 2$ MPa (±)	$t \leq 330$ °C (+)
3. Safety conditions	High requirement (−)	Conventional (+)	High requirement (−)	Medium (±)	Medium or high (−)	Conventional (+)
4. Corrosive and/or toxic media	Toxic methanol (−)	(+)	Strong acids or alkalis (±)	Amines (±)	Ammonium (±)	(+)
5. Economic aspects (including advisability of purification of the main product)	For large-scale installations (yes) (+)	For small or medium installations (no) (+)	(Yes) (±)	?	(Yes) (±)	(Yes or no) (±)
6. Product versatility	Low: DMT, EG (−)	High: oligomeric mixtures (+)	Low: TPA, EG (−)	?	Low: TPA diamide, EG (−)	Low or high (±)
Remarks	Industrially applied; large installations	Commercially applied; small or medium installations	Commercially applied high-purity TPA; large amounts of inorganic aqueous wastes	?	High-purity TPA diamide	Some may be commercially applied

Reused from Paszun and Spychaj [88] under CC-BY Permission

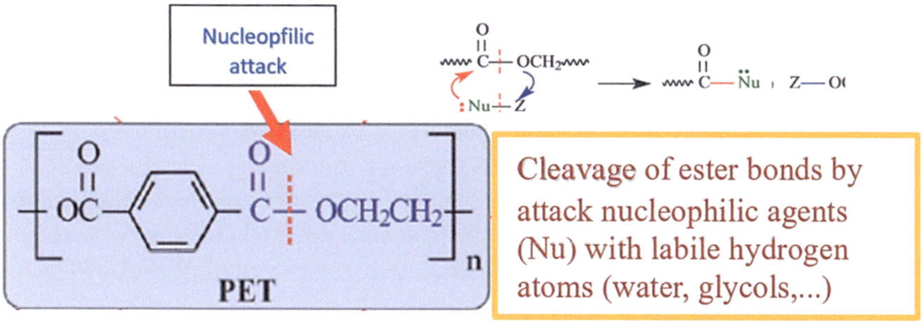

Fig. 19 Cleavage of ester bonds by nucleophilic attack

Catalytic glycolysis needs high temperature and pressure. Organometallic catalysts have a lower reaction rate and are destructive to the environment due to metal toxicity. It is one of best depolymerization method.

Ammonolysis in liquid ammonia (NH_3) at temperature range 70–180 °C under pressure 5–10 atm (low pressure can also be used) for 1–7 h with a suitable catalyst such as Zinc acetate (0.5%).

Yield terephthalamide—TDA and ethylene glycol. The TDA can be converted into p-phenylene diamine (PPD) by means of a simple reaction with chlorine and then NaOH. PPD is important as monomers of high-modulus fibers (p aramids as Kevlar).

Aminolysis needs temperature lower than 70 °C, and it is faster than ammonolysis.

Gas phase depolymerization is composed from three main steps:

1. Gas-phase treatment of PET waste in nitrating atmosphere (vaporized 12 mol/L HNO_3) at a temperature of 130–190 °C for 5–24 h.
2. Dissolving in 0.5 mol/L NaOH at room temperature for 45–60 h.
3. Ozonation of alkaline solutions for complete decomposition of the all organic compounds. PET decomposition degree is higher than 99.9%.

Catalytic-assisted depolymerization mechanisms using organo-catalysts, ionic liquids (e.g., choline-based), metal oxide/acetate/chlorides/sulfates can improve the aminolysis, glycolysis, methanolysis, and other chemical recycling processes. Zinc acetate as the best organometallic catalyst, zeolites as large surface area catalysts, and nontoxic metal salts as ionic liquid catalysts are beneficial.

In the future, the technology of depolymerization by combining physical and chemical treatments with enzymatic ones (chemoenzymatic) could be applied to textile waste [46]. The Pros/Cons of the chemical recycling method of PET is shown in Table 8.

Products of PET waste depolymerization are used for synthesizing saturated and unsaturated polyesters and polyurethanes [112]. For the production of recycled fibers, synthetic PET is used. This synthesis requires monomers of high purity and no presence of other diols or diacids able to act as co-monomers. It was shown [74], that e.g., diethylene glycol (DEG) produced during the polycondensation process of PET synthesis due to side reactions, about 1–2 wt% is mostly bonded to macromolecular chains as a modifying component. Irregular bonding of DEG in PET will adversely influence, e.g., the uniformity of dyeing.

Enzymatic depolymerization by biocatalysts capable of hydrolyzing PET is sustainable route to plastic recycling, as these environmentally benign, renewable catalysts can work under mild conditions and generate a clean product without using hazardous chemicals. Thermostable *Humicola insolens* cutinase (HiC, Novozym 51032), is suitable for enzymatic depolymerization of both crystalline and amorphous phases in PET.

Theoretically, the virgin and regenerated PET fibers have very similar behavior and properties because are chemically identical but there are many of practical reasons for obtaining differences. The main limiting factors are:

A. Quality of wastes containing PET

Quality of wastes depends on type of wastes. Post-consumer textile wastes are usually not 100% PET but are composed from other materials as different fibers types (often cotton), sewing threads, finishing agents, dyestuffs, and non-fibrous materials (zippers, buttons, strips, frills etc.). These materials can be removed only partially and special techniques should be used [14]. It is main reason for not use textile wastes for recycling in industrial scale till yet. As source waste for PET depolymerization the beverage containers (bottles) are mainly used [12]. Here recycled PET should be analyzed for benzene and limonene content. Benzene is a byproduct of PVC contamination and degradation. Limonene is a flavoring agent primarily carried over from residual citrus juices and may potentially impact the organoleptic properties of the packaged beverage. The byproduct of processing PET is also acetaldehyde [12]. Other PET based plastic packaging waste products are useful sources of monomers as well [95].

B. Pretreatment of wastes

For pretreatment of wastes the mechanical recycling is applied. This technology requires relatively little energy and resources. However, mechanical recycling causes various problems such as thermal–mechanical degradation and immiscibility of different polymers. Recyclates with a sufficient quality for subsequent chemical recycling is then very challenging [92]. Another problem is the presence of certain components such as odorous constituents, non-polymeric components and inks, often containing metals and halogens [95].

C. Depolymerization system

Depolymerization products (monomers and intermediates) quantity and quality are critically dependent on used method (see Fig. 22). Terephthalic acid (TA) and ethylene glycol (EG) are after purification used directly as raw materials for polycondensation. Other types of monomers (as diamines) will be acting as comonomers in polymeric chains changing their melting behavior, melt viscosity and arrangements in solid state (opening super molecular structure).

D. New synthesis of PET

PET synthesis is described comprehensively in the book [114]. The main problem is to select suitable method i.e. direct esterification or trans re-esterification and optimal catalytic system [68]. Especially for fibers preparation the sufficiently high molecular mass and narrow mass distribution are required. These parameters can be influenced by presence of various intermediates.

E. Preparation of fibers

Hydrolytic and thermo-oxidative degradation are also important in PET extrusion. To eliminate hydrolytic degradation, drying must be carried out before each processing operation in the melt. Properties of fibers are easily changed by drawing and thermal treatment, variability of mechanical properties are obviously not due to use of recycled fibers [68]. The presence of other organic materials in polymeric chains and different particles between chains can be reason for variation of thermal properties, sorption properties and pilling. The microplastic generation can be also different. Textile processing of recycled fibers will be influences by proper spin finish mainly. For checking of recycled PES fibers, it is therefore proposed to evaluate:

- FTIR analysis for checking presence of typical groups and components
- DSC analysis for investigation of thermal behavior
- Hot melt index for indication of melt viscosity
- Abrasion resistance
- Bending rigidity
- Dyeability testing
- Spinning ability testing
- Hand testing
- Generation of microplastics after washing.

Beneficial will be to compare results with materials from virgin PET fibres with the same geometry [79].

6 Microplastic Challenges

Common microplastics (i.e. fragments from synthetic polymers in the size range 1 μm to 5 mm) are released from wide range of thermoplastic polymers characterized by huge variation of molecular mass and morphology (different degree of degradation) having major influence on their properties including adhesion to other materials. More wide range of plastic waste fragments is shown in Fig. 20.

Majority of microplastic from industrial polymers (as abraded particles from tires) are due to their higher density deposited on bottom in calm conditions.

Primary microplastics: (constitutes 15–31% of microplastics in the oceans), for example, products in cosmetics (they form 2% of microplastics in the oceans, e.g. peeling to clean the skin) or abrasives in cleaning products.

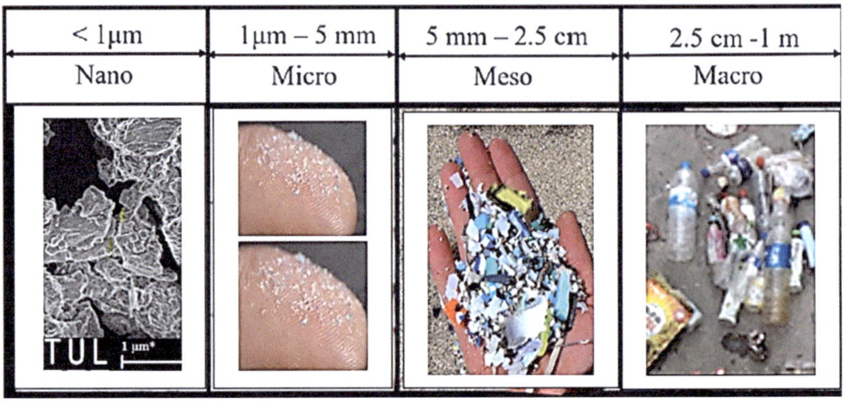

Fig. 20 Plastic fragments in different scales

Secondary microplastics: Formed by the breakdown of other products. The source is primarily the textile materials they make up 35% of all microplastics in the oceans.

Fibrous microplastics are released from textile materials during creation of fibres (preparation of polymer, filament cutting or tearing to staple fibres, texturing), production of textile fabrics (spinning, weaving, knitting and nonwovens), finishing (pretreatment, bleaching, dyeing, final treatment), creation of clothing (cutting, sewing), using of clothings (washing and drying mainly) and post-consumer waste treatment.

Generally, there are two main mechanisms of textile microplastics generation. One is due to presence of short fibres with the same diameter as regular ones which are released from yarns. Second mechanism is generation of fibrous fragments with smaller diameter than fibres as result of splitting and breaking. Textile microplastics have often needle shape (see Fig. 21).

Nanoplastics are more toxic since their smaller size allows them much more easier entering to human body, compared to microplastics. Micro- nano-plastics

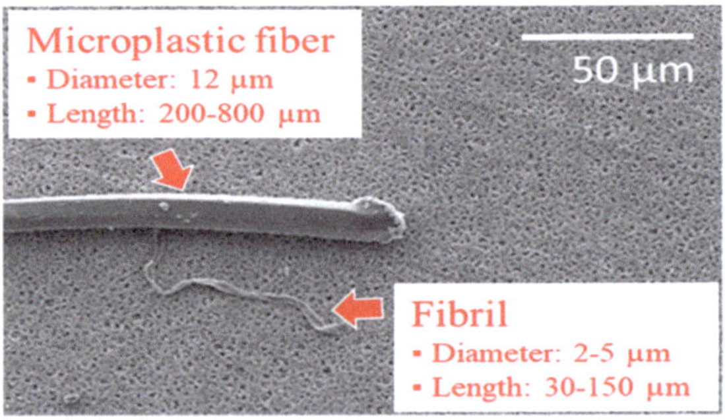

Fig. 21 Some fibrous microplastics (reused from Militký et al. [71] under CC-BY Permission)

concentrations were estimated to be about $2.4 \pm 1.3 \times 10^5$ particles per liter of bottled water, about 90% of which are nanoplastics. Tap water is lee polluted.

Microplastics are commonly colored and contain chemicals used for dyeing, chemical finishing, and functionalization of textiles. Due to the complicated chemical composition and often higher toxicity, a huge negative influence on the environment can be expected. Because the important parts of fibrous polymers are based on thermosets (as cellulose and proteins) it is in some articles, replaced term microplastic is replaced by fibrous fragments [38]. Particles, broken fibers and fragments released from synthetic fibers, and microparticles from fibers based on natural polymers are here referred to as "fibrous microplastics." The exact mechanism of origin, similar toxicity, and environmental behavior justify this extension of the term microplastics.

Fibrous microplastics release is generally dependent on fibers properties (strength and length of ultimate fibers mainly), yarn properties (content of fibrous wastes, hairiness and twist mainly), fabric construction (density of fibrous phase, fabric abrasion resistance, type of finishing) and fibrous products formation (cutting and joining). One of major source of fibrous fragments generation is abrasion during processing and using of textiles [59, 71]. Abrasion is accompanied by fibrillation especially for stiff non-thermoplastics natural fibers (see Fig. 22).

The abrasion mechanisms of synthetic thermoplastic fibers are dependent on the flexibility of polymeric chains anisotropy of mechanical properties in axial and radial direction and glass transition temperature. For rigid chains polymeric fibers with high mechanical anisotropy is fracture surface of abraded fibers following splitting mechanism. For flexible chains polymeric fibers with less mechanical anisotropy is splitting of abraded fibers result of plastic deformation accumulation [59]. SEM of abraded PET fiber is shown in Fig. 23.

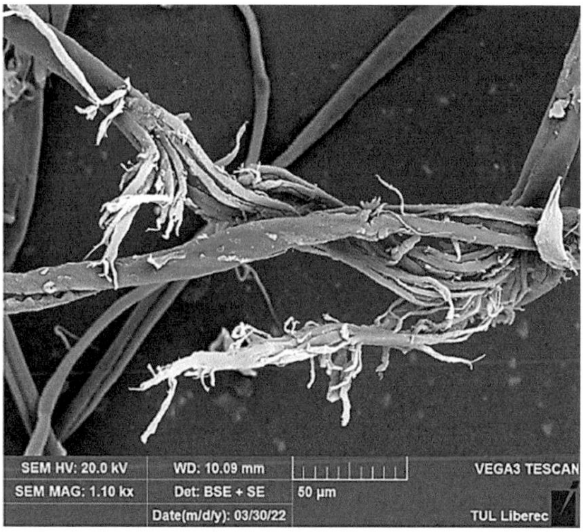

Fig. 22 SEM image of cotton fibers failure by abrasion (reused from Militký et al. [71] under CC-BY Permission)

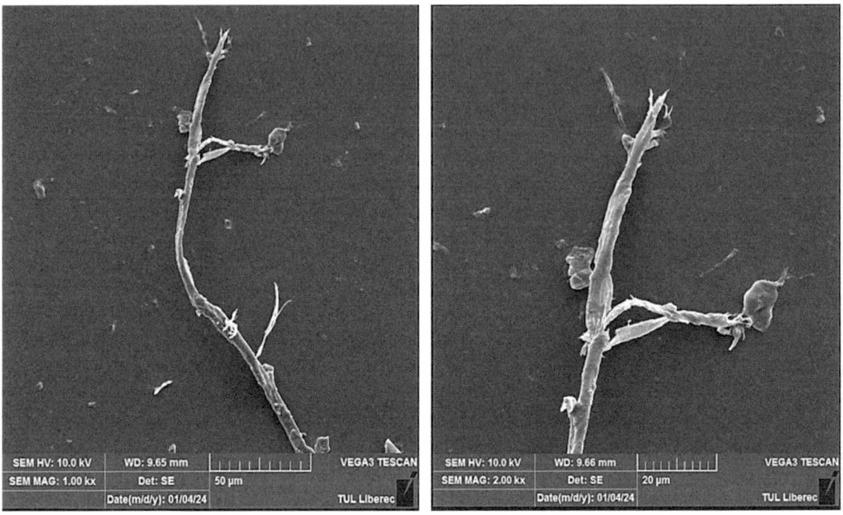

Fig. 23 SEM images of abraded PET fibers

The optimal parameters suppressing release of fibrous fragments are therefore not applicable to all types of fibers and therefore can be used only as guide for future production [38].

Conventional microplastics are often based on polyolefins, which due to their low-density float on water [6, 34]. To leads to their easy detection and relatively simple methods of capturing them. One of the largest sources of microplastics are fibrous textile products from virgin (not degraded) polymers with very high mechanical resistance.

By analysis of seashore sediments worldwide it was found that microplastic are composed mainly from polyester (56%), acrylic (23%), polypropylene (7%), polyethylene (6%), and polyamide fibers (3%). In effluents discharged from waste water were dominated polyester (67%) and acrylic (17%) fibers [17]. It is partially due to fact that polyester fibers are the most frequently used in textile products. The fiber/fiber friction coefficients of PES equal to 0.56 is over PAD (0.47) and cotton (0.22). Release of fibers from surface will therefore not so easy as in the case of other materials. Major properties of PET are excellent (see Militký [68]), but they have strong tendency to pilling (see Fig. 24).

Pills can be generated and removed from surface of textiles e.g. by washing or during wearing. Other source of microplastic generated from polyester fabrics is texturing on filament yarns and damage of polyester fibers caused by production of fabrics as spinning [69]. Local breaks and splitting due to thermal damage of polyester fibers caused by contact with metal surfaces during rotor spinning is shown in Fig. 25 [32].

Fibrous microplastics have an extremely high degree of danger, occur in the air and water, including drinking water, and immediately threaten practically the

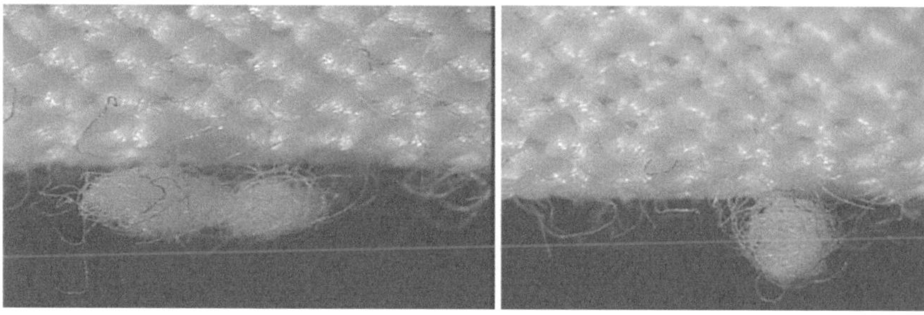

Fig. 24 Typical pills on polyester fabric surface

Fig. 25 Results of local thermal damage of PET fibers caused by rotor spinning (reused from Gharenghai [32] under CC-BY Permission)

entire human population. When textile microplastics are carried by air, they usually accumulate mainly in household and office dust and support the development of allergies, asthma and many respiratory problems. When released during washing, textile microplastics accumulate in rivers, seas, wastewater treatment plants and drinking water. Almost a third of fibrous microplastics are released mainly during the washing of garments [21, 89].

Fiber microplastics are potentially more dangerous than other types of microplastics for the following reasons:

1. They are mainly fiber fragments and are usually needle-shaped. The reason is the internal fibrillar structure of the fibers and the geometric shape of the textile fibers, which are long and thin (the typical length/diameter ratio is 10^3). The consequence of the needle shape can be easier mechanical damage to tissues and cells, mechanical disruption of cell walls, etc. [91, 103].

2. Compared to other polymers, textile fibers are usually contaminated with non-polymeric potentially toxic compounds, which can be textile dyes, finishing agents, TiO_2-based matting particles and, more recently, metal and metal oxide particles.
3. Fiber microplastics are a consequence of the everyday use, maintenance and reprocessing or mechanical recycling of textiles, so their adverse cumulative effects are serious.

The total volume of microplastic generation is related to the consumption of fibrous structures including recycling strategies. In industrialized countries, textile consumption is around 25–30 kg per person per year, and in developing countries it is about 10–15 kg per person per year. From the prediction of human population development, it is therefore possible to estimate the amount of generated microplastics in the case that there will be no modifications in the composition, construction and production technologies of fibrous structures. During the construction and recycling of textiles, it will be necessary to simultaneously solve problems related to optimizing technological processes, construction and functionalization of fibrous structures, and ecological requirements to suppress microplastic formation. Computer-aided systems are increasingly being used to develop new products. It will require the creation of software systems combining existing software focused on the classic construction of textiles (visual design) with systems predicting the properties of textiles (analogies of CAD systems in mechanical engineering), including those that contribute to the potential release of microplastics.

The formation of fibrous microplastics can be regulated by appropriate selection of fiber-forming polymers, control of supramolecular structure and surface structure of fibers, suitable construction of fibrous structures (linear and planar), and suitable methods of their maintenance and recycling. These are complicated tasks, for the solution of which both modelling and simulation, such as the abrasion, washing process, and mechanical damage during recycling, experimental results, and morphological analysis of microplastics, should be used.

7 Conclusion

Waste recycling and environmental sustainability are not a question of achieving a goal, but rather a direction and process of improvement. There is no final state that one is completely satisfied with reaching. We can always adjust to make our planet a better and safer place for all life forms. This can only be achieved through a comprehensive approach involving environmental, social, and economic aspects of problem-solving. Instead of focusing on the local growth of the economy, it is necessary to take into account a broader time horizon that also includes nature and its possibilities or limitations. This does not mean at all costs promoting either ecology at the expense of economics or economics at the expense of ecology, but comprehensively balancing the entire system of human needs integrated with nature. Industrial processes must move from linear systems where wastes are the final stage to closed-loop systems where wastes are inputs for new processes. Use of new, more efficient, types of more sustainable fibers can be a small but important step to a sustainable future.

References

1. 1987: Brundtland Report (1987)
2. Abraham, M.: Encyclopedia of Sustainable Technologies, 1st edn. Elsevier (2017)
3. Adekomaya, O., Majozi, T.: 23—Industrial and biomedical applications of fiber reinforced composites. In: Joseph, K., Oksman, K., George, G., Wilson, R., Appukuttan, S. (eds.) Fiber Reinforced Composites, pp. 753–783. Woodhead Publishing (2021). https://doi.org/10.1016/B978-0-12-821090-1.00004-1
4. Agboh, O.C., Qin, Y.: Chitin and chitosan fibers. Polym. Adv. Technol. **8**(6), 355–365 (1997). https://doi.org/10.1002/(SICI)1099-1581(199706)8:6<355::AID-PAT651>3.0.CO;2-T
5. Anastas, P.T., Warner, J.C.: Green Chemistry. Oxford University Press, Oxford (2000). https://doi.org/10.1093/oso/9780198506980.001.0001
6. Andrady, A.L.: Microplastics in the marine environment. Mar. Pollut. Bull. **62**(8), 1596–1605 (2011). https://doi.org/10.1016/j.marpolbul.2011.05.030
7. Aneja, A.P.: From textile terrific to textile terrible and back. In: 48th Textile Research Symposium (2021, Dec)
8. Arai, R., Calisto Friant, M., Vermeulen, W.J.V.: The Japanese circular economy and sound material-cycle society policies: discourse and policy analysis. Circ. Econ. Sustain. **4**(1), 619–650 (2024). https://doi.org/10.1007/s43615-023-00298-7
9. Arunpal, A.: Textile sustainability: a global imperative. XII Textile Asian Congress (2013, Nov)
10. Arunpal, A.: Textile recycling: an environmental perspective. In: 13th International Izmir Textile and Apparel Symposium (2014, Apr)
11. Batool, S.R., Ahmad, S., Nawab, Y., Hussain, M.: In: Batool, S.R., Ahmad, S., Nawab, Y., Hussain, M. (eds.) Circularity in Textiles, 1st edn. Springer Nature Switzerland (2023). https://doi.org/10.1007/978-3-031-49479-6
12. Benyathiar, P., Kumar, P., Carpenter, G., Brace, J., Mishra, D.K.: Polyethylene terephthalate (PET) bottle-to-bottle recycling for the beverage industry: a review. Polymers **14**(12), 2366 (2022). https://doi.org/10.3390/polym14122366
13. Bhambri, P., Paweloszek, I.: Digital Sustainability, 1st edn. CRC Press (2024). https://doi.org/10.1201/9781003484226
14. Bianchi, S., Bartoli, F., Bruni, C., Fernandez-Avila, C., Rodriguez-Turienzo, L., Mellado-Carretero, J., Spinelli, D., Coltelli, M.-B.: Opportunities and limitations in recycling fossil polymers from textiles. Macromol **3**(2), 120–148 (2023). https://doi.org/10.3390/macromol3020009
15. Biopolymers Facts and Statistics (2020)
16. Brine, C.J., Sandford, P.A., Zikakis, J.P.: Advances in Chitin and Chitosan. https://api.semanticscholar.org/. Corpus ID: 107237408 (1992)
17. Browne, M.A., Crump, P., Niven, S.J., Teuten, E., Tonkin, A., Galloway, T., Thompson, R.: Accumulation of microplastic on shorelines Woldwide: sources and sinks. Environ. Sci. Technol. **45**(21), 9175–9179 (2011). https://doi.org/10.1021/es201811s
18. Burgess, S.K., Karvan, O., Johnson, J.R., Kriegel, R.M., Koros, W.J.: Oxygen sorption and transport in amorphous poly(ethylene furanoate). Polymer **55**(18), 4748–4756 (2014). https://doi.org/10.1016/j.polymer.2014.07.041
19. Burgess, S.K., Kriegel, R.M., Koros, W.J.: Carbon dioxide sorption and transport in amorphous poly(ethylene furanoate). Macromolecules **48**(7), 2184–2193 (2015). https://doi.org/10.1021/acs.macromol.5b00333

20. Burgess, S.K., Mikkilineni, D.S., Yu, D.B., Kim, D.J., Mubarak, C.R., Kriegel, R.M., Koros, W.J.: Water sorption in poly(ethylene furanoate) compared to poly(ethylene terephthalate). Part 2: Kinetic sorption. Polymer **55**(26), 6870–6882 (2014). https://doi.org/10.1016/j.polymer.2014.10.065
21. Cesa, F.S., Turra, A., Checon, H.H., Leonardi, B., Baruque-Ramos, J.: Laundering and textile parameters influence fibers release in household washings. Environ. Pollut. **257**, 113553 (2020). https://doi.org/10.1016/j.envpol.2019.113553
22. Chandrashekhar, B., Mishra, S.M., Sharma, K., Dubey, S.: Bio-ethanol production from textile cotton waste via dilute acid hydrolysis and fermentation by *Saccharomyces cerevisiae*. J. Ecobiotechnol. **34**, 6–9 (2011)
23. Clark, J.H., Deswarte, F.E.I. (eds.): Introduction to Chemicals from Biomass. Wiley (2008). https://doi.org/10.1002/9780470697474
24. Coltelli, M.-B., Danti, S., De Clerck, K., Lazzeri, A., Morganti, P.: Pullulan for advanced sustainable body- and skin-contact applications. J. Funct. Biomater. **11**(1), 20 (2020). https://doi.org/10.3390/jfb11010020
25. Coltelli, M.-B., Wild, F., Bugnicourt, E., Cinelli, P., Lindner, M., Schmid, M., Weckel, V., Müller, K., Rodriguez, P., Staebler, A., Rodríguez-Turienzo, L., Lazzeri, A.: State of the art in the development and properties of protein-based films and coatings and their applicability to cellulose based products: an extensive review. Coatings **6**(1), 1 (2015). https://doi.org/10.3390/coatings6010001
26. Cui, J., Fu, D., Mi, L., Li, L., Liu, Y., Wang, C., He, C., Zhang, H., Chen, Y., Wang, Q.: Effects of thermal treatment on the mechanical properties of bamboo fiber bundles. Materials **16**(3), 1239 (2023). https://doi.org/10.3390/ma16031239
27. Ebringerová, A., Hromádková, Z., Heinze, T.: Hemicellulose. In: Polysaccharides I, pp. 1–67. Springer. https://doi.org/10.1007/b136816
28. Elnashar, M.M.: In: Elnashar, M. (ed.) Biopolymers. Sciyo (2010). https://doi.org/10.5772/286
29. Fei, X., Wang, J., Zhang, X., Jia, Z., Jiang, Y., Liu, X.: Recent progress on bio-based polyesters derived from 2,5-furandicarbonxylic acid (FDCA). Polymers **14**(3), 625 (2022). https://doi.org/10.3390/polym14030625
30. Fernandes, M., Padrão, J., Ribeiro, A.I., Fernandes, R.D.V., Melro, L., Nicolau, T., Mehravani, B., Alves, C., Rodrigues, R., Zille, A.: Polysaccharides and metal nanoparticles for functional textiles: a review. Nano **12**(6), 1006 (2022). https://doi.org/10.3390/nano12061006
31. Flores-Nieves, M.M., Castellanos-Espinoza, R., Estevez, M., Baldenegro-Pérez, L.A., Trejo, J.F.G., García, M.E., Cano, B.M., Soto-Zarazúa, G.M., España-Sánchez, B.L.: Electrospun casein fibers obtained from revalued milk with mechanical and antibacterial properties. Arab. J. Chem. **15**(11), 104201 (2022). https://doi.org/10.1016/j.arabjc.2022.104201
32. Gharenghai, A.A.: Thermal damage of PET fibres in OE rotor spinning. In: Proceedings of the International Conference on "Textile Science", pp. 1–6 (2008)
33. Gholamzad, E., Karimi, K., Masoomi, M.: Effective conversion of waste polyester–cotton textile to ethanol and recovery of polyester by alkaline pretreatment. Chem. Eng. J. **253**, 40–45 (2014). https://doi.org/10.1016/j.cej.2014.04.109
34. Gigault, J., Halle, A., Baudrimont, M., Pascal, P.-Y., Gauffre, F., Phi, T.-L., El Hadri, H., Grassl, B., Reynaud, S.: Current opinion: what is a nanoplastic? Environ. Pollut. **235**, 1030–1034 (2018). https://doi.org/10.1016/j.envpol.2018.01.024
35. Gonzalez, V., Lou, X., Chi, T.: Evaluating environmental impact of natural and synthetic fibers: a life cycle assessment approach. Sustainability **15**(9), 7670 (2023). https://doi.org/10.3390/su15097670
36. Gupta, B., Revagade, N., Anjum, N., Atthoff, B., Hilborn, J.: Preparation of poly(lactic acid) fiber by dry-jet-wet-spinning. I. Influence of draw ratio on fiber properties. J. Appl. Polym. Sci. **100**(2), 1239–1246 (2006). https://doi.org/10.1002/app.23497
37. Gürses, A.: 2—Sustainable colorants. In: Shahid-ul-Islam, Butola, B.S. (eds.) The Impact and Prospects of Green Chemistry for Textile Technology, pp. 21–55. Woodhead Publishing (2019). https://doi.org/10.1016/B978-0-08-102491-1.00002-2

38. Han, J., McQueen, R.H., Batcheller, J.C.: From fabric to fallout: a systematic review of the impact of textile parameters on fibre fragment release. Textiles **4**(4), 459–492 (2024). https://doi.org/10.3390/textiles4040027
39. Hanuš, J., Dvořáková, E.: Possibilities of using cryogenic technologies to recycle textiles and shoes (2008)
40. Hasan, A., Rabbi, M.S., Maruf Billah, M.: Making the lignocellulosic fibers chemically compatible for composite: a comprehensive review. Cleaner Mater. **4**, 100078 (2022). https://doi.org/10.1016/j.clema.2022.100078
41. Heijungs, R.: Probability, Statistics and Life Cycle Assessment. Springer International Publishing (2024). https://doi.org/10.1007/978-3-031-49317-1
42. Horrocks, A.R. (ed.): Recycling Textile and Plastic Waste. Woodhead Publishing (1996)
43. Ibrahim, M.A., Alhalafi, M.H., Emam, E.-A.M., Ibrahim, H., Mosaad, R.M.: A review of chitosan and chitosan nanofiber: preparation, characterization, and its potential applications. Polymers **15**(13), 2820 (2023). https://doi.org/10.3390/polym15132820
44. Ibrahim, N.: Green Chemistry for Sustainable Textiles. Elsevier (2021). https://doi.org/10.1016/C2020-0-01624-5
45. Jawahir, I.S., Bradley, R.: Technological elements of circular economy and the principles of 6R-based closed-loop material flow in sustainable manufacturing. Procedia CIRP **40**, 103–108 (2016). https://doi.org/10.1016/j.procir.2016.01.067
46. Jönsson, C., Wei, R., Biundo, A., Landberg, J., Schwarz Bour, L., Pezzotti, F., Toca, A., Jacques, L.M., Bornscheuer, U.T., Syrén, P.: Biocatalysis in the recycling landscape for synthetic polymers and plastics towards circular textiles. ChemSusChem. **14**(19), 4028–4040 (2021). https://doi.org/10.1002/cssc.202002666
47. Joziasse, C.A.P., Grijpma, D.W., Bergsma, J.E., Cordewener, F.W., Bos, R.R.M., Pennings, A.J.: The influence of morphology on the (hydrolytic degradation of as-polymerized and hot-drawn poly(L-lactide)). Colloid Polym. Sci. **276**(11), 968–975 (1998). https://doi.org/10.1007/s003960050335
48. Kaifu, K., Nishi, N., Komai, T., Tokura, S., Somorin, O.: Studies on chitin. V. Formylation, propionylation, and butyrylation of chitin. Polym. J. **13**(3), 241–245 (1981). https://doi.org/10.1295/polymj.13.241
49. Karimah, A., Ridho, M.R., Munawar, S.S., Ismadi, Amin, Y., Damayanti, R., Lubis, M.A.R., Wulandari, A.P., Nurindah, Iswanto, A.H., Fudholi, A., Asrofi, M., Saedah, E., Sari, N.H., Pratama, B.R., Fatriasari, W., Nawawi, D.S., Rangappa, S.M., Siengchin, S.: A comprehensive review on natural fibers: technological and socio-economical aspects. Polymers **13**(24), 4280 (2021). https://doi.org/10.3390/polym13244280
50. Kicińska-Jakubowska, A., Bogacz, E., Zimniewska, M.: Review of natural fibers. Part I—Vegetable fibers. J. Nat. Fibers **9**(3), 150–167 (2012). https://doi.org/10.1080/15440478.2012.703370
51. Kim, T., Kim, D., Park, Y.: Recent progress in regenerated fibers for "green" textile products. J. Clean. Prod. **376**, 134226 (2022). https://doi.org/10.1016/j.jclepro.2022.134226
52. Kopf, S., Åkesson, D., Skrifvars, M.: Textile fiber production of biopolymers—a review of spinning techniques for polyhydroxyalkanoates in biomedical applications. Polym. Rev. **63**(1), 200–245 (2023). https://doi.org/10.1080/15583724.2022.2076693
53. Kosan, B., Schwikal, K., Meister, F.: Solution states of cellulose in selected direct dissolution agents. Cellulose **17**(3), 495–506 (2010). https://doi.org/10.1007/s10570-010-9402-1
54. Kozlowski, R., Barriga, J.: Renewable natural fibrous raw materials, production and application. In: 3rd International Conference of Textile Research Division, pp. 597–612 (2006)
55. Křemenáková, D.: Properties of rotor yarns from reprocessed cotton fibers. In: Proceedings of the Beltwide Cotton Conference (2008, Jan 8)
56. Lein, J.K.: Futures Research and Environmental Sustainability. CRC Press (2016). https://doi.org/10.9774/gleaf.9781315371139
57. Lendelein, A., Sisson, A.: In: Lendlein, A., Sisson, A. (eds.) Handbook of Biodegradable Polymers. Wiley (2011). https://doi.org/10.1002/9783527635818

58. Liang, J., Fu, J., Lin, H., Chen, J., Peng, S., Sun, Y., Xu, Y., Kang, S.: Valorization of polyethylene terephthalate wastes to terephthalamide via catalyst-free ammonolysis. J. Ind. Eng. Chem. **132**, 578–587 (2024). https://doi.org/10.1016/j.jiec.2023.11.053
59. Li, H., He, G., Zhao, Z., Liu, Q., Wang, J., Yin, Y., Cui, Y., Ning, X., Ning, F.: Abrasion performance and failure mechanism of fiber yarns based on molecular segmental differences. J. Eng. Fibers Fabr. **19** (2024). https://doi.org/10.1177/15589250241228263
60. Lorenzetti, C., Manaresi, P., Berti, C., Barbiroli, G.: Chemical recovery of useful chemicals from polyester (PET) waste for resource conservation: a survey of state of the art. J. Polym. Environ. **14**(1), 89–101 (2006). https://doi.org/10.1007/s10924-005-8711-1
61. Lunt, J.: Large-scale production, properties and commercial applications of polylactic acid polymers. Polym. Degrad. Stab. **59**(1–3), 145–152 (1998). https://doi.org/10.1016/S0141-3910(97)00148-1
62. Malucelli, G.: 9—Textile finishing with biomacromolecules: a low environmental impact approach in flame retardancy. In: Shahid-ul-Islam, Butola, B.S. (eds.) The Impact and Prospects of Green Chemistry for Textile Technology, pp. 251–279. Woodhead Publishing (2019). https://doi.org/10.1016/B978-0-08-102491-1.00009-5
63. Mansor, M.R., Nurfaizey, A.H., Tamaldin, N., Nordin, M.N.A.: Natural fiber polymer composites. In: Biomass, Biopolymer-Based Materials, and Bioenergy, pp. 203–224. Elsevier (2019). https://doi.org/10.1016/B978-0-08-102426-3.00011-4
64. Markevičiūtė, Z., Varžinskas, V.: Plant-origin feedstock applications in fully green food packaging: the potential for tree-free paper and plant-origin bio-plastics in the Baltic Sea region. Sustainability **14**(12), 7393 (2022). https://doi.org/10.3390/su14127393
65. Materials Market Report (2024)
66. Mehta, N., Shah, K.J., Lin, Y.-I., Sun, Y., Pan, S.-Y.: Advances in circular bioeconomy technologies: from agricultural wastewater to value-added resources. Environments **8**(3), 20 (2021). https://doi.org/10.3390/environments8030020
67. Militký, J.: Trends in the Development of the Textile and Clothing Industry, Implementation Action Plan of Czech Technology Platform (CTPT) (2011)
68. Militký, J.: 13—Tensile failure of polyester fibers. In: Bunsell, A.R. (ed.) Handbook of Properties of Textile and Technical Fibres, 2nd edn, pp. 421–514. Woodhead Publishing (2018). https://doi.org/10.1016/B978-0-08-101272-7.00013-4
69. Militký, J., Ibrahim, S.: Effect of textile processing on fatigue. In: Miraftab, M. (ed.) Fatigue Failure of Textile Fibres. Woodhead Publishing (2009)
70. Militký, J., Křemenáková, D., Václavík, M., Klička, V., Dídek, S.: Textile branch and main breakthroughs of The Czech Republic in the field of textile machinery: an illustrated review. Textiles **1**(3), 466–482 (2021). https://doi.org/10.3390/textiles1030024
71. Militký, J., Kremenakova, D., Venkataraman, M.: Microplastics and fibrous fragments generated during the production and maintenance of textiles (2024)
72. Militký, J., Křemenáková, D., Venkataraman, M., Aneja, A.P.: Towards sustainable textiles for a safer planet: main topics. Sustainability (Switzerland) **16**(13) (2024). https://doi.org/10.3390/su16135344
73. Militký, J., Kryštůfek, J., Hartych, V.: Modified Polyester Fibres. Elsevier Science Publishers (1991)
74. Militký, J., Vaníček, J., Dostál, J., Jansa, J., Čáp, J.: Influence of diethylene glycol content on behavior of poly(ethylene terephthalate) fibers. J. Appl. Polym. Sci. **25**(6), 1195–1208 (1980). https://doi.org/10.1002/app.1980.070250621
75. Militký, J., Venkataraman, M., Periyasamy, A.P.: In: Militký, J., Venkataraman, M., Periyasamy, A.P. (eds.) Fibrous Structures and Their Impact on Textile Design, 1st edn. Springer Nature Singapore (2023). https://doi.org/10.1007/978-981-19-4827-5
76. Mistra Future Fashion: Mistra, Foundation for Environmental Strategic Research (2017). https://mistra.org/program/mistra-future-fashion/
77. Mora-Pale, M., Meli, L., Doherty, T.V., Linhardt, R.J., Dordick, J.S.: Room temperature ionic liquids as emerging solvents for the pretreatment of lignocellulosic biomass. Biotechnol. Bioeng. **108**(6), 1229–1245 (2011). https://doi.org/10.1002/bit.23108

78. Munasinghe, P., Druckman, A., Dissanayake, D.G.K.: A systematic review of the life cycle inventory of clothing. J. Clean. Prod. **320**, 128852 (2021). https://doi.org/10.1016/j.jclepro.2021.128852
79. Muthu, S.S. (ed.): Recycled Polyester. Springer Singapore (2020). https://doi.org/10.1007/978-981-32-9559-9
80. Muthu, S.S.: In: Muthu, S.S. (ed.) Sustainable Approaches in Textiles and Fashion, 1st edn. Springer Nature Singapore (2022). https://doi.org/10.1007/978-981-19-0874-3
81. Muthu, S.S.: Life Cycle Assessment & Circular Economy. Springer (2023)
82. Muthu, S.S.: In: Muthu, S.S. (ed.) Sustainable Manufacturing Practices in the Textiles and Fashion Sector, 1st edn. Springer Nature Switzerland (2024). https://doi.org/10.1007/978-3-031-51362-6
83. Muzzarelli, R.A.A., Dimitriu, S.: Polymeric Biomaterials. Marcel Dekker, Inc. (1993)
84. Naser, A.Z., Deiab, I., Darras, B.M.: Poly(lactic acid) (PLA) and polyhydroxyalkanoates (PHAs), green alternatives to petroleum-based plastics: a review. RSC Adv. **11**(28), 17151–17196 (2021). https://doi.org/10.1039/D1RA02390J
85. Niinimäki, K., Peters, G., Dahlbo, H., Perry, P., Rissanen, T., Gwilt, A.: Author correction: the environmental price of fast fashion. Nat. Rev. Earth Environ. **1**(5), 278–278 (2020). https://doi.org/10.1038/s43017-020-0054-x
86. Noritake, A., Hori, M., Shigematsu, M., Tanahashi, M.: Recycling of polyethylene terephthalate using high-pressure steam treatment. Polym. J. **40**(6), 498–502 (2008). https://doi.org/10.1295/polymj.PJ2007237
87. Novel packaging films and textiles with tailored end of life and performance based on bio-based copolymers and coatings (2019)
88. Paszun, D., Spychaj, T.: Chemical recycling of poly(ethylene terephthalate). Ind. Eng. Chem. Res. **36**(4), 1373–1383 (1997). https://doi.org/10.1021/ie960563c
89. Periyasamy, A.P., Tehrani-Bagha, A.: A review on microplastic emission from textile materials and its reduction techniques. Polym. Degrad. Stab. **199**, 109901 (2022). https://doi.org/10.1016/j.polymdegradstab.2022.109901
90. Phanthong, P., Miyoshi, Y., Yao, S.: Development of tensile properties and crystalline conformation of recycled polypropylene by re-extrusion using a twin-screw extruder with an additional molten resin reservoir unit. Appl. Sci. **11**(4), 1707 (2021). https://doi.org/10.3390/app11041707
91. Prata, J.C., da Costa, J.P., Lopes, I., Duarte, A.C., Rocha-Santos, T.: Environmental exposure to microplastics: an overview on possible human health effects. Sci. Total Environ. **702**, 134455 (2020). https://doi.org/10.1016/j.scitotenv.2019.134455
92. Ragaert, K., Delva, L., Van Geem, K.: Mechanical and chemical recycling of solid plastic waste. Waste Manag. **69**, 24–58 (2017). https://doi.org/10.1016/j.wasman.2017.07.044
93. Rani, R.: Management of biodegradable waste. J. Biotechnol. Biomater. **11**(6) (2011)
94. Regulation (EU) 2024/1781 of the European Parliament and of the Council, Pub. L. No. 2024/1781 (2024)
95. Roosen, M., Mys, N., Kusenberg, M., Billen, P., Dumoulin, A., Dewulf, J., Van Geem, K.M., Ragaert, K., De Meester, S.: Detailed analysis of the composition of selected plastic packaging waste products and its implications for mechanical and thermochemical recycling. Environ. Sci. Technol. **54**(20), 13282–13293 (2020). https://doi.org/10.1021/acs.est.0c03371
96. Rouhani Shirvan, A., Shakeri, M., Bashari, A.: Recent advances in application of chitosan and its derivatives in functional finishing of textiles. In: The Impact and Prospects of Green Chemistry for Textile Technology, pp. 107–133. Elsevier (2019). https://doi.org/10.1016/B978-0-08-102491-1.00005-8
97. Sandin, G., Roos, S., Johansson, M.: Environmental impact of textile fibres—what we know and what we don't know. Fiber Bible Part 2 (2019)
98. Scheffer, M., Aneja, A.: Biofibers—an exploration in the prediction of innovation diffusion. In: Autex 10th World Textile Conference (2010)
99. Schlöesser, T.P.: Natural fiber reinforced automotive parts. In: Natural Fibers, Plastics and Composites, pp. 275–285. Springer US (2004). https://doi.org/10.1007/978-1-4419-9050-1_15

100. Scribante, A., Vallittu, P.K., Özcan, M.: Fiber-reinforced composites for dental applications. Biomed. Res. Int. **2018**, 1–2 (2018). https://doi.org/10.1155/2018/4734986
101. Seifali Abbas-Abadi, M., Tomme, B., Goshayeshi, B., Mynko, O., Wang, Y., Roy, S., Kumar, R., Baruah, B., De Clerck, K., De Meester, S., D'hooge, D.R., Van Geem, K.M.: Advancing textile waste recycling: challenges and opportunities across polymer and non-polymer fiber types. Polymers **17**(5), 628 (2025). https://doi.org/10.3390/polym17050628
102. Siddiqui, M.N., Redhwi, H.H., Al-Arfaj, A.A., Achilias, D.S.: Chemical recycling of PET in the presence of the bio-based polymers, PLA, PHB and PEF: a review. Sustainability **13**(19), 10528 (2021). https://doi.org/10.3390/su131910528
103. Song, S., van Dijk, F., Vasse, G.F., Liu, Q., Gosselink, I.F., Weltjens, E., Remels, A.H.V., de Jager, M.H., Bos, S., Li, C., Stoeger, T., Rehberg, M., Kutschke, D., van Eck, G.W.A., Wu, X., Willems, S.H., Boom, D.H.A., Kooter, I.M., Spierings, D., Wardenaar, R., Cole, M., Nawijn, M.C., Salvati, A., Gosens, R., Melgert, B.N.: Inhalable textile microplastic fibers impair airway epithelial differentiation. Am. J. Respir. Crit. Care Med. **209**(4), 427–443 (2024). https://doi.org/10.1164/rccm.202211-2099OC
104. Sousa, A.F., Patrício, R., Terzopoulou, Z., Bikiaris, D.N., Stern, T., Wenger, J., Loos, K., Lotti, N., Siracusa, V., Szymczyk, A., Paszkiewicz, S., Triantafyllidis, K.S., Zamboulis, A., Nikolic, M.S., Spasojevic, P., Thiyagarajan, S., van Es, D.S., Guigo, N.: Recommendations for replacing PET on packaging, fiber, and film materials with biobased counterparts. Green Chem. **23**(22), 8795–8820 (2021). https://doi.org/10.1039/D1GC02082J
105. Sun, L., Wang, J., Mahmud, S., Jiang, Y., Zhu, J., Liu, X.: New insight into the mechanism for the excellent gas properties of poly(ethylene 2,5-furandicarboxylate) (PEF): role of furan ring's polarity. Eur. Polym. J. **118**, 642–650 (2019). https://doi.org/10.1016/j.eurpolymj.2019.06.033
106. Swatloski, R.P., Spear, S.K., Holbrey, J.D., Rogers, R.D.: Dissolution of cellose with ionic liquids. J. Am. Chem. Soc. **124**(18), 4974–4975 (2002). https://doi.org/10.1021/ja025790m
107. Szosland, L.: A simple method for the production of chitin materials from the chitin ester derivatives. Adv. Chitin Sci. **I**, 297–302 (1996)
108. Szosland, L., Janowska, G.: The method of preparation of dibutyrylchitin (Patent 169077 B1) (1996)
109. Takarada, W., Sugimoto, K., Nakajima, H., Visser, H.A., Gruter, G.-J.M., Kikutani, T.: Melt-spun fibers from bio-based polyester–fiber structure development in high-speed melt spinning of poly(ethylene 2,5-furandicarboxylate) (PEF). Materials **14**(5), 1172 (2021). https://doi.org/10.3390/ma14051172
110. Textiles and the Environment: The Role of Design in Europe's Circular Economy. European Environment Agency (2022)
111. Tokiwa, Y., Calabia, B.P.: Biodegradability and biodegradation of poly(lactide). Appl. Microbiol. Biotechnol. **72**(2), 244–251 (2006). https://doi.org/10.1007/s00253-006-0488-1
112. Vaidya, U.R., Nadkarni, V.M.: Polyester polyols from glycolyzed PET waste: effect of glycol type on kinetics of polyesterification. J. Appl. Polym. Sci. **38**(6), 1179–1190 (1989). https://doi.org/10.1002/app.1989.070380615
113. Vancov, T., Alston, A.-S., Brown, T., McIntosh, S.: Use of ionic liquids in converting ligno-cellulosic material to biofuels. Renew. Energy. **45**, 1–6 (2012). https://doi.org/10.1016/j.renene.2012.02.033
114. Veit, D.: Fibers: History, Production, Properties, Market. Springer International Publishing (2022). https://doi.org/10.1007/978-3-031-15309-9
115. Wang, H., Gurau, G., Rogers, R.D.: Ionic liquid processing of cellulose. Chem. Soc. Rev. **41**(4), 1519 (2012). https://doi.org/10.1039/c2cs15311d
116. Wei, Y.C., Hudson, S.M., Mayer, J.M., Kaplan, D.L.: The crosslinking of chitosan fibers. J. Polym. Sci. A Polym. Chem. **30**(10), 2187–2193 (1992). https://doi.org/10.1002/pola.1992.080301013
117. Winnacker, M., Rieger, B.: Biobased polyamides: recent advances in basic and applied research. Macromol. Rapid Commun. **37**(17), 1391–1413 (2016). https://doi.org/10.1002/marc.201600181

118. Xiao, J., Li, P., Zhang, X., Wang, X.: Study on preparation of regenerated cellulose fiber from biomass based on mixed solvents. Materials **17**(4), 819 (2024). https://doi.org/10.3390/ma17040819
119. Youjiang, W. (ed.): Recycling in Textiles, 1st edn. Woodhead Publishing (2006)
120. Zahmani, B.: Carbon Footprint and Energy Use of Textile Recycling Techniques (2012)

Coir Fibers—Production, Properties and Applications

Mohanapriya Venkataraman, Mahalakshmi Subramanian, Jiří Militký, and Yan Wang

1 Introduction

Natural fibers have gained tremendous attention in the field of materials science due to their renewability, biodegradability, low cost, and acceptable mechanical properties. Among various plant fibers, coir fiber is a fruit lignocellulosic fiber, extracted from the husk of coconut (*Cocos nucifera*), stands out because of its unique combination of mechanical strength, high lignin content, and resistance to microbial degradation [11, 53].

There are two types of natural fibers, purpose-grown and produced as a by-product. Grown for the purpose of producing technical products (cotton, flax, hemp, kenaf, etc.). Grown for the purpose of food, and their waste and by-products are used for other applications (coconut (coir), sugarcane, banana, and pineapple).

The term "coir" was first derived from the Tamil word "kayiru," which means "cord or rope," and then from the Malayalam word "kayar," which means "rope" (traditionally, a type of rope is manufactured from the coconut fiber).

M. Venkataraman (✉) · J. Militký
Department of Material Engineering, Faculty of Textile Engineering, Technical University of Liberec, Liberec, Czech Republic
e-mail: mohanapriya.venkataraman@tul.cz

J. Militký
e-mail: jiri.militky@tul.cz

M. Subramanian
NIL Textile, s. r. o., Ostrava, Czech Republic

Y. Wang
College of Textile Science and Engineering, Zhejiang Sci-Tech University, Hangzhou, China

Xiangshan Knitting Research Institute of Zhejiang Sci-Tech University, Xiangshan, China

Y. Wang
e-mail: yan.wang@zstu.edu.cn

J. Militký and M. Venkataraman (eds.), *Fibrous Structures for Sustainable Future*, Engineering Materials, https://doi.org/10.1007/978-981-95-1382-6_2

Coconut waste is a very common biomass resource globally, with an estimated 64 million tons produced annually by over 90 countries. To put that into perspective, it's equivalent to about 10.9 Great Pyramids of Giza in terms of volume. This significant amount of waste presents both challenges and opportunities, particularly regarding environmental sustainability and resource utilization [68]. With annual production of 17.5 Mt., 14.7 Mt., and 11 Mt. of coconuts produced in harvest areas of 2.8, 3.6, and 2.2 million ha, respectively, Indonesia, the Philippines, and India are the top producers.

This chapter offers a comprehensive examination of coir fibers, discussing their sources, physical and chemical characteristics, thermal properties, mechanical behavior, surface treatments, and future possibilities for advanced uses [3, 17, 39, 45, 70].

2 Biology of Coconut Palm

The coconut palm (*Cocos nucifera* L.), a member of the Arecaceae family, is one of the most versatile and economically valuable cultivated tropical tree species. Cultivated for more than 4000 years, coconut palm trees are among the oldest plant species in the world.

Indigenous to the coastal regions of Southeast Asia and the Pacific, the coconut tree thrives in tropical climates and plays a central role in the agricultural systems and economies of many countries, particularly in Asia-Pacific, which accounts for over 80% of global production [7, 36].

Botanically, the coconut palm is a monocotyledon with a single, unbranched stem that can reach heights of 80–100 feet.

Coconut varieties are broadly grouped into tall, dwarf, and hybrid types as in Table 1 and Fig. 1:

- Tall varieties are cross-pollinated and exhibit considerable genetic diversity. They are known for their long lifespan and higher overall yield, although they take longer to bear fruit [12, 46].
- Dwarf varieties are self-pollinating and typically begin fruiting within three years of planting. Their shorter stature makes them easier to manage, and they are mainly cultivated for tender nut water [23].
- Hybrids, resulting from cross-breeding dwarf and tall types, are developed to combine early bearing with high yield and improved resistance to disease and environmental stress [33, 41].

The fruit, a drupe, comprises three layers: the outer smooth exocarp, a fibrous mesocarp, and a hard endocarp protecting the edible endosperm (see Fig. 2). The coconut wastes can be categorized into coco husk, coco peat, coco coir fiber, coco husk pieces, and coco shells (endocarp). This natural architecture not only protects the seed but also enables water dispersal, an adaptation typical of wild coconut types [7].

Table 1 Characteristics of coconut tree types

Characteristics	Typical (tall)	Nana (dwarf)	Hybrid
Height	30 m	10–12 m	20 m
Life cycle	60–80 years	30 years	50 years
Growth	Fast	Slow	Intermediate
Fluorescence time	5–7 years	2–3 years	3–4 years
Applications	Agro-industry/food preparation	Water consumption	Water consumption/agro-industry/food preparation

Reused from Vieira et al. [68] under CC-BY Permission

Fig. 1 Tall and dwarf coconut trees

The soil and environmental conditions are crucial to the successful growth and productivity of coconut palms. They thrive in well-drained soils such as red sandy loam, lateritic, and coastal alluvial types, ideally with a pH between 5.2 and 8.0 and a depth of at least 1.2 m [63]. Adequate moisture through evenly distributed rainfall or irrigation is essential, while poor drainage and waterlogging can lead to diseases like root wilt, which severely affect yield [61].

Fig. 2 Parts of coconut and its wastes (reused from Lédo et al. [34] under CC-BY Permission)

Temperature sensitivity is another critical factor. Coconut palms cannot withstand frost and are adversely affected by prolonged exposure to cold. As such, cultivation is limited to frost-free tropical regions. Additionally, nutrient availability, particularly nitrogen, plays a role in flowering and fruit development [31].

3 Coir Fibers

Coir fibers come from the dense mesocarp layer of the coconut fruit, making up about 35% of the entire fruit's weight [66]. Based on the coconut's ripeness during extraction, the fibers are typically divided into two categories: white coir and brown coir (see Fig. 3). White coir is derived from unripe green coconuts and is noted for its delicate, smooth texture. It is mainly utilized in the production of higher-quality items like brushes and upholstery cushioning. In comparison, brown coir is sourced from fully grown coconuts and has a coarse, rigid, and robust nature, which makes it suitable for load-bearing uses and composite strengthening [54, 58]. The varieties analyzed were the Chowghat Orange Dwarf variety and the Indian Country Coconut variety. Both are categorized as types that produce brown coir, appropriate for applications in structural and material science [63].

Brown coir fibers have a cellulose content of 32–47% and a lignin content of 40–46%. Coir fibers have good resiliency, toughness, and abrasion resistance due to their low cellulose and hemi-cellulose content, and high lignin content. Surface morphological examination through Scanning Electron Microscopy (SEM) showed that untreated coir fibers have a coarse surface laden with impurities, pits, and tyloses, as shown in Fig. 4. Figure 5 shows the cross-section of coir fibers with the hollow structure, cell wall, and lumens. Coir fibers contain around 100–300 elementary fibers.

Technical coconut fibers are usually 14–18 cm long and 130–280 μm thick. The fibers are highly porous and multicellular. They are composed of parallelly arranged short (length from 0.25 to 1.4 mm) and fine (7–19 μm) tubular elementary fibers with a

Fig. 3 Types of coir fibers [63]

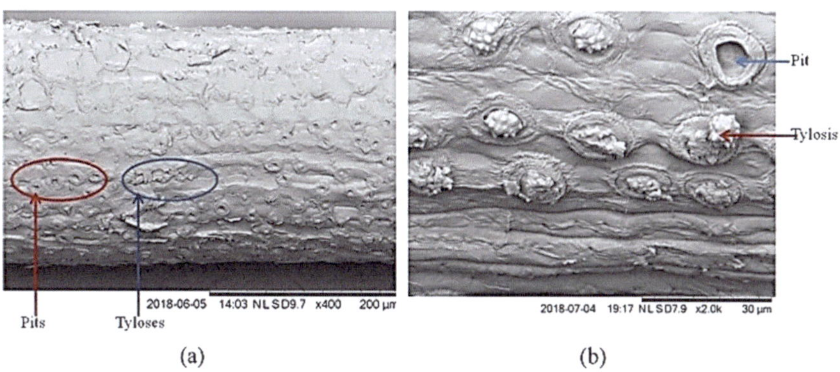

Fig. 4 SEM micrographs surface showing pits and tyloses (reused from Madueke et al. [37] under CC-BY Permission)

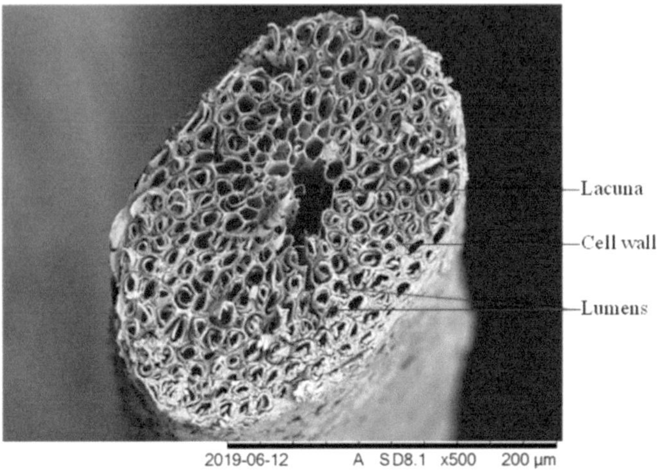

Fig. 5 SEM micrograph of cross-section of coir fiber (reused from Madueke et al. [37] under CC-BY Permission)

wall thickness of about 0.9–1.5 μm and a polygonal cross section embedded in a non-fibrous matrix formed by hemicelluloses and lignin. In the longitudinal direction, the fibers are highly uneven, with waxes on the surface and frequent circular protrusions (termed bulges, tyloses, phytoliths, or volcanoes) and grooves with an outer diameter of about 9–14 μm (see Fig. 4) [72].

There is usually a large lumen in the middle of technical coir. The tubular fiber is formed by a right-handed spiral crystalline ribbon laterally cohered by amorphous cellulose matrix. This is the main reason for the lower modulus of the tubular fiber and also of the coir fibers. The length of the crystalline ribbon can reach 3.58–19.98 mm and their thickness is about 0.64–1.06 μm (see Fig. 6).

4 Extraction of Coir Fibers

Coir fibers are extracted from the fibrous husk that surrounds the coconut fruit, specifically from the mesocarp layer. This husk, comprising up to 35% of the coconut's total mass, contains a mix of long fibers and non-fibrous pith. The maturity of the coconut determines the classification of the fiber: mature coconuts yield coarse brown fibers, while immature green coconuts produce finer white fibers [66] (Fig. 7).

Traditional extraction involves retting, a microbial process where the husks are soaked in water for several weeks to loosen the fibrous material. Although effective, this method results in inconsistent fiber quality and high environmental load due to water contamination [54]. Mechanical decortication has emerged as an alternative that reduces processing time and environmental impact. However, it often results in broken fibers and reduced aspect ratio, impacting reinforcement quality

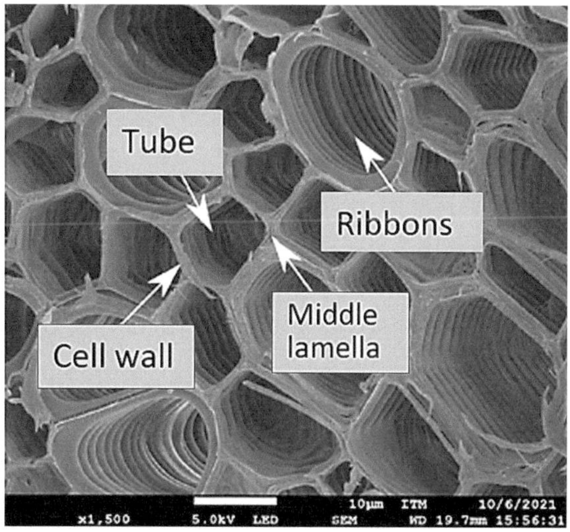

Fig. 6 Parts of coir fiber cross section after alkali treatment (reused from Barrera-Fajardo et al. [5] under CC-BY Permission)

Traditional	**Mechanical**	**Green Decortication and Microbial Treatments**
• Retting techniques are generally used in ponds of brackish waters (3-6 months). • Requires 10-12 months of anaerobic (bacterial) fermentation. • Laborious and time-consuming process. • Retting softens the husks and allows decortication. The fibre is extracted by beating by hand. • After hacking, washing, and drying (in the shade), the fibres are loosened manually and cleaned. • Traditional practices of this kind yield the highest quality of (white) fibre for spinning and weaving.	• Mechanical processes using either defibering or decorticating equipment after only five days of immersion in water tanks. • The stronger fibres are washed, cleaned, dried, hackled, and combed.	• Novel environmentally friendly methods for fibre productionMethod 1: Using microbial enzymes (biotechnological approach) to reduce the retting time (5 days) and increase quality. • Method 2: Specific (microbial) ligninolytic enzymes (laccase/phenoloxidase) can bleach (or activate) the fibre surface to react more easily with the dyes. • **Outputs**: Green and improved production, Improved fiber quality, Enhanced surface properties and porosity, and environmentally safe (Khan et al., 2022).

Fig. 7 Extraction methods of coir fibers

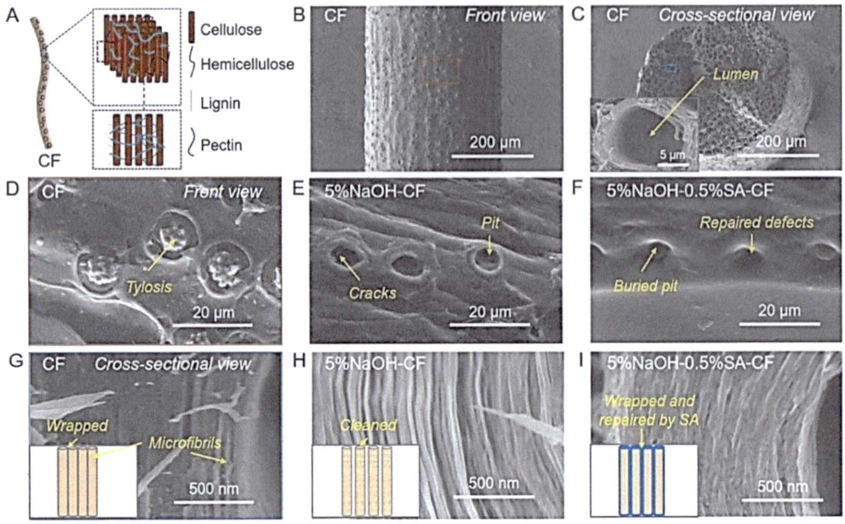

Fig. 8 SEM micrograph of cross-section of coir fiber processed in alkalis (reused from Xiong et al. [69] under CC-BY Permission)

in composites [24]. Green decortication of coir and microbial treatments offer environmentally friendly methods for producing high-quality fibers. This involves using specific microbial enzymes to break down the non-fibrous components, reducing retting time and producing smoother, more porous fibers [29] (see Fig. 8).

Post-extraction processing steps such as combing, grading, and drying are crucial to achieve uniformity and reduce microbial contamination. These steps improve fiber orientation and length distribution, which directly influence the mechanical behavior of fiber-reinforced composites [8].

The effectiveness of the extraction process also affects chemical composition. Poor retting or incomplete decortication may leave behind lignin, pectin, or hemicellulose, hindering interfacial bonding in composite matrices. Hence, optimizing extraction protocols is essential to achieve desired performance characteristics.

The final step of fiber extraction is use of alkalis for partial removal of hemicellulose and lignin, activating surface and extending interfacial area. Surface morphology is a key factor governing interfacial adhesion in fiber-reinforced systems. Scanning Electron Microscopy (SEM) analysis has revealed that untreated coir fibers generally have a rough surface covered with waxes and impurities, which reduce bonding efficiency. George et al. [18] observed that alkali-treated coir fibers show significant improvement in surface quality, with exposed fibrils and fewer impurities, thereby enhancing mechanical interlocking with matrix materials shown in Fig. 8.

Supporting this, Sgriccia et al. [59] emphasized that surface purification increases fiber consistency and fiber–matrix interaction. Further, Ramesh et al. [51] found that coir fibers that underwent size homogenization and chemical cleaning demonstrated better mechanical integrity and interfacial strength when used in thermoset composites.

5 Physical Properties

The mechanical performance of fibers used as reinforcements is directly affected by their physical structure and morphology. Coir fibers display considerable diversity in diameter, density, and surface traits based on their origin and processing circumstances (see Table 2). The typical diameter of Chowghat Orange Dwarf fibers was recorded at 0.25 mm, whereas Indian Country Coconut fibers showed a marginally smaller diameter of 0.23 mm [63]. Elevated aspect ratios, surpassing 580 in each instance, are especially beneficial for composite reinforcement, as they facilitate efficient load distribution between the matrix and the fiber [58].

These surface characteristics impede successful adhesion to polymer matrices. After being treated with alkali solutions such as sodium hydroxide (NaOH) and sodium sulfite (Na_2SO_3), the fibers showed notably smoother and cleaner surfaces, revealing cellulose microfibrils and improving interfacial adhesion in composites [8, 18, 38, 59].

According to Bledzki and Gassan [9], the density of coir fibers typically ranges between 1.15 and 1.25 g/cm^3. This relatively low density, combined with their high lignin content (~ 40–45%), makes them highly suitable for lightweight structural applications and provides good resistance to moisture and microbial attack. The average fiber diameter varies based on variety, maturity, and processing technique, ranging from 100 to over 450 μm. Satyanarayana et al. [58] reported that this variability directly impacts fiber performance and uniformity in composites. High aspect ratios, often exceeding 500, enable more effective load transfer from matrix to fiber, enhancing composite strength and stiffness.

Table 2 Physical properties of coir fiber

Physical properties	Coir fiber
Density (g/cm^3)	1.1–1.5
Microfibrillar angle of secondary cell wall (°)	30–49
Diameter (μm)	50–300
Lumen width (μm)	11.75–13.59
Tensile strength (MPa)	105–593
Initial modulus (GPa)	2–8
Elongation at break (%)	3–40
Aspect ratio (L/D)	Between 500 and 1380

In summary, coir fibers offer promising physical characteristics including low density, acceptable strength-to-weight ratio, and large aspect ratio, all of which are desirable in bio-composite manufacturing. However, to achieve consistent performance, especially in structural applications, fiber quality control and surface modification are essential to reduce variability and optimize adhesion.

6 Chemical Composition

Coir fibers mainly consist of lignocellulosic substances, specifically cellulose, hemicellulose, and lignin [9], as well as minor components like pectin, ash, surface waxes, traces of silica, and water, as shown in Table 3. The high lignin content contributes to the fiber's stiffness, longevity, and defense against microbial assaults, setting coir apart from other natural fibers like jute or flax.

Fourier Transform Infrared Spectroscopy (FTIR) assessment provides important information regarding the chemical composition of coir fibers (see Fig. 9). Notable spectral peaks identified consist of a wide absorption near 3340 cm^{-1} linked to O–H stretching vibrations from hydroxyl groups in cellulose and hemicellulose. Peaks around 1735 cm^{-1} are associated with the C=O stretching vibrations of ester linkages primarily derived from hemicellulose. The peaks detected around 1515 cm^{-1} are associated with aromatic skeletal vibrations that signify the presence of lignin [63]. Following chemical treatment, there was a decrease in the intensity of bands associated with hemicellulose, indicating the successful elimination of amorphous elements and the refinement of the cellulose structure [1, 56].

The chemical processing of lignocellulosic materials enhances their surface characteristics, alters surface chemical groups, and leads to structural changes. These techniques typically rely on the application of reagent functional groups that can interact with lignocellulose structures and alter their composition [42, 43, 60].

According to Narendar and Priya Dasan [43], during sulfuric acid treatment, hydrogen ions infiltrate the amorphous structure of lignocelluloses, aiding in the breakdown of glycosidic bonds and releasing the crystallites. Extended exposure to

Table 3 Chemical composition of coir fiber

Chemical compositions	Values (%)
Cellulose content	27–45
Hemicellulose content	0.15–2.5
Lignin content	40–50
Pectin	Around 3
Ash	Approx. 2.22
Moisture content	9–11
Surface waxes	1.3
Silica oxide	Traces

Fig. 9 FT-IR analysis of untreated coir fiber (UCF) and treated coir fiber (OCF) (reused from Ru et al. [56] under CC-BY Permission)

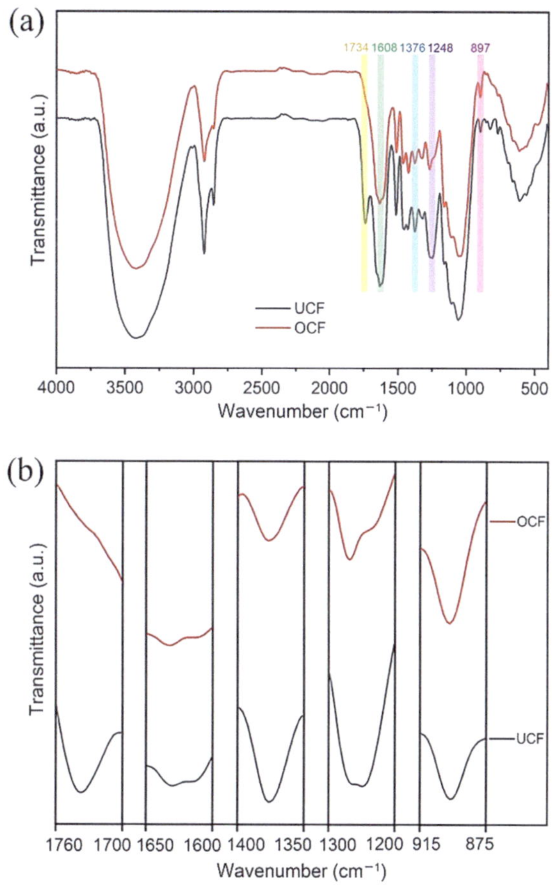

acid leads to the breakdown of amorphous parts and the partial disruption of the crystalline areas in lignocelluloses [53].

Ru et al. demonstrated (see Fig. 10) that the crystalline structure of cellulose I in coir fiber remained unchanged after alkali treatment. Furthermore, the treated coir fiber's XRD spectrum peak is more noticeable, indicating that its crystallinity index is higher than that of the untreated coir fiber. Untreated and treated coir fibers had crystallinity indices of 35.85% and 40.40%, respectively. This shift may be explained by the fact that untreated coir fibers contain high levels of lignin and hemicellulose, while treated coir fibers have some amorphous components like lignin and hemicellulose removed. Cellulose crystals accumulate, and molecules rearrange as a result of this rise in the relative concentration of crystalline cellulose. Thus, compared to untreated coir fibers, treated coir fibers have a higher crystallinity index, indicating that they have superior mechanical qualities [56].

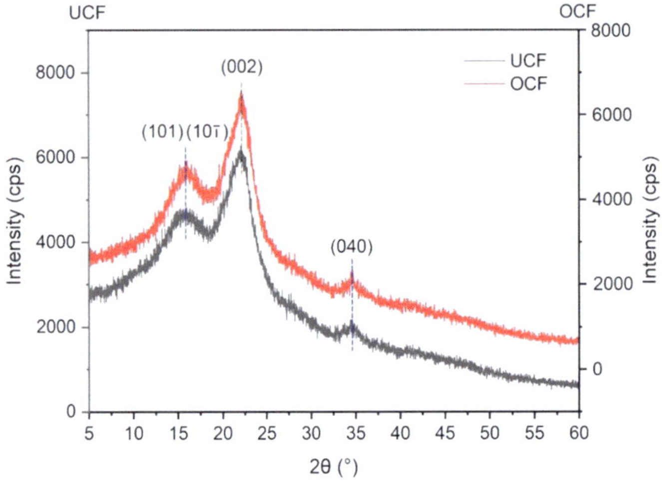

Fig. 10 XRD spectra of untreated (black) and treated (red) coir fibers (reused from Ru et al. [56] under CC-BY Permission)

7 Mechanical Properties

The mechanical properties of coir fibers are essential for their effectiveness as reinforcement in composite uses. Tensile tests performed on both untreated and treated fibers showed a notable effect of surface alteration on the fiber strength and stiffness. The tensile strength and modulus of coir fibers have huge variation based on findings of various authors due to the unknown procedure of the cross-section area calculation, complications in setting a cross-section size of the sample, and no details about the measurement of cross section [67].

The untreated Chowghat Orange Dwarf fibers displayed an average tensile strength of 112 MPa, whereas Indian Country Coconut fibers demonstrated a slightly reduced strength of 105 MPa. Following chemical treatment, the tensile strength rose to about 148 MPa for Chowghat fibers and 139 MPa for Indian Country fibers shown in Fig. 11 [63].

The rise in tensile strength and Young's modulus after chemical treatment can be linked to various reasons, such as the elimination of surface contaminants, improved alignment of cellulose microfibrils, and stronger fiber-matrix adhesion [26]. Treated fibers showed decreased elongation at break, suggesting enhanced stiffness, which is beneficial for structural composite applications that need high modulus materials [1, 9, 26, 40, 48, 52, 55].

Natural fibers such as coir naturally have a hydrophilic quality because of their hydroxyl groups, whereas most polymer matrices tend to be hydrophobic. Their surface area is not sufficient for interfacial bonding with the matrix. This mismatch frequently causes weak fiber-matrix bonding, resulting in subpar mechanical performance in composites. Techniques for the enhancement of accessible surface area by

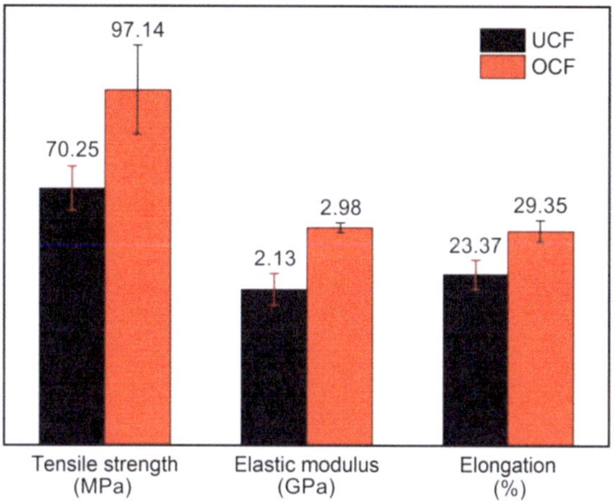

Fig. 11 Mechanical properties of (black) and treated (red) coir fibers (reused from Ru et al. [56] under CC-BY Permission)

surface modification are therefore used to improve the bonding properties of coir fibers. Sodium hydroxide (NaOH) alkali treatment is one of the most widely used techniques for modifying fiber surfaces. The process generally consists of soaking the fibers in a 5–10% NaOH solution at room temperature for a few hours. This procedure results in the elimination of surface contaminants, waxes, pectin, and hemicellulose, which increases surface roughness and improves mechanical interlocking with the matrix [18, 24, 35].

Furthermore, sodium sulfite (Na_2SO_3) treatment was utilized to enhance the purification of the fibers and partially delignify them while preserving the main cellulose structure. Modified fibers also demonstrated a significant enhancement in wettability and tensile strength, essential for the advancement of high-performance composites [47].

The mechanical properties of coir fibers are significantly influenced by their chemical composition, surface morphology, and treatment history. In their raw state, coir fibers exhibit moderate tensile strength, high elongation, and relatively low stiffness due to the presence of amorphous components such as hemicellulose, pectin, and surface waxes. Typical tensile strengths for untreated fibers range between 100–150 MPa, with initial modulus values in the range of 4–6 GPa and elongation at break exceeding 25% [9]. These properties reflect coir's natural toughness and ductility, largely attributed to its high lignin content, which can account for up to 45% of its mass.

Chemical surface treatments, particularly alkaline treatment using sodium hydroxide (NaOH), have been widely employed to enhance the mechanical performance of coir fibers. Alkali treatment induces partial delignification and removes surface impurities, waxes, and hemicellulose, thereby increasing surface roughness and improving fiber–matrix adhesion.

According to George et al. [18], tensile strength increased by 25–45% after 5% NaOH treatment, while initial modulus also improved significantly. The removal of non-cellulosic components aligns the microfibrils and increases crystallinity, which in turn enhances stiffness [35]. Post-treatment tensile strengths as high as 180–220 MPa and modulus values above 6 GPa have been reported for brown coir fibers [25]. However, this improved strength and stiffness are often accompanied by a reduction in elongation at break, sometimes dropping to as low as 10%, due to increased brittleness caused by higher crystallinity and microfibril alignment.

Beyond simple alkali treatment, combined chemical methods have also been explored. Dewaxing treatments using ethanol-toluene solutions before alkali exposure remove the superficial waxy layer more effectively, allowing deeper penetration of alkaline agents. Joseph et al. [26] observed that this dual-step treatment enhanced tensile strength by up to 40% and increased modulus by nearly 60%, while also improving fiber–matrix bonding. Similarly, Sgriccia et al. [59] reported that dewaxed and treated coir fibers showed reduced interfacial failure under load in composite applications.

Partial bleaching and sodium sulfite (Na_2SO_3) treatments have been investigated to selectively reduce lignin without fully compromising the fiber integrity. These treatments can enhance flexibility while still moderately improving mechanical performance. Rout et al. [54] reported that lightly sulfite-treated fibers showed tensile strengths of up to 170 MPa and better stiffness compared to fully delignified samples. However, excessive bleaching may deteriorate the cellulose network, leading to decreased tensile performance. Pickering et al. [47] emphasized that maintaining an optimal balance between lignin reduction and cellulose preservation is crucial for retaining mechanical integrity during treatment.

Advanced surface modifications incorporating silane coupling agents post-alkali treatment have shown further enhancement in mechanical characteristics. These silanes form covalent bonds between hydroxyl groups on the fiber and reactive groups in the matrix, thus improving stress transfer across the interface. John and Thomas [24] observed that tensile strength could be further increased by 30–35% with silane treatment, especially when used in epoxy or polyester matrices. Sreekumar et al. [62] additionally demonstrated that silane-treated fibers exhibited better moisture resistance and dimensional stability, leading to higher long-term retention of mechanical performance in humid environments.

Overall, the mechanical behavior of coir fibers can be substantially optimized through chemical surface modification, with alkali and hybrid treatments consistently showing marked improvements in tensile strength and modulus. However, treatment severity, time, and concentration must be carefully controlled to avoid degradation of the fiber's structural integrity.

The variation in results across different studies also reflects the influence of coir source, fiber maturity, and environmental conditions, necessitating standardized protocols for consistent performance evaluation.

8 Thermal Characteristics and Stability

Thermal stability is an essential factor for fiber-reinforced composites, particularly when they are anticipated to experience higher temperatures during use.

8.1 *Thermogravimetric Assessment (TGA)*

Thermogravimetric analysis (TGA) has consistently been used to characterize the degradation stages of natural fibers, including coir. Thermogravimetric analysis indicated that the initial weight reduction below 100 °C is mainly caused by the evaporation of moisture. Significant thermal degradation of hemicellulose and cellulose took place between 220 and 350 °C. A subsequent stage above 400 °C was associated with the breakdown of the more thermally stable lignin component [63]. Treated fibers exhibited a greater onset degradation temperature than untreated fibers, suggesting enhanced thermal resistance due to the purification and stabilization of the fiber structure (see Fig. 12) [30, 49, 71].

These findings were corroborated by Yang et al. [71], where they emphasized the wide degradation range of lignin, contributing to the broad and gradual mass loss of lignocellulosic materials.

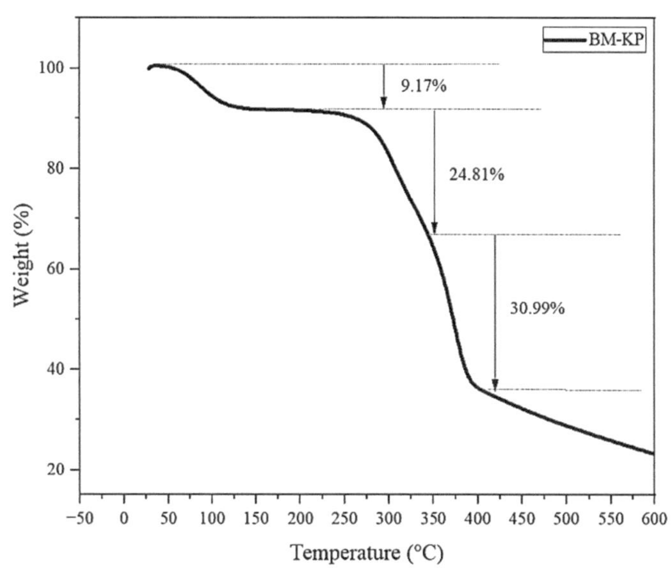

Fig. 12 Mass loss of coir fibers [63]

8.2 Differential Scanning Calorimetry (DSC)

Differential Scanning Calorimetry (DSC) complements TGA by revealing enthalpic changes associated with thermal events. Subramanian [63] had results, showing endothermic peaks associated with moisture loss occurring below 100 °C and exothermic degradation transitions occurring between 220 and 400 °C. Treated fibers displayed a shift of decomposition peaks to higher temperatures, indicating improved thermal stability from the chemical treatments.

George et al. [18] reported endothermic peaks at temperatures below 100 °C, consistent with water desorption, and exothermic transitions between 250 and 400 °C corresponding to the decomposition of cellulose and lignin components. Importantly, they observed that alkali-treated fibers exhibited higher thermal stability than untreated ones, as reflected in delayed onset of exothermic peaks.

Khedari et al. [30] demonstrated that alkali-treated coir-based insulation panels retained their structural form and thermal resistivity up to 300 °C, making them promising candidates for non-structural thermal insulation applications.

A deeper examination of fiber structure and crystallinity conducted by Poletto et al. [48], revealed that chemically treated coir fibers possess enhanced thermal resistance due to increased cellulose crystallinity and removal of amorphous fractions. Their findings affirm the notion that surface and bulk chemical composition strongly influences thermal degradation behavior.

In conclusion, coir fibers exhibit distinct and predictable thermal degradation profiles that are strongly influenced by chemical treatment. Alkali and bleaching processes not only improve thermal resistance but also increase the structural stability of the fiber during composite processing. However, overtreatment can lead to fiber embrittlement or degradation of the cellulose backbone. Therefore, optimizing treatment parameters is essential to preserve the delicate balance between thermal enhancement and mechanical integrity in coir-reinforced composite systems.

9 Sound Absorption of Materials Based on Coir

Materials that have pores are commonly utilized as sound-absorbing agents in noise management engineering. They are made up of multiple synthetic substances such as glass wool, minerals, and agricultural-based foams and fibers. Delany and Bazley [13] examined the impedance and wave propagation characteristics of fibrous substances. They standardized these characteristics into dimensionless categories and created a straightforward empirical method that remains an effective and quick approximation to the solution. Dunn and Davern [14] employed the same technique to establish empirical connections between the characteristic impedance and propagation constant of porous foam substances and multilayer absorbers. Qunli [50] collected an extensive set of experimental data and computed those relationships for foam materials exhibiting a broader spectrum of flow resistivity.

The effective functioning of sound absorbers and the attainment of acoustic comfort rely on a significant and efficient characteristic known as "material porosity." This inherent characteristic arrangement in fibrous sound absorbers (whether natural or synthetic) can convert sound energy into heat and scatter sound waves. The exceptional characteristics of natural lignocellulosic fibers, including biodegradability, lightweight, low density, affordability, and non-toxicity, along with their use as substitutes for synthetic fiber sound absorbers, have drawn considerable attention from researchers in acoustics [2, 6, 64, 65].

Coconut coir fiber possesses a high lignin content, which renders the fibers more rigid and durable, offers an air porosity of 95%, acts as a heat retardant, is biodegradable, and is viewed as a renewable resource. The absorption of sound by porous materials is grounded in the principle of converting sound energy into thermal energy. Typically, common porous materials like fibrous mineral wool and glass fiber primarily achieve acoustic absorption through viscous losses as air flows through their pores. Porous substances can offer greater absorption when situated in areas where the sound wave's particle velocity is high. In porous materials, the speed of sound is less than in air, and the wavelength gets reduced. Typically, the absorption coefficient is small at low frequencies, whereas at high frequencies, it increases, especially at frequencies where the material thickness is equal to $\lambda/4$. The greater the material's thickness, the more effective the absorption at lower frequencies as shown in Fig. 13. To achieve better absorption at low frequencies, it is necessary to include an air space. Every substance absorbs sound to a certain degree. Nonetheless, when sound is viewed as wave motion, it becomes essential to apply the concept of acoustic impedance [64]. Fibrous sound-absorbing materials, based on their composition, contain various open and closed cells (channels) that facilitate sound absorption across different frequency ranges [15, 19, 64].

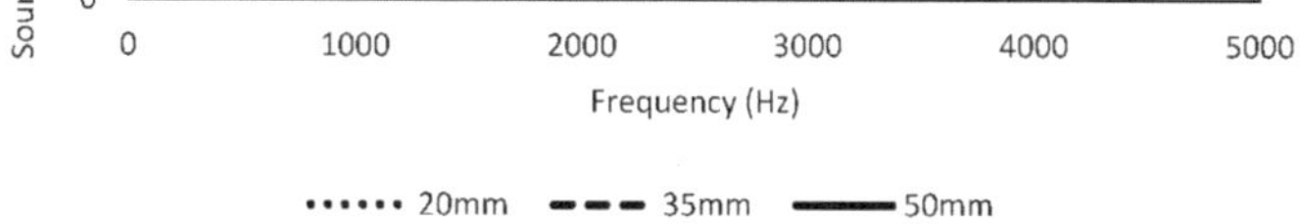

Fig. 13 Sound absorption coefficient of coir fiber according to thickness (reused from Jang [20] under CC-BY Permission)

10 Other Applications of Coconut Products

The unique characteristics render coir fibers ideal for various applications, from conventional ones like ropes and mats to modern uses like reinforcements in composite materials. As environmental issues grow and the focus on sustainable development intensifies, coir fibers are becoming a viable substitute for synthetic fibers.

Coir fibers, because of their well-rounded mechanical characteristics and ecological advantages, have been widely utilized in creating composite materials for the automotive, construction, and packaging sectors. In polymeric composites, coir fibers serve as reinforcements in matrices like polyester, polypropylene, epoxy, and more recently, in geopolymer matrices. Their elevated aspect ratio and adaptability enable the production of lightweight composites that offer improved impact resistance and acoustic insulation [54].

Composites reinforced with coir fiber display excellent flexural strength, durability, and energy absorption properties, rendering them appropriate for non-structural and semi-structural applications like door panels, roofing materials, insulation boards, and lightweight partitions. Recent developments have investigated the incorporation of chemically modified coir fibers in geopolymer composites for eco-friendly construction purposes [10, 16, 22].

Figure 14 displays a flowchart of the main products that can be made from green coconut shells. After the shells have been crushed and pressed, Products that use simply the fibers or both the fiber and coconut powder can be made. Briquettes, art supplies, and gardening items are examples of goods that use both the fiber and the powder.

The fruit's mesocarp and endocarp serve as the fiber and powder that make up the green coconut shell. Crushing and pressing shells can extract a liquid known as Green Coconut Shell Liquid (LCCV). This liquid can be utilized for fertilizing crops, fermentative processes, phenolic resin formation, and biogas generation [45]. When LCCV is not employed in other processes, it can be securely disposed of by electrocoagulation. This method is less expensive and more efficient than the conventional physical-chemical treatment, costing 0.36 (R$/L) as opposed to 1.58 (R$/L) for chemical treatments [45].

The coconut shells are torn, dried, and sieved to extract the fiber and powder of the green coconut. Due to its numerous applications, affordability, renewable supply, and customers' current concerns about environmentally friendly products, coconut fiber and powder are becoming more and more popular in both domestic and foreign markets [45].

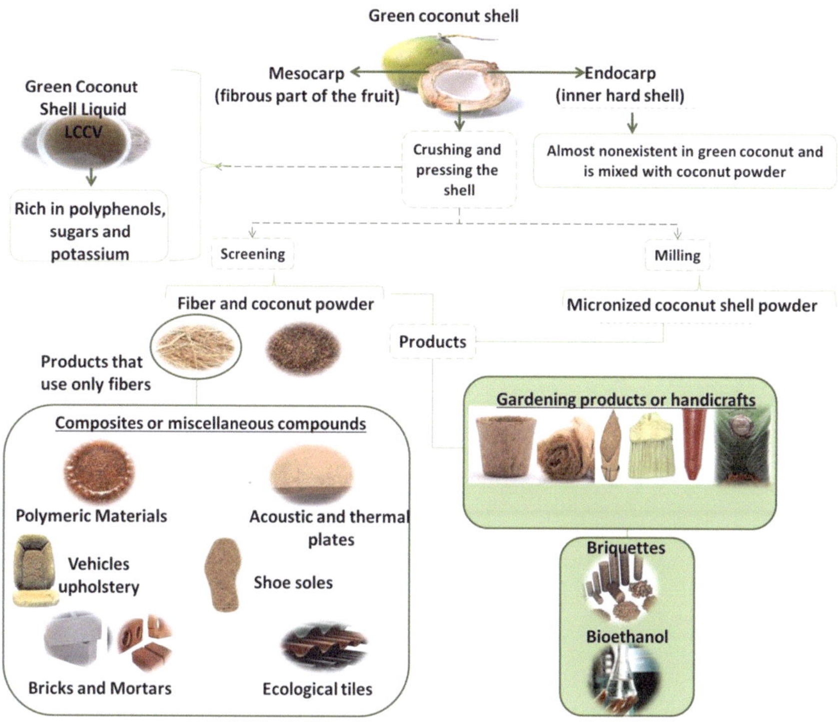

Fig. 14 Application of coconut parts (reused from Nunes et al. [45] under CC-BY Permission)

11 Ecological Effects and Sustainability

The rising need for sustainable and eco-friendly materials has made coir fibers an appealing choice for reinforcement. Coir manufacturing is an energy-efficient, low-emission procedure mainly performed in tropical nations with limited industrial resources. In contrast to synthetic fibers, coir fibers are fully biodegradable and do not present any long-term environmental risks when discarded. Moreover, the conversion of agricultural waste like coconut husks into valuable fiber products greatly aids in waste minimization, circular economy frameworks, and job creation in rural areas. The carbon footprint linked to coir fiber processing is significantly smaller than that of synthetic fiber manufacturing, positioning coir as an essential element in creating environmentally friendly composites. Even with these benefits, issues like fluctuations in fiber quality, sensitivity to moisture, and restricted thermal stability in extreme conditions continue to be topics for future study and advancement [27, 32, 40, 44].

12 Challenges and Research Gaps

While coir fibers present many benefits, there are various obstacles that need to be overcome to fully utilize their capabilities in advanced uses. Main obstacles consist of:

- Moisture Absorption: Due to their hydrophilic characteristics, coir fibers typically soak up moisture, resulting in dimensional instability and a decline in mechanical properties as time progresses.
- Variability: The inherent differences in fiber size, strength, and chemical makeup influence the uniformity of composite characteristics.
- Thermal Stability: While coir fibers exhibit greater thermal stability compared to numerous other plant fibers, their application remains restricted in high-temperature settings.
- Fiber-Matrix Compatibility: Additional studies on surface modification methods and the creation of compatibilizers are essential to improve interfacial adhesion [4, 21, 28, 57].

13 Conclusion

Coir fibers, characterized by their unique combination of mechanical strength, high lignin content, biodegradability, and affordability, are a promising resource for sustainable composite creation. Chemical and physical processes greatly improve their characteristics, rendering them appropriate for various industrial uses.

A thorough understanding of their physical, chemical, thermal, and mechanical properties, as outlined in this chapter, lays the groundwork for deeper investigation into advanced composites, especially in eco-friendly industries. Through continuous research and technological progress, coir fibers are set to take on a more substantial role in the future of sustainable materials science.

References

1. Abraham, E., Deepa, B., Pothan, L., Thomas, S., John, M., Anandjiwala, R., Cvelbar, U.: Extraction of nanocellulose fibrils from lignocellulosic fibres: a novel approach. Carbohydr. Polym. **86**, 1468–1475 (2011). https://doi.org/10.1016/j.carbpol.2011.06.034
2. Amares, S., Sujatmika, E., Hong, T.W., Durairaj, R., Hamid, H.S.H.B.: A review: characteristics of noise absorption material. J. Phys. Conf. Ser. **908**(1), 012005 (2017). https://doi.org/10.1088/1742-6596/908/1/012005
3. Ardanuy, M., Claramunt, J., Toledo Filho, R.D.: Cellulosic fiber reinforced cement-based composites: a review of recent research. Constr. Build. Mater. **79**, 115–128 (2015). https://doi.org/10.1016/j.conbuildmat.2015.01.035
4. Ashori, A.: Wood–plastic composites as promising green-composites for automotive industries! Bioresour. Technol. **99**(11), 4661–4667 (2008). https://doi.org/10.1016/j.biortech.2007.09.043

5. Barrera-Fajardo, I., Rivero-Romero, O., Unfried-Silgado, J.: Investigation of the effect of chemical treatment on the properties of colombian banana and coir fibers and their adhesion behavior on polylactic acid and unsaturated polyester matrices. Fibers **12**(1), 6 (2024)
6. Berardi, U., Iannace, G.: Acoustic characterization of natural fibers for sound absorption applications. Build. Environ. **94**, 840–852 (2015). https://doi.org/10.1016/j.buildenv.2015.05.029
7. Beveridge, F.C., Kalaipandian, S., Yang, C., Adkins, S.W.: Fruit biology of coconut (*Cocos nucifera* L.). Plants **11**(23), 3293 (2022)
8. Bismarck, A., Lampke, T.: Plant fibers as reinforcement for green composites. Nat. Fibers Bioploymers Biocomposites (2005). https://doi.org/10.1201/9780203508206.ch2
9. Bledzki, A.K., Gassan, J.: Composites reinforced with cellulose based fibres. Prog. Polym. Sci. **24**(2), 221–274 (1999). https://doi.org/10.1016/S0079-6700(98)00018-5
10. Bledzki, A.K., Mamun, A.A., Faruk, O.: Abaca fibre reinforced PP composites and comparison with jute and flax fibre PP composites. Express Polym. Lett. **1**(11), 755–762 (2007)
11. Brígida, A.I.S., Calado, V.M.A., Gonçalves, L.R.B., Coelho, M.A.Z.: Effect of chemical treatments on properties of green coconut fiber. Carbohydr. Polym. **79**(4), 832–838 (2010). https://doi.org/10.1016/j.carbpol.2009.10.005
12. Chan, E., Elevitch, C.R.: *Cocos nucifera* (coconut). Species Profiles Pacific Island Agrofor. **2**(1), 1–27 (2006)
13. Delany, M.E., Bazley, E.N.: Acoustical properties of fibrous absorbent materials. Appl. Acoust. **3**(2), 105–116 (1970). https://doi.org/10.1016/0003-682X(70)90031-9
14. Dunn, I.P., Davern, W.A.: Calculation of acoustic impedance of multi-layer absorbers. Appl. Acoust. **19**(5), 321–334 (1986). https://doi.org/10.1016/0003-682X(86)90044-7
15. Elwaleed, A.K., Nikabdullah, N., Nor, M.J.M., Tahir, M.F.M., Zulkifli, R.: Experimental investigation of sound absorption properties of perforated date palm fibers panel. IOP Conf. Ser. Mater. Sci. Eng. **46**(1), 012027 (2013). https://doi.org/10.1088/1757-899X/46/1/012027
16. Faruk, O., Bledzki, A., Fink, H., Sain, M.: Progress report on natural fiber reinforced composites. Macromol. Mater. Eng. **299** (2014). https://doi.org/10.1002/mame.201300008
17. Faruk, O., Bledzki, A.K., Fink, H.-P., Sain, M.: Biocomposites reinforced with natural fibers: 2000–2010. Prog. Polym. Sci. **37**(11), 1552–1596 (2012). https://doi.org/10.1016/j.progpolymsci.2012.04.003
18. George, J., Sreekala, M.S., Thomas, S.: A review on interface modification and characterization of natural fiber reinforced plastic composites. Polym. Eng. Sci. **41**(9), 1471–1485 (2001)
19. Hosseini Fouladi, M., Ayub, M., Jailani Mohd Nor, M.: Analysis of coir fiber acoustical characteristics. Appl. Acoust. **72**(1), 35–42 (2011). https://doi.org/10.1016/j.apacoust.2010.09.007
20. Jang, E.-S.: Sound absorbing properties of selected green material—a review. Forests **14**(7), 1366 (2023)
21. Jawaid, M., Abdul Khalil, H.P.S., Abu Bakar, A.: Woven hybrid composites: tensile and flexural properties of oil palm-woven jute fibres based epoxy composites. Mater. Sci. Eng. A **528**(15), 5190–5195 (2011). https://doi.org/10.1016/j.msea.2011.03.047
22. Jawaid, M.H.P.S., Khalil, H.A.: Cellulosic/synthetic fibre reinforced polymer hybrid composites: a review. Carbohydr. Polym. **86**, 1–18 (2011). https://doi.org/10.1016/j.carbpol.2011.04.043
23. Jayasekara, C., Amarasinghe, N.: Coir–coconut cultivation, extraction and processing of coir. In: Industrial Applications of Natural Fibres: Structure, Properties and Technical Applications, pp. 197–217 (2010)
24. John, M.J., Thomas, S.: Biofibres and biocomposites. Carbohydr. Polym. **71**(3), 343–364 (2008). https://doi.org/10.1016/j.carbpol.2007.05.040
25. Joseph, K., Thomas, S., Pavithran, C.: Effect of chemical treatment on the tensile properties of short sisal fibre-reinforced polyethylene composites. Polymer **37**(23), 5139–5149 (1996)
26. Joseph, P.V., Joseph, K., Thomas, S.: Effect of processing variables on the mechanical properties of sisal-fiber-reinforced polypropylene composites. Compos. Sci. Technol. **59**(11), 1625–1640 (1999). https://doi.org/10.1016/S0266-3538(99)00024-X

27. Joshi, S.V., Drzal, L.T., Mohanty, A.K., Arora, S.: Are natural fiber composites environmentally superior to glass fiber reinforced composites? Compos. A Appl. Sci. Manuf. **35**(3), 371–376 (2004). https://doi.org/10.1016/j.compositesa.2003.09.016
28. Kabir, M.M., Wang, H., Lau, K.T., Cardona, F.: Chemical treatments on plant-based natural fibre reinforced polymer composites: an overview. Compos. Part B **43**(7), 2883–2892 (2012). https://doi.org/10.1016/j.compositesb.2012.04.053
29. Khan, T., Singh, B., Ahmad, K.A., Pai, R.: Extraction of coir fibers by different methods. In: Coir Fiber and Its Composites, pp. 19–42. Elsevier (2022)
30. Khedari, J., Charoenvai, S., Hirunlabh, J.: New insulating particleboards from durian peel and coconut coir. Build. Environ. **38**(3), 435–441 (2003)
31. Kumar, S.N., Bai, K.K., Rajagopal, V., Aggarwal, P.K.: Simulating coconut growth, development and yield with the InfoCrop-coconut model. Tree Physiol. **28**(7), 1049–1058 (2008)
32. La Mantia, F.P., Morreale, M.: Green composites: a brief review. Compos. A Appl. Sci. Manuf. **42**(6), 579–588 (2011). https://doi.org/10.1016/j.compositesa.2011.01.017
33. Lantican, D.V., Strickler, S.R., Canama, A.O., Gardoce, R.R., Mueller, L.A., Galvez, H.F.: De novo genome sequence assembly of dwarf coconut (*Cocos nucifera* L.'Catigan green dwarf') provides insights into genomic variation between coconut types and related palm species. G3: Genes, Genomes Genet. **9**(8), 2377–2393 (2019)
34. Lédo, A.D.S., Passos, E.E.M., Fontes, H.R., Ferreira, J.M.S., Talamini, V., Vendrame, W.A.: Advances in coconut palm propagation. Rev. Bras. Frutic. **41**, e-159 (2019)
35. Li, X., Tabil, L.G., Panigrahi, S.: Chemical treatments of natural fiber for use in natural fiber-reinforced composites: a review. J. Polym. Environ. **15**, 25–33 (2007)
36. Louis, K.J., Raloul, S.S., Pierre, N.G.S., Thierry, L.T., Kouassi, A.L.L.O.U.: Assessment of vegetative growth and production of new improved coconut (*Cocos nucifera* L.) hybrids. J. Appl. Biosci. **26**, 1664–1674 (2010)
37. Madueke, C.I., Kolawole, F., Tile, J.: Property evaluations of coir fibres for use as reinforcement in composites. SN Appl. Sci. **3**, 1–11 (2021)
38. Mbeche, S.M., Wambua, P.M., Githinji, D.N.: Mechanical properties of sisal/cattail hybrid-reinforced polyester composites. Adv. Mater. Sci. Eng. **2020**(1), 6290480 (2020)
39. Mohanty, A.K., Misra, M., Drzal, L.T.: Sustainable bio-composites from renewable resources: opportunities and challenges in the green materials world. J. Polym. Environ. **10**(1), 19–26 (2002). https://doi.org/10.1023/A:1021013921916
40. Mohanty, A.K., Misra, M., Drzal, L.T. (eds.): Natural Fibers, Biopolymers, and Biocomposites. CRC Press (2005). https://doi.org/10.1201/9780203508206
41. Nair, K.P., Nair, K.P.: Technological Advancements in coconut, arecanut and cocoa research: a century of service to the global farming community by the Central Plantation Crops Research Institute, Kasaragod, Kerala State, India. In: Tree Crops: Harvesting Cash from the World's Important Cash Crops, pp. 377–536 (2021)
42. Nam, T.H., Ogihara, S., Tung, N.H., Kobayashi, S.: Effect of alkali treatment on interfacial and mechanical properties of coir fiber reinforced poly(butylene succinate) biodegradable composites. Compos. Part B **42**(6), 1648–1656 (2011). https://doi.org/10.1016/j.compositesb.2011.04.001
43. Narendar, R., Priya Dasan, K.: Chemical treatments of coir pith: morphology, chemical composition, thermal and water retention behavior. Compos. Part B **56**, 770–779 (2014). https://doi.org/10.1016/j.compositesb.2013.09.028
44. Netravali, A.N., Chabba, S.: Composites get greener. Mater. Today **6**(4), 22–29 (2003). https://doi.org/10.1016/S1369-7021(03)00427-9
45. Nunes, L.A., Silva, M.L., Gerber, J.Z., Kalid, R. de A.: Waste green coconut shells: diagnosis of the disposal and applications for use in other products. J. Clean. Prod. **255**, 120169 (2020)
46. Perera, L., Russell, J.R., Provan, J., Powell, W.: Studying genetic relationships among coconut varieties/populations using microsatellite markers. Euphytica **132**, 121–128 (2003)
47. Pickering, K.L., Efendy, M.A., Le, T.M.: A review of recent developments in natural fibre composites and their mechanical performance. Compos. A: Appl. Sci. Manuf. **83**, 98–112 (2016)

48. Poletto, M., Ornaghi Junior, H.L., Zattera, A.J.: Native cellulose: structure, characterization and thermal properties. Materials **7**(9), 6105–6119 (2014)
49. Poletto, M., Pistor, V., Zattera, A.J.: Thermal properties of native cellulose. Cellul.: Fundam. Asp. **45** (2013)
50. Qunli, W.: Empirical relations between acoustical properties and flow resistivity of porous plastic open-cell foam. Appl. Acoust. **25**(3), 141–148 (1988). https://doi.org/10.1016/0003-682X(88)90090-4
51. Ramesh, M., Palanikumar, K., Reddy, K.H.: Mechanical property evaluation of sisal–jute–glass fiber reinforced polyester composites. Compos. Part B **48**, 1–9 (2013)
52. Reddy, N., Yang, Y.: Biofibers from agricultural byproducts for industrial applications. Trends Biotechnol. **23**(1), 22–27 (2005). https://doi.org/10.1016/j.tibtech.2004.11.002
53. Rosa, M.F., Medeiros, E.S., Malmonge, J.A., Gregorski, K.S., Wood, D.F., Mattoso, L.H.C., Glenn, G., Orts, W.J., Imam, S.H.: Cellulose nanowhiskers from coconut husk fibers: effect of preparation conditions on their thermal and morphological behavior. Carbohydr. Polym. **81**(1), 83–92 (2010). https://doi.org/10.1016/j.carbpol.2010.01.059
54. Rout, J., Misra, M., Tripathy, S.S., Nayak, S.K., Mohanty, A.K.: The influence of fibre treatment on the performance of coir-polyester composites. Compos. Sci. Technol. **61**(9), 1303–1310 (2001). https://doi.org/10.1016/S0266-3538(01)00021-5
55. Rowell, R.M.: Characterization and factors effecting fiber properties. In: Natural Polymers and Agrofibers Based Composites (2000)
56. Ru, S., Zhao, C., Yang, S.: Multi-objective optimization and analysis of mechanical properties of coir fiber from coconut forest waste. Forests **13**(12), 2033 (2022)
57. Saheb, D.N., Jog, J.P.: Natural fiber polymer composites: a review. Adv. Polym. Technol. **18**(4), 351–363 (1999). https://doi.org/10.1002/(SICI)1098-2329(199924)18:4<351::AID-ADV6>3.0.CO;2-X
58. Satyanarayana, K.G., Guimarães, J.L., Wypych, F.: Studies on lignocellulosic fibers of Brazil. Part I: Source, production, morphology, properties and applications. Compos. A: Appl. Sci. Manuf. **38**(7), 1694–1709 (2007). https://doi.org/10.1016/j.compositesa.2007.02.006
59. Sgriccia, N., Hawley, M.C., Misra, M.: Characterization of natural fiber surfaces and natural fiber composites. Compos. A: Appl. Sci. Manuf. **39**(10), 1632–1637 (2008). https://doi.org/10.1016/j.compositesa.2008.07.007
60. Shih, Y.-F., Cai, J.-X., Kuan, C.-S., Hsieh, C.-F.: Plant fibers and wasted fiber/epoxy green composites. Compos. Part B **43**(7), 2817–2821 (2012). https://doi.org/10.1016/j.compositesb.2012.04.044
61. Solomon, J.J., Geetha, L.: Phytoplasma diseases of coconut in India-root (wilt) and tatipaka diseases. CORD **20**(01), 34–34 (2004)
62. Sreekumar, P.A., Gopalakrishnan, P., Leblanc, N., Saiter, J.M.: Effect of glycerol and short sisal fibers on the viscoelastic behavior of wheat flour based thermoplastic. Compos. A: Appl. Sci. Manuf. **41**(8), 991–996 (2010)
63. Subramanian, M.: Characterization and Modification of Coir Fibers (n.d.)
64. Taban, E., Tajpoor, A., Faridan, M., Samaei, S.E., Beheshti, M.H.: Acoustic absorption characterization and prediction of natural coir fibers. Acoust. Aust. **47**(1), 67–77 (2019). https://doi.org/10.1007/s40857-019-00151-8
65. Tang, X., Yan, X.: Acoustic energy absorption properties of fibrous materials: a review. Compos. A: Appl. Sci. Manuf. **101**, 360–380 (2017). https://doi.org/10.1016/j.compositesa.2017.07.002
66. Thomas, S., Paul, S.A., Pothan, L.A., Deepa, B.: Natural fibres: structure, properties and applications. In: Cellulose Fibers: Bio- and Nano-polymer Composites: Green Chemistry and Technology, pp. 3–42. Springer (2011)
67. Valášek, P., Müller, M., Šleger, V., Kolář, V., Hromasová, M., D'Amato, R., Ruggiero, A.: Influence of alkali treatment on the microstructure and mechanical properties of coir and abaca fibers. Materials **14**(10), 2636 (2021)
68. Vieira, F., Santana, H.E., Jesus, M., Santos, J., Pires, P., Vaz-Velho, M., Silva, D.P., Ruzene, D.S.: Coconut waste: discovering sustainable approaches to advance a circular economy. Sustainability **16**(7), 3066 (2024)

69. Xiong, T., Wang, H., Sun, H., Zhao, J., Li, M., Du, Q., Li, D., Nan, Y., Zhang, Z., Sun, D.: Simultaneously enhancing strength and toughness of coir fiber through NaOH and sodium alginate treatment. Ind. Crop. Prod. **222**, 119668 (2024)
70. Yan, L., Chouw, N., Jayaraman, K.: Flax fibre and its composites—a review. Compos. Part B **56**, 296–317 (2014). https://doi.org/10.1016/j.compositesb.2013.08.014
71. Yang, H., Yan, R., Chen, H., Lee, D.H., Zheng, C.: Characteristics of hemicellulose, cellulose and lignin pyrolysis. Fuel **86**(12), 1781–1788 (2007). https://doi.org/10.1016/j.fuel.2006.12.013
72. Yu, W., Xie, W., Du, Z.: Structure of the right-handed helical crystal ribbon and multilevel fibrils in a tube fiber from a coir fiber. Cellulose **23**, 2841–2852 (2016)

Geopolymer Coir Composites

Mahalakshmi Subramanian, Mohanapriya Venkataraman, Jakub Wiener, and Jiří Militký

1 Introduction

The growing demand for sustainable and eco-friendly materials has led to significant advancements in composite development, particularly in the construction and manufacturing industries. Among various composite materials, geopolymer coir composites have become popular due to their environmental benefits, mechanical strength, and ease of production. Geopolymers are inorganic polymers that serve as a sustainable alternative to traditional cement-based materials by using industrial byproducts like fly ash or slag, which helps reduce carbon emissions and conserve natural resources. Compared to ordinary Portland cement (OPC), geopolymers offer significant advantages: superior fire and heat resistance, excellent chemical durability, low water permeability, and a substantially lower carbon footprint, especially when industrial byproducts are used as precursors [6]. These properties position geopolymers as promising eco-friendly materials for applications in sustainable construction, hazardous waste encapsulation, and high-temperature ceramics. However, geopolymers exhibit inherent brittleness, with limited tensile and flexural strength, a common issue for inorganic binders [31]. To overcome this limitation, incorporation of fibers,

M. Subramanian
NIL Textile, s. r. o., Ostrava, Czech Republic
e-mail: mahalakshmisubramanian00@gmail.com

M. Venkataraman (✉) · J. Wiener · J. Militký
Department of Material Engineering, Faculty of Textile Engineering, Technical University of Liberec, Liberec, Czech Republic
e-mail: mohanapriya.venkataraman@tul.cz

J. Wiener
e-mail: jakub.wiener@tul.cz

J. Militký
e-mail: jiri.militky@tul.cz

J. Militký and M. Venkataraman (eds.), *Fibrous Structures for Sustainable Future*, Engineering Materials, https://doi.org/10.1007/978-981-95-1382-6_3

synthetic or natural, can restrain crack propagation and improve mechanical performance. Among natural fibers, coir (coconut husk fiber) stands out due to its renewable nature, biodegradability, and mechanical toughness thanks to high lignin content [9].

Coir, sourced from coconut husks, is widely studied and used as a reinforcing agent in composites thanks to its availability, biodegradability, and favorable mechanical properties. When coir fibers are incorporated into geopolymer matrices, they enhance the overall performance of the composite, improving strength, durability, and resistance to environmental degradation.

Improving fiber-matrix adhesion by means of impurity removal from the shells of natural fibres helps to strengthen the mechanical interlocking and bonding contacts between the fibres hydroxyl groups and chemicals including dyes and some resins. Lignocellulosic materials have several well-known chemical treatment techniques. Kabir et al. [20] has highlighted various studies on how chemical treatments affect the shape and characteristics of fibres and their composites. Alkaline treatment, especially with sodium hydroxide, is one of the most common ways to chemically treat lignocellulosic materials. This process removes non-cellulosic components such as hemicelluloses, lignin, and waxes [4, 7, 21, 36] and enhances interfacial surface area. Figure 1 shows geopolymer coir composites with different coir fiber quantities.

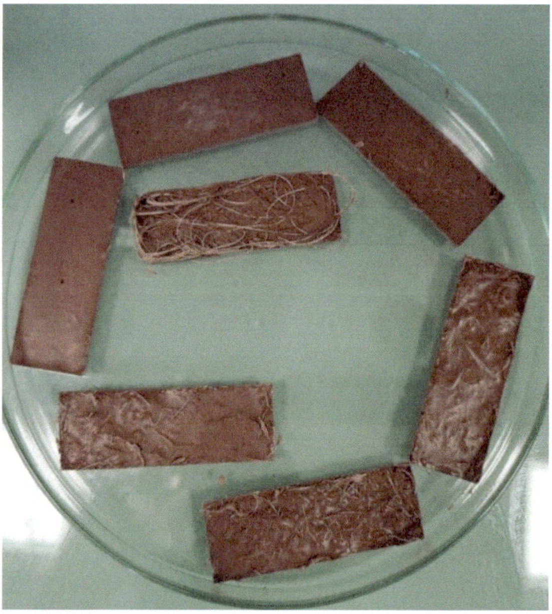

Fig. 1 Geopolymer coir composites

1.1 Geopolymer Materials and Their Characteristics

Geopolymers are synthetic polymeric compounds constructed from aluminosilicate precursors, which undergo a chemical process resulting in a solid, three-dimensional network. The primary raw materials utilized in the synthesis of geopolymers are typically those high in silica (SiO_2) and alumina (Al_2O_3), such as fly ash, slag, metakaolin, or naturally occurring minerals. The geopolymerization process is initiated by activating these materials with alkaline solutions, which facilitates the formation of a hardened matrix [12].

A significant advantage of geopolymers compared to conventional cement-based materials is their considerably lower carbon emissions during production. The manufacturing of Portland cement is characterized by its high energy consumption and substantial contribution to global CO_2 emissions; conversely, geopolymers present a more sustainable alternative. Furthermore, geopolymers are known for their enhanced durability, high resistance to chemical degradation, and exceptional fire resistance, making them suitable for a wide range of applications, including construction and environmental remediation. The geopolymerization process commences with the activation of aluminosilicate materials using alkaline solutions, such as sodium hydroxide (NaOH) or potassium hydroxide (KOH). This activation leads to the creation of a gel-like phase that subsequently solidifies over time, akin to the hydration process observed with Portland cement. However, geopolymers demonstrate greater versatility, as variations in the silica-to-alumina ratio or the selection of alkaline activators can be employed to tailor the final properties of the material [3, 50].

Additionally, the incorporation of coir fibers into the geopolymer matrix has the potential to enhance specific mechanical properties of the resulting composite. Coir fibers, recognized for their high cellulose content and tensile strength, can effectively improve the stiffness, toughness, and load-bearing capacity of the composite material [46].

2 Coir Fibers and Their Role in Composite Materials

Coir fibers, derived from the husk of coconuts, are naturally occurring fibers recognized for their exceptional tensile strength, durability, and biodegradability. These properties render coir an appealing choice for reinforcing composite materials. Historically, coir fibers have found applications in the production of ropes, mats, and brushes; however, their utilization in contemporary composite materials has garnered considerable interest due to their environmental advantages. The composition of coir fibers primarily encompasses cellulose, hemicellulose, and lignin, which confer strength and rigidity to the fibers. The mechanical properties of coir fibers are influenced by multiple factors, including the source of the coir, the extraction

Fig. 2 Brown coir fibers

methodology, and the environmental conditions under which the fibers are cultivated. Coir fibers are categorized into two distinct types: brown coir, sourced from mature coconuts, and white coir, extracted from younger coconuts. Generally, brown coir fibers exhibit superior strength and durability compared to their white counterparts, thus making them better suited for incorporation into composite materials [15, 48]. Brown coir fibers are shown in Fig. 2.

Incorporating coir fibers into geopolymer matrices has been demonstrated to enhance the mechanical properties of the resulting composites. Specifically, the addition of coir fibers can improve tensile strength, flexural strength, and impact resistance. Furthermore, these fibers function as reinforcing agents, effectively transferring loads and stresses, which mitigates crack propagation and bolsters the overall toughness of the composite material. The interaction between coir fibers and the geopolymer matrix is pivotal for achieving optimal mechanical performance. Various techniques, including fiber treatment and surface modification, have been utilized to enhance the adhesion between the fibers and the matrix, thereby improving the mechanical integrity of the composite [43].

3 Material Preparation and Formulation of Geopolymer Coir Composites

The fabrication of geopolymer coir composites encompasses several essential stages, notably the selection and processing of raw materials, the formulation of the geopolymer matrix, and the integration of coir fibers into that matrix. The subsequent sections will delineate the materials utilized, their properties, and the methodology employed in the creation of geopolymer coir composites.

3.1 Selection of Geopolymer Matrix Materials

The foundational components of a geopolymer matrix are aluminosilicate materials, which predominantly consist of fly ash, metakaolin, and ground granulated blast-furnace slag (GGBFS). These materials are characterized by their high silicon (Si) and aluminum (Al) content, essential for establishing the geopolymeric network upon activation with alkaline solutions. The selection of raw materials for the geopolymer matrix is crucial, as it significantly influences the resultant properties of the composite. Figure 3 shows different types of geopolymer sources.

Fly Ash: Fly ash is frequently utilized as a primary raw material in geopolymer production due to its widespread availability and cost-effectiveness. Its chemical composition, typically rich in silica and alumina, renders it an excellent candidate for the geopolymerization process [38].

Metakaolin: As a dehydroxylated variant of kaolin clay, metakaolin is also a commonly employed source for geopolymers. It possesses superior reactivity compared to fly ash, thereby facilitating a more expedient geopolymerization process [16, 17]. Table 1 shows the chemical composition of kaolin and metakaolin.

Slag: Ground granulated blast-furnace slag (GGBFS), a byproduct of steel manufacturing, is abundantly rich in silica and alumina. Its application in geopolymer production has been well-documented, providing enhanced durability and mechanical properties [11].

An additional critical consideration in the geopolymerization process is the choice of activator, which is generally an alkaline solution such as sodium hydroxide (NaOH) or potassium hydroxide (KOH). The concentration of the activator plays a pivotal role, influencing the rate of polymerization and the ultimate strength characteristics of the geopolymer matrix [13].

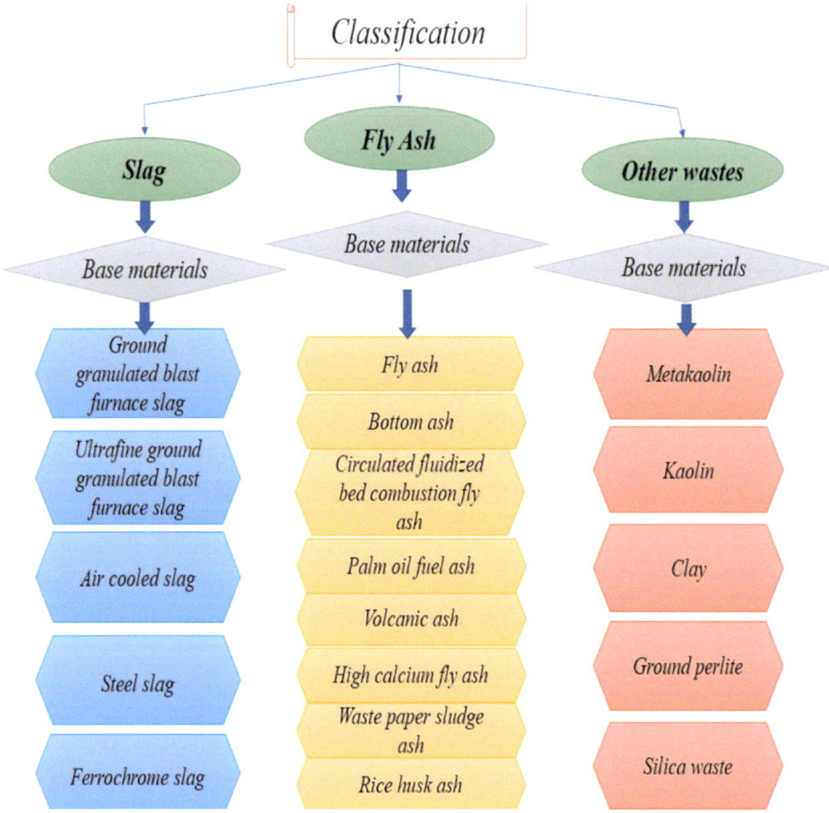

Fig. 3 Geopolymer sources (reused from Saeed et al. [41] under CC-BY Permission)

Table 1 Chemical composition of kaolin and metakaolin

Mass ratio (%)	SiO_2	Al_2O_3	Fe_2O_3	TiO_2	K_2O	MnO	ZrO_2	CaO	BaO	LOI
Kaolin	60.09	26.15	1.46	0.27	0.70	0.20	0.01	0.06	0.21	10.85
MK	60.50	34.30	2.41	0.15	0.20	0.15	0.09	0.04	0.01	2.15

Reused from Ayeni et al. [6] under CC-BY Permission
LOI Loss on ignition

3.2 Integration of Coir Fibers to the Geopolymer Matrix

Coir fibers are integrated into the geopolymer matrix to serve as reinforcements, thereby enhancing the mechanical properties of the composite material. The methodology employed for fiber incorporation significantly influences the interaction

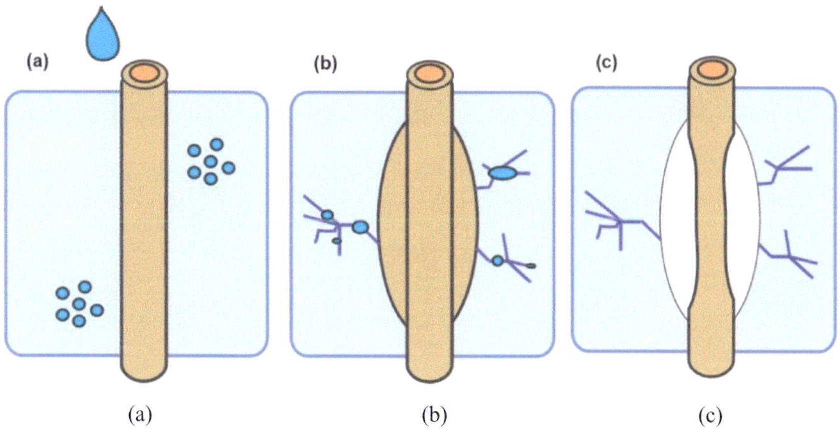

Fig. 4 Effect of water on alkali-treated coir fiber in geocomposite **a** water diffusion into the geocomposite; **b** swelling of fibers after moisture adsorption, creating microcracks; **c** fiber–matrix debonding (reused from Camargo et al. [9] under CC-BY Permission)

between the fibers and the matrix, ultimately affecting the composite's overall characteristics. To enhance the bond between coir fibers and the geopolymer matrix, the fibers may undergo various treatments. Common methods include alkali treatment, utilizing sodium hydroxide or potassium hydroxide to eliminate lignin and other contaminants, as well as surface modification with silane coupling agents. These treatments improve the surface roughness of the fibers, facilitating stronger adhesion to the geopolymer matrix [1, 49].

The main problem is enhanced water diffusion in the coir fibers treated by alkalis because they promote swelling and debonding, as shown in Fig. 4.

It is indicated that alkaline treatment improves the mechanical properties of fibres, augmenting tensile strength and elongation at break, while concurrently reducing the fiber's diameter and hydrophilicity. According to Rosa et al. [40], alkaline treatment of coir fibres strengthened fiber-matrix adhesion and increased tensile strength by 53% when compared to untreated coir fibre composites. Haseena et al. [18] conducted a thorough investigation into the influence of chemical treatments on coir/sisal/natural rubber hybrid composites. Their findings indicated that the treated composites exhibited significant improvements in abrasion resistance, tear strength, and hardness when compared to untreated coir/sisal fiber composites. Similarly, Kostic et al. [26] reported that chemically modified sisal fiber composites demonstrated enhanced fiber fineness and flexibility as a result of the treatment. Cordeiro et al. [10] emphasized that chemical modifications can substantially enhance the properties of agro fibers relative to wood fibers. The chemical treatment of surface fibers is known to improve wettability and modify key characteristics such as microstructure, surface roughness (see Fig. 5), chemical functionalities, and tensile strength.

These modifications can beneficially influence the interactions within composite materials, including those between the matrix and polymer, between adsorbents

Fig. 5 Increased interface interlocking due to higher surface roughness of alkali-treated coir fibers (reused from Camargo et al. [9] under CC-BY Permission)

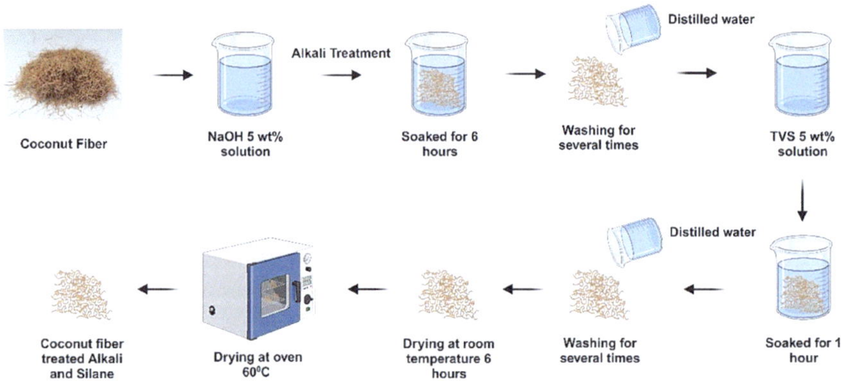

Fig. 6 Fabrication of alkali and silane treated coir fibers (reused from Mulana et al. [34] under CC-BY Permission)

and adsorbates, which is relevant for metal and pigment adsorption, and between enzymes/support and functionalizing agents/support, particularly in the context of immobilization processes [4, 8, 36]. Figure 6 shows the process of treatment of coir fibers with silane and alkali.

Yan et al. [52] describe how alkaline (NaOH) treatment removes surface waxes, lignin, and hemicellulose, resulting in rougher fiber surfaces, reduced diameter, increased aspect ratio, and exposure of cellulose fibrils, factors that significantly improve mechanical interlocking and stress transfer in composites.

Faridul Hasan et al. [15] detail various treatments (e.g., benzoylation, $Cr_2(SO_4)_3$, $NaHCO_3$) that enhance initial modulus by reducing lignin content while maintaining adequate tensile strength. Siakeng et al. [44], through alkali and silane treatments, demonstrated higher interfacial shear strength and lower water uptake. A broader review by Sahu and Gupta [42], confirms that alkali treatment consistently enhances tensile strength, crystallinity, and thermal stability in coir- and other lignocellulosic-fiber composites. Tan et al. [47], revealed that 5% NaOH treatment significantly improves bonding with polymer matrices by increasing fiber roughness (interfacial area) and decreasing diameter. The alkali-treated fibers show higher crystallinity and better fiber–matrix interaction. Lv and Liu [29] emphasizes benefits such as dimensional stability, and enhanced interfacial micro-mechanical bonding through mechanical interlocking and hydrogen bond modification. Figure 7 shows the preparation steps of coir fibers to integrate in the geopolymer composites.

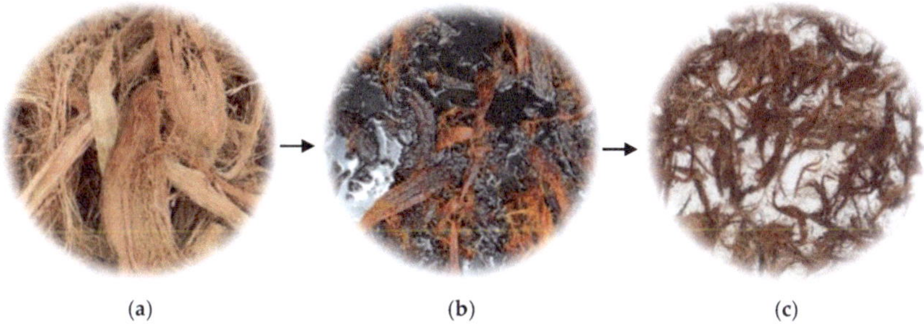

Fig. 7 a Dehusked coir fibers b coir fibers pretreated with 5% NaOH c air dried pretreated coir fibers (reused from Ayeni et al. [6] under CC-BY Permission)

Taken together, these sources demonstrate that alkali and silane treatments enhance fiber–matrix cohesion via chemical cleaning, increased surface roughness, and exposure of cellulose microfibrils. These changes significantly reduce debonding and fiber pull-out in geopolymer composites.

The quantity of coir fibers introduced into the geopolymer matrix is a critical determinant of the composites mechanical properties. Research indicates that increasing fiber content can enhance both tensile strength and impact resistance; however, it may also compromise the workability of the mixture. Furthermore, the length of the fibers is influential in terms of mechanical performance, with longer fibers typically offering superior reinforcement. Achieving an even distribution of coir fibers within the geopolymer matrix is essential for ensuring consistent mechanical properties. Inadequate dispersion can result in weak points and diminished performance of the composite. In practical applications, fibers are either directly mixed into the geopolymer paste or oriented in specific alignments, depending on the desired application of the composite material [6, 14].

Ayeni et al. [6] highlighted that fibers worn inside the matrix carry tensile stress via interfacial shear. Both 0.5 and 1.0 wt% coir loading allowed effective stress transmission from the brittle matrix into the higher modulus fibers. SEM analysis from Ayeni et al. [6] showed multiple fine microcracks instead of single macro-fractures in 0.5–1.0 wt% composites, due to coir's micro-scale bridging effect. This corresponds to enhanced toughness and delayed brittle failure, reinforcing the importance of high-aspect-ratio natural fiber reinforcement. Figure 8 shows the SEM images of coir fibers before and after alkali chemical modifications.

X-ray diffraction (XRD) analysis of untreated and treated coir fibers revealed characteristic cellulose peaks. A prominent peak at 22.5° 2θ (assigned to the cellulose (002) plane) indicated the crystalline nature of the fiber. Additional lower-intensity peaks at around 17° and 35° 2θ (corresponding to cellulose (101) and (004)) suggested the presence of amorphous components like lignin and hemicellulose. Post-treatment, the sharper peak at 22.5° 2θ confirmed enhanced crystallinity due to increased cellulose content, while quartz-related peaks visible in the untreated sample were no

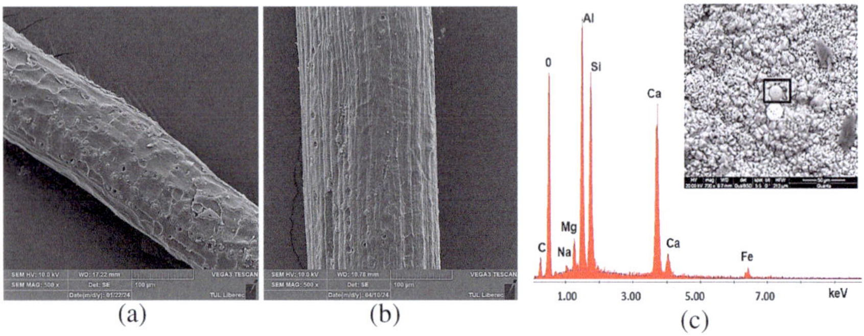

Fig. 8 SEM images of coir fiber **a** before alkali modification **b** after alkali modification **c** EDX analysis geopolymer (reused from Saeed et al. [41] under CC-BY Permission)

Fig. 9 XRD of treated and untreated coir fibers (reused from Ayeni et al. [6] under CC-BY Permission)

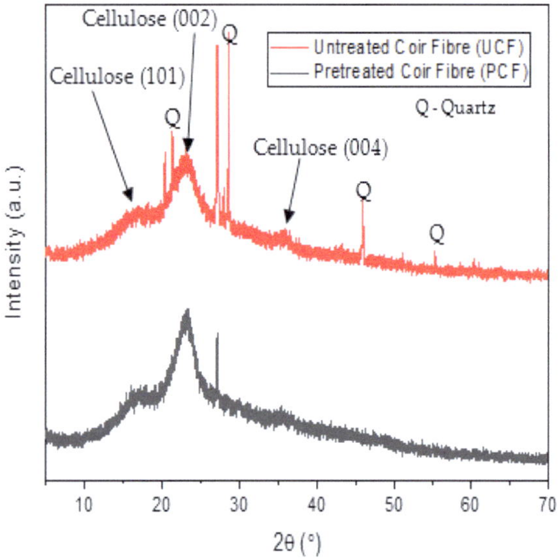

longer present [6]. Figure 9 shows the XRD graph of treated and untreated coir fibers.

3.3 *Fabrication of Geopolymer Composites*

The sample was prepared using Baucis LBNa. This aluminosilicate binder has 2 parts—metakaolin (part A), activated by liquid alkaline (part B). The reinforced fibers were classified into 3 sizes. The geopolymer mixture was combined with fibres in a silicone mould and dried in the oven at 70 °C for an hour. The silicon mould

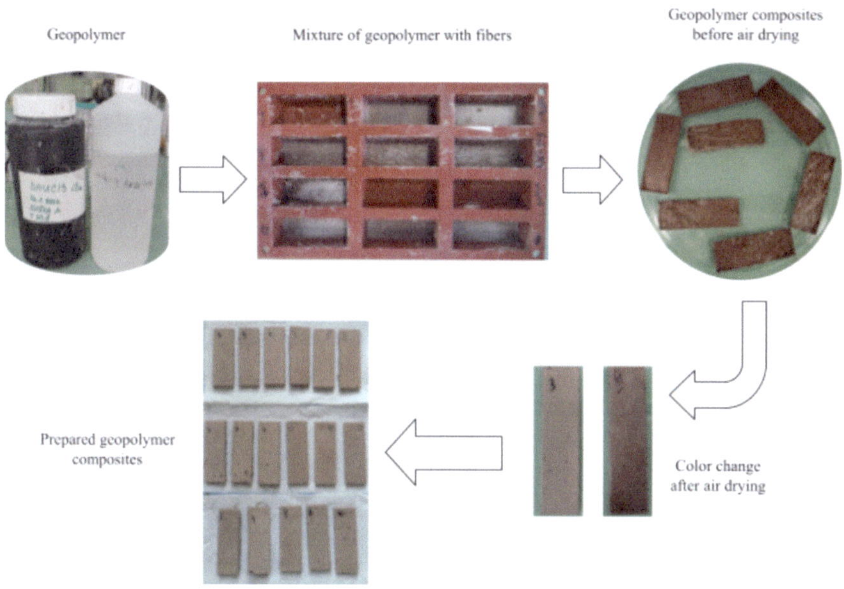

Fig. 10 Fabrication of geopolymer composites [46]

was entirely sealed to prevent evaporation. The geopolymer composites were then wrapped separately and left undisturbed for one day. This phase helps to identify any subsequent reactions that may arise. The composites were then allowed to air dry at room temperature for one day to ensure that any remaining water is completely evaporated. This method considerably improves the strength and stability of composites [46]. Figure 10 shows this fabrication process of geopolymer composites.

3.4 *Curing Conditions and Hardening Process*

The curing methodology for geopolymer coir composites distinctly contrasts with that of conventional Portland cement-based materials. In geopolymer systems, the hardening mechanism does not rely on water for hydration; instead, a chemical reaction referred to as geopolymerization facilitates the solidification of the matrix. The parameters of curing temperature and duration are pivotal in influencing the ultimate properties of the composite. Geopolymer coir composites are typically subjected to curing temperatures ranging from 30 to 80 °C. Elevated curing temperatures can significantly expedite the geopolymerization process, thereby enhancing compressive strength. However, it is crucial to note that excessively high temperatures may result in the degradation of coir fibers, which could detract from their reinforcing capabilities. The curing duration for geopolymer coir composites is contingent upon both the temperature and the specific types of raw materials employed. Generally, these composites are allowed to cure for a period of 24–48 h under ambient conditions, followed by a post-curing phase at elevated temperatures to further augment strength [6, 19, 35].

While geopolymer materials do not necessitate water for the hydration process, moisture remains a critical factor in the geopolymerization reaction. An excess of moisture can dilute the alkaline activator, subsequently reducing the efficiency of the chemical reaction and, as a result, leading to diminished strength in the final composite product [19].

4 Mechanical Properties of Geopolymer Coir Composites

The integration of coir fibers into geopolymer matrices has been extensively studied as a way to improve the mechanical behavior of otherwise brittle inorganic systems. Geopolymers, which are synthesized from aluminosilicate sources such as fly ash or slag and activated with alkaline solutions, are known for their high compressive strength, low shrinkage, and excellent fire resistance. However, their inherent brittleness limits their application in structures requiring tensile or flexural resilience. The addition of coir fibers introduces a ductile phase that enhances the toughness and post-crack energy absorption capacity of geopolymer composites. This observation is supported by the work of Assaedi et al. [5], where they demonstrated that natural fiber reinforcement significantly improved the fracture resistance of geopolymer matrices.

The tensile strength of geopolymer coir composites represents a fundamental parameter for their utilization in scenarios involving tensile or pulling forces. The addition of coir fibers to geopolymer matrices markedly improves flexural performance and toughness compared to plain geopolymer systems. Coir fiber additions (0.5–1.0 wt%) increased flexural strength from approximately 3.1 MPa (for unreinforced mortar) to 5.3–6.6 MPa [30]. This significantly enhances tensile strength by facilitating the distribution of stresses throughout the matrix and augmenting the fiber-matrix adhesion. Nevertheless, the ultimate tensile strength is influenced by various factors, including fiber content, fiber length, and the effectiveness of the fiber-matrix interaction. Compressive strength is a vital mechanical property for construction materials, particularly in the manufacture of building components such as blocks and panels. Geopolymer coir composites typically demonstrate lower compressive strength compared to conventional concrete. Nonetheless, the integration of coir fibers aids in mitigating brittle failure and enhances the material's capacity to withstand compressive forces. The flexural strength of a composite material quantifies its resistance to bending. Coir fibers play a crucial role in augmenting the bending resistance of geopolymer composites by functioning as reinforcements that inhibit excessive deformation. Key factors affecting the flexural strength include the optimal fiber content and the uniform distribution of fibers within the geopolymer matrix [3, 12, 24, 33].

An experimental study by Ayeni et al. [6], investigated metakaolin-based geopolymers with 0.5, 1.0, 1.5 and 2.0 wt% coir fibers. They found optimal compressive strength (~ 21.25 MPa) at 0.5 wt% coir and peak flexural strength (~ 10.39 MPa) at 1.0 wt%, noting that higher fiber loads reduced bulk density and introduced porosity. Similarly, Korniejenko et al. [25] demonstrated that adding 2 wt% coir fibers increased compressive strength by 25–56% and improved flexural strength by ~ 5%, indicating consistent strength gains across fiber types and dosages.

In Fig. 11, the flexural properties of geopolymer coir composites with different percentages of coir fiber are shown. Figure 12 depicts the mode of failure of geopolymer composites with no and varying lengths of coir fibres.

According to the study by Korniejenko et al. [25], geopolymer composites reinforced with coir fibers showed a clear improvement in compressive strength with curing time, reaching maximum performance at 28 days. Among the tested dosages, a 2% coir fiber addition resulted in the highest strength gain, approximately 43.3%

Fig. 11 Flexural properties of geopolymer coir composite samples (metakaoline geopolymer for different percentages of coir fiber) (reused from Ayeni et al. [6] under CC-BY Permission)

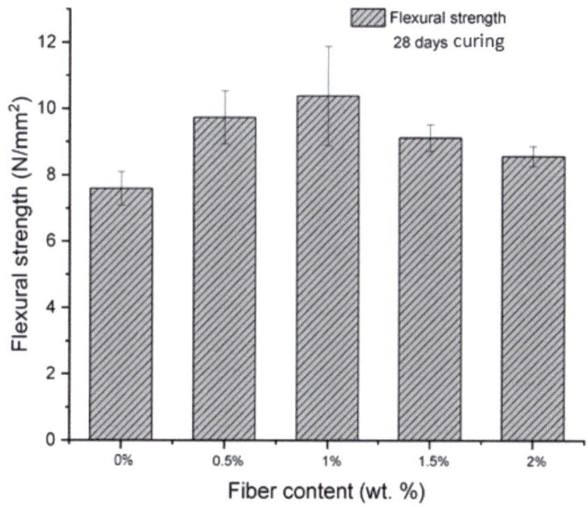

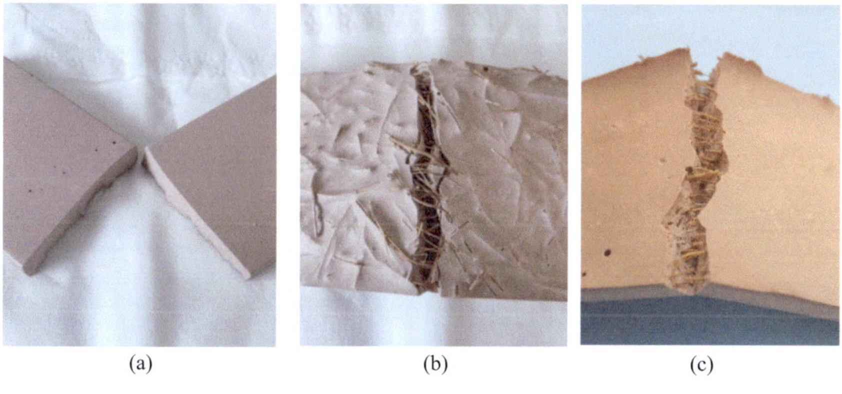

Fig. 12 Mode of failure of geopolymer composites with **a** no fibers **b** 0.5–1.2 cm long coir fibers **c** 0.2–0.5 cm long coir fibers [46]

greater than the unreinforced matrix. Lower and higher additions, specifically 1% and 5%, also enhanced strength by 27.7% and 31.5%, respectively, though to a lesser extent. In addition to strength enhancement, long-term testing revealed a reduction in variability, with standard deviations decreasing significantly by 28 days. This indicates more reliable and predictable mechanical behavior, which is critical for structural applications. Table 2 shows these results.

Compressive strength is one of the most critical indicators used to evaluate the mechanical behavior of construction materials. As illustrated in Fig. 13, the compressive strength of metakaolin-based geopolymer composites varies with different coir fiber additions ranging from 0.5 to 2%. A notable reduction in strength was observed at 1% fiber content, likely due to poor dispersion of the coir fibers within the matrix. This uneven distribution may have caused fiber agglomeration, which can initiate microcracks and weaken the material. When the fiber content was increased to 1.5%, the compressive strength reached approximately 16.94 N/mm^2, comparable to the unreinforced geopolymer composites, indicating that the added fibers did not significantly enhance the composite's strength at this level. At 2% coir fiber addition, a further decline in compressive strength was recorded. This suggests that higher fiber loadings reduce composite density, hinder matrix compaction, and increase internal voids, all of which negatively affect the overall strength [6, 24]. So according to Ayeni et al. [6], 0.5% of coir fibers is the optimal quantity to improve the compressive strength.

The incorporation of coir fibers significantly bolsters the impact resistance of geopolymer composites by absorbing energy during impact events and hindering crack propagation. This attribute is particularly advantageous for applications where materials are exposed to abrupt shocks or loads. Alkali treatment of coir fibers improves the adhesion between the fibers and the geopolymer matrix, which results in better impact resistance.

Table 2 Compressive and flexural strengths test after 28 days [24]

Sample	Compressive strength (MPa)	Standard deviation of recorded values of strength	Flexural strength (MPa)	Standard deviation of recorded values of strength
Geopolymer (matrix)	24.78	1.89	5.55	0.72
Geopolymer with coir fibers (1%)	31.36	10.10	5.25	0.57

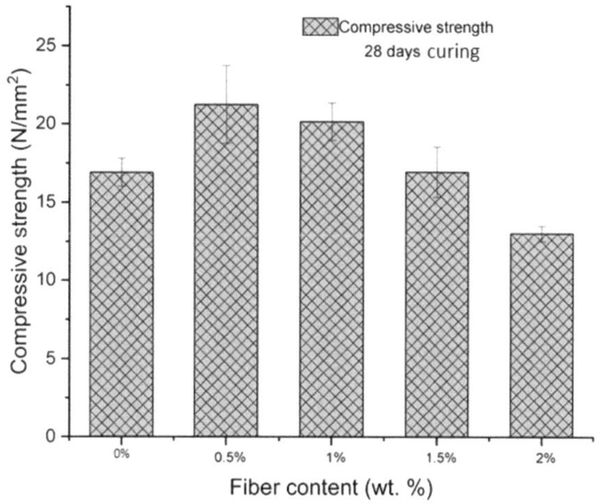

Fig. 13 Compressive strength of geopolymer coir composite samples (metakaoline geopolymer for different percentages of coir fiber) (reused from Ayeni et al. [6] under CC-BY Permission)

5 Thermal Behavior of Coir Fibers and Geopolymer Coir Composites

The thermal analysis of coir fibers reveals a characteristic three-stage degradation pattern, consistently reported across multiple studies. TGA data show an initial mass loss below ~ 100 °C due to moisture evaporation (typically 2–9%), followed by major losses between ~ 200 and 350 °C, which correspond to hemicellulose and cellulose decomposition, and a final gradual decline beyond ~ 350 °C linked to lignin degradation. Coir fiber composites exhibit enhanced thermal stability compared to untreated fibers. In coir PP composites, effective inclusion of coir can raise decomposition temperatures from the ~ 274 °C range typical for pure PP up to ~ 292–479 °C depending on fiber load [23]. This increase has been attributed to the high lignin content in coir, which promotes char formation and improves thermal resistance (see Fig. 14) [32].

Proctoring this data to the coir–geopolymer system, natural fiber matrices typically benefit from coir's char formation and moisture retention properties. Alkaline chemical treatments (e.g., NaOH or sodium sulfite) of coir fibers improve thermal stability by reducing hemicellulose and surface impurities, where treated fibers lost less mass (~ 6% less) around 380 °C compared to untreated samples [37].

Coir fibers and their composites display three-stage thermal degradation transitions, with chemically treated fibers and composites showing higher decomposition thresholds and char yield. This enhanced thermal behavior, facilitated by intrinsic lignin content and post-treatment stabilization, makes coir fibers attractive additives for geopolymer composites, provided operational temperatures remain below ~ 300 °C to maintain structural integrity.

During fire resistance evaluations, geopolymer foams maintained their structural integrity with minimal macroscopic damage after being subjected to high temperatures. The duration of fire resistance improved with the thickness of the

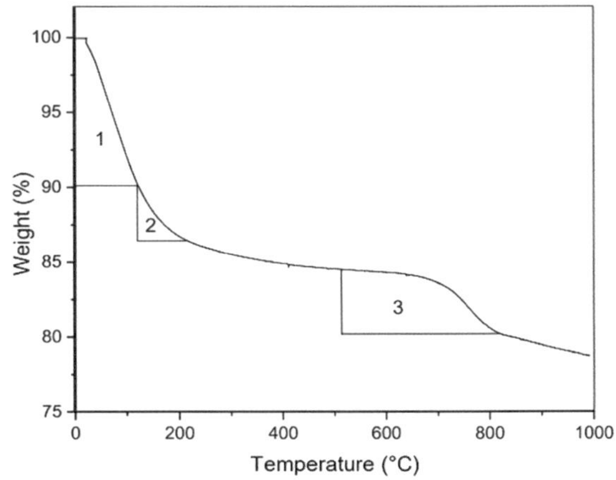

Fig. 14 Thermogravimetric analysis (TGA) of geopolymer composites (reused from Saeed et al. [41] under CC-BY Permission)

geopolymer layer, with a 20 mm thick layer providing up to 99 min of fire protection on wood substrates. The mechanical properties of geopolymer composites can be adversely affected by repeated thermal cycling. A study involving fly ash-based geopolymer composites indicated a 56% reduction in mechanical strength after exposure to ten thermal cycles at 700 °C, which was attributed to increased porosity and microstructural changes occurring within the material [6, 21, 27, 28, 48–50, 53].

6 Durability and Long-Term Performance of Geopolymer Coir Composites

Geopolymer matrices display a high level of alkalinity due to the incorporation of alkali activators such as sodium hydroxide or potassium hydroxide. While this alkaline nature enhances the overall strength of the matrix, it poses a risk for natural fibers like coir. The elevated pH levels can lead to the degradation of lignin and hemicellulose within coir fibers, thereby weakening the fiber-matrix interface and ultimately diminishing the mechanical properties of the composite over time. The exposure of coir fibers to moisture and repeated wet-dry cycles can induce swelling and shrinkage, resulting in microcracking within the matrix. However, research indicates that the incorporation of coir fibers can improve the toughness of the composite, effectively delaying crack propagation. Notably, after undergoing 20 wetting/drying cycles, geopolymer composites exhibited enhanced performance, demonstrating increased ductility attributed to superior fiber-matrix bonding. Geopolymer coir composites have shown promising resistance to acidic conditions. In a specific study, a straw-based geopolymer retained a compressive strength of approximately 36 MPa when subjected to acid rain leaching, with the pH maintained above 3, suggesting robust acid resistance [24, 33, 38, 45].

The research by Agrawal and Malviya [2] identified an optimal mix incorporating 0.25% coir fiber and a sodium silicate to sodium hydroxide (Na_2SiO_3/NaOH) ratio of 2.0, which achieved a peak compressive strength of 33.7 MPa and flexural strength of 7.0 MPa. The addition of coir fibers enhanced fiber–matrix cohesion, contributing to improved crack control and mechanical integrity through a bridging effect that limited crack propagation. Durability testing revealed that CFR-GPC maintained its structural integrity under aggressive environmental conditions, including sulfate and acidic exposure, with minimal mass loss and strength degradation. The authors also applied a Random Forest Regression model to accurately predict compressive strength with an R^2 value of 0.956, demonstrating the feasibility of data-driven optimization in geopolymer mix design.

These results positions coir fiber-reinforced geopolymer composites as a viable, sustainable alternative to ordinary Portland cement-based systems, particularly for infrastructure components like pavements, precast elements, and coastal structures.

7 Potential Applications of Geopolymer Coir Composites

Geopolymer composites reinforced with coir fibers have shown great potential in various sustainable construction applications due to their combined benefits of mechanical resilience, thermal stability, and environmental friendliness. These composites are particularly suitable for non-load-bearing structural elements, such as precast wall panels, partition boards, and pavement blocks, where enhanced ductility and crack resistance are essential [6, 30]. The fire-resistant nature of the geopolymer matrix, combined with the natural insulating properties of coir, makes them ideal for thermal and acoustic insulation panels, especially in low-rise or eco-friendly buildings [25]. Additionally, their resistance to chemical attack and moisture positions them well for use in repair mortars, overlays, and durable protective coatings in aggressive environments [9]. Given that coir is a renewable agricultural byproduct, its incorporation into geopolymer composites also contributes to waste valorization and circular economy goals, making it a low-cost, bio-based reinforcement option with increasing relevance in green building technologies [12]. Potential applications of geopolymer coir composites are illustrated in Fig. 15.

8 Conclusion

The integration of coir fibers into geopolymer matrices presents a sustainable alternative to conventional composites by enhancing toughness, reducing environmental impact, and improving ductility without compromising the fundamental strength of the material. Studies have shown that treated coir fibers improve the flexural and tensile performance of fly ash and metakaolin-based geopolymers, although care must be taken to address their degradation in highly alkaline environments. Durability

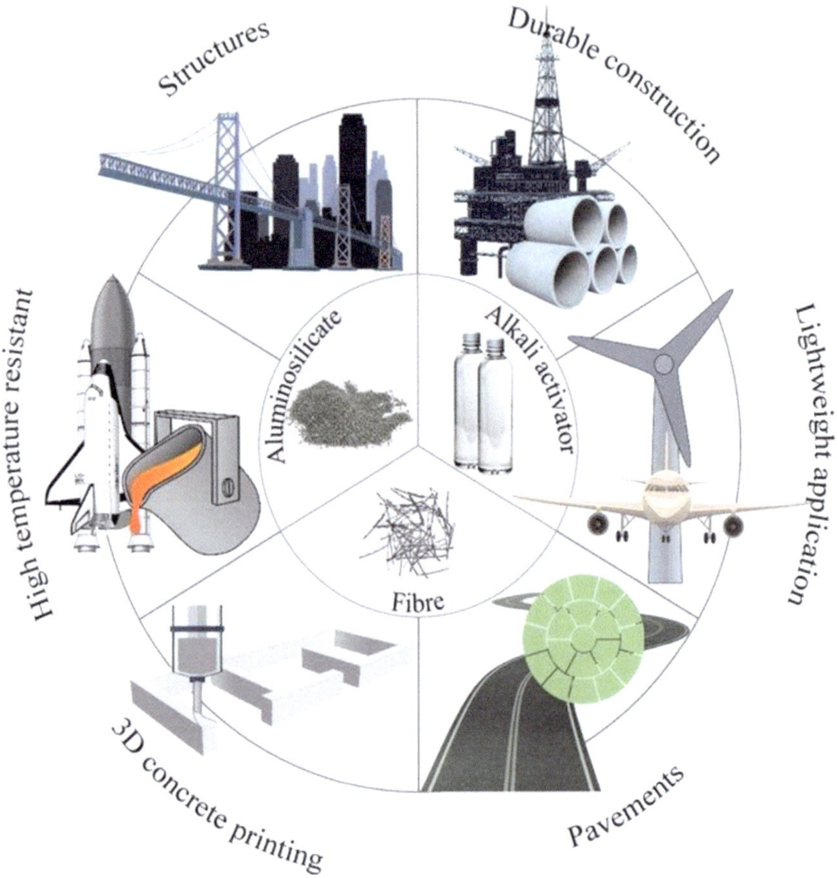

Fig. 15 Potential applications of geopolymer coir composites (reused from Ranjbar and Zhang [39] under CC-BY Permission)

under thermal and moisture cycling remains promising, and SEM analysis confirms effective fiber–matrix bonding when proper treatment is applied. Furthermore, life cycle assessments show that these composites can significantly lower CO_2 emissions compared to traditional Portland cement materials [20]. Thus, geopolymer coir composites offer both technical viability and ecological value for future construction applications [12, 22, 48, 51].

References

1. Abdalla, J.A., Hawileh, R.A., Bahurudeen, A., Jyothsna, G., Sofi, A., Shanmugam, V., Thomas, B.S.: A comprehensive review on the use of natural fibers in cement/geopolymer concrete: a

step towards sustainability. Case Stud. Constr. Mater. **19**, e02244 (2023). https://doi.org/10.1016/j.cscm.2023.e02244

2. Agrawal, A., Malviya, N.: Advanced geopolymer concrete with coconut fiber reinforcement: optimizing strength, durability, and predictive modelling for sustainable construction. Archit. Struct. Constr. **5**(1), 1–16 (2025)
3. Aleem, M.I.A., Arumairaj, P.D.: Geopolymer concrete—a review. Int. J. Eng. Sci. Emerg. Technol. **1**(2), 118–122 (2012). https://doi.org/10.7323/ijeset/v1_i2_14
4. Arrakhiz, F.Z., Malha, M., Bouhfid, R., Benmoussa, K., Qaiss, A.: Tensile, flexural and torsional properties of chemically treated alfa, coir and bagasse reinforced polypropylene. Compos. B Eng. **47**, 35–41 (2013). https://doi.org/10.1016/j.compositesb.2012.10.046
5. Assaedi, H., Alomayri, T., Shaikh, F.U.A., Low, I.M.: Advances in geopolymer composites with natural reinforcement. In: Advances in Ceramic Matrix Composites, pp. 461–474. Elsevier (2018)
6. Ayeni, O., Mahamat, A.A., Bih, N.L., Stanislas, T.T., Isah, I., Savastano Junior, H., Boakye, E., Onwualu, A.P.: Effect of coir fiber reinforcement on properties of metakaolin-based geopolymer composite. Appl. Sci. **12**(11), 5478 (2022)
7. Boopathi, L., Sampath, P.S., Mylsamy, K.: Investigation of physical, chemical and mechanical properties of raw and alkali treated *Borassus* fruit fiber. Compos. B Eng. **43**(8), 3044–3052 (2012). https://doi.org/10.1016/j.compositesb.2012.05.002
8. Brígida, A.I.S., Calado, V.M.A., Gonçalves, L.R.B., Coelho, M.A.Z.: Effect of chemical treatments on properties of green coconut fiber. Carbohyd. Polym. **79**(4), 832–838 (2010). https://doi.org/10.1016/j.carbpol.2009.10.005
9. Camargo, M.M., Adefrs Taye, E., Roether, J.A., Tilahun Redda, D., Boccaccini, A.R.: A review on natural fiber-reinforced geopolymer and cement-based composites. Materials **13**(20), 4603 (2020)
10. Cordeiro, N., Ornelas, M., Ashori, A., Sheshmani, S., Norouzi, H.: Investigation on the surface properties of chemically modified natural fibers using inverse gas chromatography. Carbohyd. Polym. **87**(4), 2367–2375 (2012). https://doi.org/10.1016/j.carbpol.2011.11.001
11. Das, S.K., Mishra, J., Singh, S.K., Mustakim, S.M., Patel, A., Das, S.K., Behera, U.: Characterization and utilization of rice husk ash (RHA) in fly ash—blast furnace slag based geopolymer concrete for sustainable future. Mater. Today Proc. **33**, 5162–5167 (2020). https://doi.org/10.1016/j.matpr.2020.02.870
12. Davidovits, J.: Geopolymers (2005). https://doi.org/10.1007/bf01912193
13. Fan, X.: A brief introduction on the research status and future prospects on geopolymer concrete. IOP Conf. Ser. Earth Environ. Sci. **508**(1), 012124 (2020). https://doi.org/10.1088/1755-1315/508/1/012124
14. Farhan, K.Z., Johari, M.A.M., Demirboğa, R.: Impact of fiber reinforcements on properties of geopolymer composites: a review. J. Build. Eng. **44**, 102628 (2021). https://doi.org/10.1016/j.jobe.2021.102628
15. Faridul Hasan, K.M., György Horváth, P., Bak, M., Alpár, T.: A state-of-the-art review on coir fiber-reinforced biocomposites. RSC Adv. **11**(18), 10548–10571 (2021). https://doi.org/10.1039/D1RA00231G
16. Görhan, G., Aslaner, R., Şinik, O.: The effect of curing on the properties of metakaolin and fly ash-based geopolymer paste. Compos. B Eng. **97**, 329–335 (2016). https://doi.org/10.1016/j.compositesb.2016.05.019
17. Han, X., Zhang, P., Zheng, Y., Wang, J.: Utilization of municipal solid waste incineration fly ash with coal fly ash/metakaolin for geopolymer composites preparation. Constr. Build. Mater. **403**, 133060 (2023). https://doi.org/10.1016/j.conbuildmat.2023.133060
18. Haseena, A.P., Dasan, K.P., Unnikrishnan, G., Thomas, S.: Mechanical properties of sisal/coir hybrid fibre reinforced natural rubber. Prog. Rubber Plast. Recycl. Technol. **21**(3), 155–181 (2005)
19. Jumrat, S., Muangprathub, J., Wongpanich, J., Klaiklay, S., Chotikhun, A., Karrila, S., Pianroj, Y.: Enhancing geopolymer mortar with coir and pineapple leaf fibers through various curing methods. Sci. Technol. Asia 260–270 (2025). https://ph02.tci-thaijo.org/index.php/SciTechAsia/article/view/254919

20. Kabir, M.M., Wang, H., Lau, K.T., Cardona, F.: Chemical treatments on plant-based natural fibre reinforced polymer composites: an overview. Compos. B Eng. **43**(7), 2883–2892 (2012). https://doi.org/10.1016/j.compositesb.2012.04.053
21. Kaewkuk, S., Sutapun, W., Jarukumjorn, K.: Effects of interfacial modification and fiber content on physical properties of sisal fiber/polypropylene composites. Compos. B Eng. **45**(1), 544–549 (2013). https://doi.org/10.1016/j.compositesb.2012.07.036
22. Kamaruddin, F.A., Anggraini, V., Kim Huat, B., Nahazanan, H.: Wetting/drying behavior of lime and alkaline activation stabilized marine clay reinforced with modified coir fiber. Materials **13**(12), Article 12 (2020). https://doi.org/10.3390/ma13122753
23. Khuntia, T., Biswas, S.: An investigation on the flammability and dynamic mechanical behavior of coir fibers reinforced polymer composites. J. Ind. Text. **51**(10), 1616–1640 (2022)
24. Korniejenko, K., Frączek, E., Pytlak, E., Adamski, M.: Mechanical properties of geopolymer composites reinforced with natural fibers. Procedia Eng. **151**, 388–393 (2016). https://doi.org/10.1016/j.proeng.2016.07.395
25. Korniejenko, K., Łach, M., Mikuła, J.: The influence of short coir, glass and carbon fibers on the properties of composites with geopolymer matrix. Materials **14**(16), 4599 (2021)
26. Kostic, M., Pejic, B., Skundric, P.: Quality of chemically modified hemp fibers. Biores. Technol. **99**(1), 94–99 (2008). https://doi.org/10.1016/j.biortech.2006.11.050
27. Li, L., Wang, J., Peng, T.: The influence of pH environments on the long-term durability of coir fiber-reinforced epoxy resin composites. Sustainability **17**(1), Article 1 (2025). https://doi.org/10.3390/su17010364
28. Liu, J., Lv, C.: Durability of cellulosic-fiber-reinforced geopolymers: a review. Molecules **27**(3), Article 3 (2022). https://doi.org/10.3390/molecules27030796
29. Lv, C., Liu, J.: Alkaline degradation of plant fiber reinforcements in geopolymer: a review. Molecules **28**(4), 1868 (2023)
30. Lv, C., Liu, J., Guo, G., Zhang, Y.: The mechanical properties of plant fiber-reinforced geopolymers: a review. Polymers **14**(19), 4134 (2022)
31. Lv, C., Wu, D., Guo, G., Zhang, Y., Liu, S., Qu, E., Liu, J.: Effect of plant fiber on early properties of geopolymer. Molecules **28**(12), 4710 (2023)
32. Mahmud, Md.A., Abir, N., Anannya, F.R., Nabi Khan, A., Rahman, A.N.M.M., Jamine, N.: Coir fiber as thermal insulator and its performance as reinforcing material in biocomposite production. Heliyon **9**(5), e15597 (2023). https://doi.org/10.1016/j.heliyon.2023.e15597
33. Moujoud, Z., Sair, S., Ait Ousaleh, H., Ayouch, I., El Bouari, A., Tanane, O.: Geopolymer composites reinforced with natural fibers: a review of recent advances in processing and properties. Constr. Build. Mater. **388**, 131666 (2023). https://doi.org/10.1016/j.conbuildmat.2023.131666
34. Mulana, F., Aulia, M.P., Aprilia, S.: Coconut fiber and fly ash polymer hybrid composite treated silane coupling agent: Study on morphology, physical, mechanical, and thermal properties. S. Afr. J. Chem. Eng. **50**, 10–19 (2024)
35. Nagral, M.R., Ostwal, T., Chitawadagi, M.V.: Effect of curing temperature and curing hours on the properties of geo-polymer concrete. Int. J. Comput. Eng. Res. **4**(9), 1–11 (2014)
36. Narendar, R., Priya Dasan, K.: Chemical treatments of coir pith: morphology, chemical composition, thermal and water retention behavior. Compos. B Eng. **56**, 770–779 (2014). https://doi.org/10.1016/j.compositesb.2013.09.028
37. Olonade, K.A., Savastano Junior, H.: Performance evaluation of treated coconut fibre in cementitious system. SN Appl. Sci. **5**(8), 218 (2023). https://doi.org/10.1007/s42452-023-05444-2
38. Rangan, B.V.: Fly Ash-Based Geopolymer Concrete. Curtin University of Technology, 7 (2008). https://www.academia.edu/download/73106243/Fly_Ash-Based_Geopolymer_Concrete20211018-19768-1t468ih.pdf
39. Ranjbar, N., Zhang, M.: Fiber-reinforced geopolymer composites: a review. Cement Concr. Compos. **107**, 103498 (2020)
40. Rosa, M.F., Medeiros, E.S., Malmonge, J.A., Gregorski, K.S., Wood, D.F., Mattoso, L.H.C., Glenn, G., Orts, W.J., Imam, S.H.: Cellulose nanowhiskers from coconut husk fibers: effect of

preparation conditions on their thermal and morphological behavior. Carbohyd. Polym. **81**(1), 83–92 (2010). https://doi.org/10.1016/j.carbpol.2010.01.059
41. Saeed, A., Najm, H.M., Hassan, A., Sabri, M.M.S., Qaidi, S., Mashaan, N.S., Ansari, K.: Properties and applications of geopolymer composites: a review study of mechanical and microstructural properties. Materials **15**(22), 8250 (2022)
42. Sahu, P., Gupta, M.K.: A review on the properties of natural fibres and its bio-composites: effect of alkali treatment. Proc. Inst. Mech. Eng. Part L: J. Mater. Des. Appl. **234**(1), 198–217 (2020)
43. Sathishkumar, T.P., Navaneethakrishnan, P., Shankar, S., Rajasekar, R., Rajini, N.: Characterization of natural fiber and composites—a review. J. Reinf. Plast. Compos. **32**(19), 1457–1476 (2013)
44. Siakeng, R., Jawaid, M., Ariffin, H., Salit, M.S.: Effects of surface treatments on tensile, thermal and fibre-matrix bond strength of coir and pineapple leaf fibres with poly lactic acid. J. Bionic Eng. **15**, 1035–1046 (2018)
45. Siddique, R., Khan, M.I.: Supplementary Cementing Materials. Springer Science & Business Media (2011)
46. Subramanian, M.: Characterization and Modification of Coir Fibers (n.d.)
47. Tan, H.S., Yu, Y.Z., Liu, L.L., Xing, L.X.: Effect of alkali treatment of coir fiber on its morphology and performance of the fiber/LLDPE bio-composites. Adv. Mater. Res. **139**, 348–351 (2010)
48. Wang, B., Yan, L., Kasal, B.: A review of coir fibre and coir fibre reinforced cement-based composite materials (2000–2021). J. Clean. Prod. **338**, 130676 (2022). https://doi.org/10.1016/j.jclepro.2022.130676
49. Wang, T., Fan, X., Gao, C., Qu, C., Liu, J., Yu, G.: The influence of fiber on the mechanical properties of geopolymer concrete: a review. Polymers **15**(4), 827 (2023)
50. Wong, L.S.: Durability performance of geopolymer concrete: a review. Polymers **14**(5), 868 (2022)
51. Xie, T., Ozbakkaloglu, T.: Behavior of low-calcium fly and bottom ash-based geopolymer concrete cured at ambient temperature. Ceram. Int. **41**(4), 5945–5958 (2015). https://doi.org/10.1016/j.ceramint.2015.01.031
52. Yan, L., Chouw, N., Huang, L., Kasal, B.: Effect of alkali treatment on microstructure and mechanical properties of coir fibres, coir fibre reinforced-polymer composites and reinforced-cementitious composites. Constr. Build. Mater. **112**, 168–182 (2016)
53. Zhang, H.Y., Kodur, V., Qi, S.L., Cao, L., Wu, B.: Development of metakaolin–fly ash based geopolymers for fire resistance applications. Constr. Build. Mater. **55**, 38–45 (2014). https://doi.org/10.1016/j.conbuildmat.2014.01.040

Banana Fibers and Their Advanced Applications

S. Pavithra, J. Madhumitha, R. B. Shivani, R. Vanathi Vijayalakshmi, Mohanapriya Venkataraman, and Jiří Militký

1 Introduction

Over the past years, researchers have placed a high emphasis on renewable, recyclable, and sustainable products. This is mainly due to the rising concern of environmental issues like global warming, the depletion of non-renewable resources, and forest conservation. The textile industry, which has been subjected to fashion trends worldwide, is responsible for producing 5% of landfill waste. The disposal of textile waste leads to many environmental risks [1, 2]. Hence, the textile industry has focused its research on plant-based natural fibers composed of cellulose, given their numerous advantages, such as eco-friendliness, recyclability, and cost-effectiveness [3]. Among them, banana, a bast fiber obtained from the pseudo-stem of the banana plant, has gained attention due to its promising physical, thermal, and mechanical properties, making it a valuable resource for sustainable development.

Banana plantations are mainly cultivated for fruits and are an abundant natural resource in many Asian countries. Almost every part of a banana plantation is harnessed for various purposes, while the trunk, known as the pseudo-stem, is often left behind as agro-waste residue once the fruit is harvested [4]. They are often discarded as waste or burned, resulting in environmental pollution (mainly CO_2 and

S. Pavithra · J. Madhumitha · R. Vanathi Vijayalakshmi (✉)
Department of Physics, Queen Mary's College (Autonomous), Chennai, Tamil Nadu, India
e-mail: r.vanathivijayalakshmi@queenmaryscollege.edu.in

R. B. Shivani
Department of Physics, Presidency College (Autonomous), Chennai, Tamil Nadu, India

M. Venkataraman (✉) · J. Militký
Department of Material Engineering, Faculty of Textile Engineering, Technical University of Liberec, Liberec, Czech Republic
e-mail: mohanapriya.venkataraman@tul.cz

J. Militký
e-mail: jiri.militky@tul.cz

J. Militký and M. Venkataraman (eds.), *Fibrous Structures for Sustainable Future*, Engineering Materials, https://doi.org/10.1007/978-981-95-1382-6_4

CO) [5]. Thus, the usage of this agricultural waste can promote sustainability and contribute to economic growth. As a natural fiber, banana fiber plays a significant role in decomposition due to the presence of cellulose and hemicellulose compounds. Unlike synthetic fiber, which can persist in the ecological community for decades, banana fiber contributes to soil enrichment during decomposition, supporting its sustainable and zero-waste practices.

The chemical composition of natural fibers consists of cellulose, hemicellulose, lignin, pectin, ash, and wax [6], which plays its role in determining fiber strength, flexibility, water repellence, biodegradability, etc. Table 1 illustrates the chemical composition of natural fibers. From Table 1, banana fiber, compared with other fibers, has relatively high cellulose content for the fiber strength and suitability for composites.

The fiber of the banana pseudo-stem has shown good physical properties, such as good absorption capacity and tensile strength. It has also been reported that banana fiber possesses Young's modulus which can be used in industries where durability and lightweight materials are crucial [9].

One of the main reasons for developing natural fibers is to achieve sustainability and reduce reliance on synthetic fibers. Although synthetic fiber usage cannot be

Table 1 Chemical composition of some natural fibers

Fiber	Cellulose (%)	Hemicellulose (%)	Lignin (%)	Pectin (%)	Waxes (%)	References
Banana	62.5	12.5	7.5	4	–	Mejía Osorio et al. [7]
Bamboo	34.5	20.5	26			Mejía Osorio et al. [7]
Cotton	89	4	0.75	6	0.6	Mejía Osorio et al. [7]
Sisal	60	11.5	8	1.2	–	Mejía Osorio et al. [7], Ramesh [8]
Jute	67	16	9	0.2	0.5	Mejía Osorio et al. [7], Ramesh [8]
Coir	46	0.3	45	4	–	Mejía Osorio et al. [7], Ramesh [8]
Hemp	81	20	4	0.9	0.8	Mejía Osorio et al. [7], Ramesh [8]
Flax	70.5	16.5	2.5	0.9	–	Mejía Osorio et al. [7]
Ramie	72	14	0.8	1.95	–	Mejía Osorio et al. [7], Ramesh [8]

entirely replaced, it can be reduced by substituting them with natural alternatives. Therefore, many researchers have focused on optimizing the performance of natural fiber-reinforced composites to reduce the usage of synthetic materials. Banana fibers have demonstrated their potential in various technical applications. They are notably utilized as eco-friendly reinforcement in composite materials, marking a progressive step toward reducing the dependency on synthetic materials in the production of automobile components [10]. They are also employed with chemical surface modification and tailored as banana fiber–reinforced composites for transportation applications [11]. Woven banana fiber composites also exhibit good stability and mechanical strength, with remarkable stress and Young's modulus, demonstrating their capability to withstand substantial loads and deformations, making them ideal for producing fine-quality household products.

The thermal behaviour of banana fiber has confirmed that the fiber has acceptable thermal properties suitable for producing polymer composites [12]. Banana-fiber-reinforced composites exhibit reliable thermal stability, making them suitable for applications [13, 14]. Research indicates that banana fiber-reinforced composite materials enhance fiber strength, and the overall performance of natural fiber composites.

By gaining deeper insights, banana fiber can be innovatively utilized to address environmental challenges posed by waste biomass.

This study aims to investigate the fibers morphology, as well as its flexural, physical, mechanical, and thermal properties, contributing to eco-friendly applications.

2 Morphology of Banana Pseudo-stem and Fibers

Banana plant is producing about 25 of pseudo stems being mature at different times. The banana pseudo-stem is barrel shaped form composed from soft central core and tightly wrapped leaf tails (up to 25) see Fig. 1. Fresh Banana pseudo-stem contains approximately 96% of water. Their main components are relatively high portion (about 73%) of holocellulose (combination of hemicelluloses and cellulose) and low portion (about 10%) of lignin. Fibrous bundles in banana pseudo-stem are composed from elementary fibers and narrow fibers (diameter 3–4 μm) embedded by a non-cellulose membrane. Banana pith is without elementary fibers [15].

Banana pseudo-stem can be source for preparation of super absorbent hydrogel [17] and pseudo stem sap can be used for improvement of flame retardancy of cellulosic fibers [18]. Banana fiber is a side-product of banana plants often treated as waste and not appropriately used [19]. The fiber bundles approximately 5–8 cm long are separated from the remaining parts mechanically by Tuxing following by retting or degumming. For removal of nonfibrous gums composed from arabans and xylans mostly not soluble in water is the degumming obviously realized in alkaline solutions. Banana pseudo stems are used for various practical applications but banana fibers are only about 2% [16] (see Fig. 2).

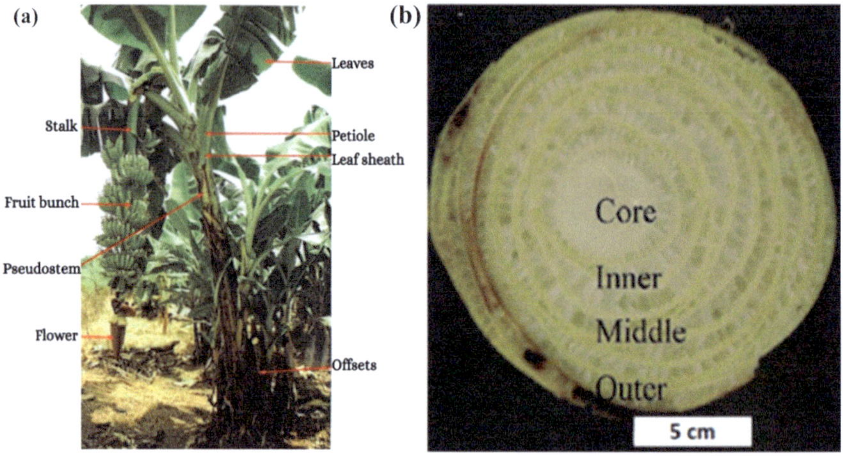

Fig. 1 Banana pseudo-stem **a** longitudinal **b** cross section (reused from Subagyo and Chafidz [16] under CC-BY Permission)

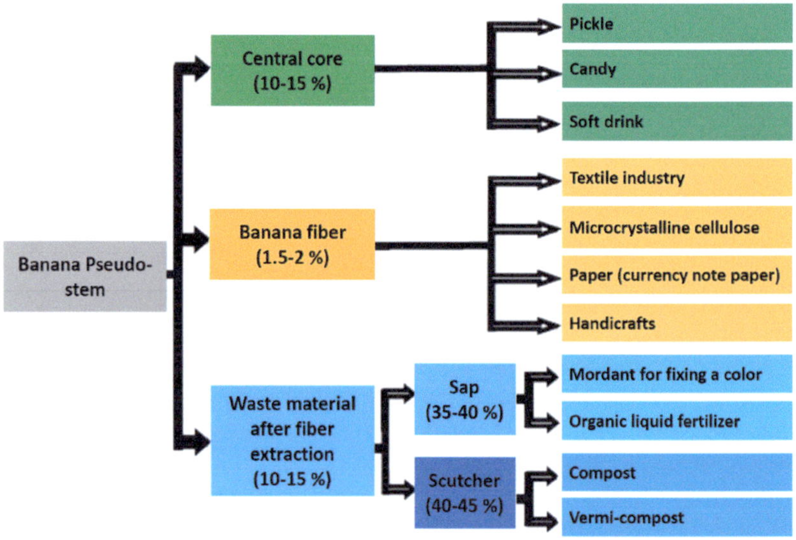

Fig. 2 Applications the banana pseudo-stem components (reused from Subagyo and Chafidz [16] under CC-BY Permission)

Banana plants contain different antimicrobial substances, including dopamine, gentisic acid, ferulic acid, lupeol, and 3-carene and their extracts such as peel, pseudo-stems, and flowers have possible antimicrobial activity [20]. It was found that also banana cellulose fiber have strong antibacterial activity indicated by elevated inhibitory zones [21]. It is in contradiction with other results where antimicrobial

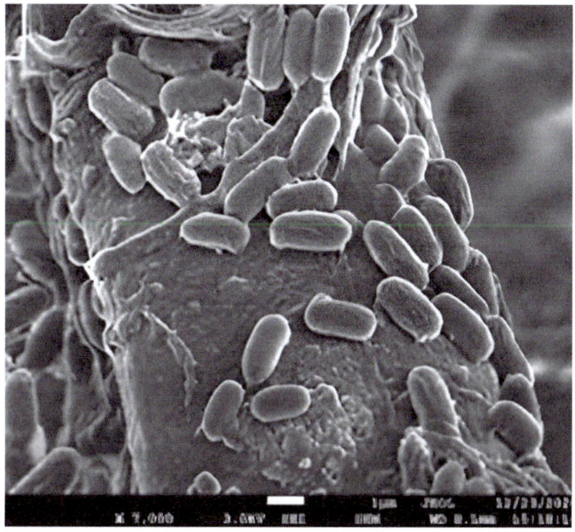

Fig. 3 SEM images of untreated banana fabric showing adhesion/growth of *Escherichia coli* bacteria (reused from Vajpayee et al. [22] under CC-BY Permission)

activity of fabrics from banana fibers were obtained only when proper antimicrobial finishing was used and untreated fabrics were no active (see Fig. 3) [22].

Despite some publications promoting antimicrobial functions of some plant fibers, it was quantitatively found that, e.g., bamboo, flax, and jute fibers have no antibacterial properties and their bacteriostatic action is zero [23].

Morphologically, banana fiber is a multicellular structure composed of very short ultimate (elementary) fibers, like most bast natural fibers. Ultimate (elementary) banana fibers have circular hollow with an outer diameter in the range 10–30 μm and cell wall thicknesses of about 1.25 μm. The ultimate fiber lengths range from 2.7 to 5.5 mm, which is the main limitation for textile use. Banana pseudo-stem fiber has a scaly and fibrillar structure as is shown in Fig. 4 [16].

The cross-section of the pseudo-stem fiber confirms the hollow structure of the ultimate fibers and their relatively thin walls [16].

Microfibrils of banana fibers (diameter 1–2 μm and the length of few mm) are embedded in a matrix of hemicellulose and lignin, forming a complex network. Their

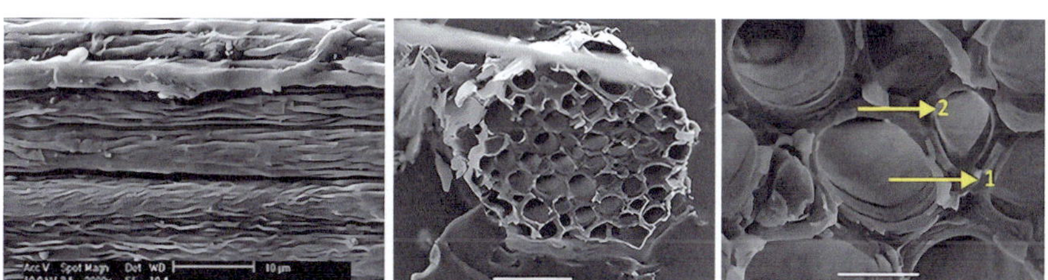

Fig. 4 SEM images of banana fibers (reused from Subagyo and Chafidz [16] under CC-BY Permission)

orientation, whether aligned longitudinally or in a cross-linked pattern, affects the fibers rigidity and stiffness. The cellulose microfibrils present in banana pseudo-stem fibers are arranged in a spiral configuration, wrapping around vascular bundles. This helicoidal orientation provides good tensile strength and allows the fiber to withstand mechanical stresses [22]. Rough surface of banana fibers may enhance bonding with other materials, which is beneficial for creating reinforced composites [24].

3 Processing Fibers from Pseudo-stem

Fiber quality depends on their placement in a banana pseudo-stem, so before extraction, they can be cut into parts for further processing (see Fig. 5).

The extraction of banana fiber is a fundamental step involving the separation of fibrous material from non-cellulosic components such as pith, gums, and lignin. Extraction techniques are broadly classified into manual, mechanical, chemical, and biological methods [25] (see Fig. 6).

Manual extraction, though precise, is labour-intensive and inefficient for large-scale applications. Mechanical extraction, using decorticator machines equipped with rollers, motors, and pulleys, is significantly more efficient—up to 25 times faster than manual methods—but requires immediate post-harvest processing due to the reduced efficacy on dried sheaths. Mechanical extraction involves physically separating banana fibers using methods like decortication, scraping, or crushing. In this method, Cellulose Crystallinity Strong β-glycosidic bonds provide fiber integrity, making mechanical extraction challenging if crystalline cellulose remains intact. Excess lignin makes fibers rigid and brittle, affecting mechanical separation [26]. Hemicellulose and Pectin partial removal through retting softens fibers, making extraction smoother.

Fig. 5 Main parts of banana pseudo stem (reused from [54] under CC-BY Permission)

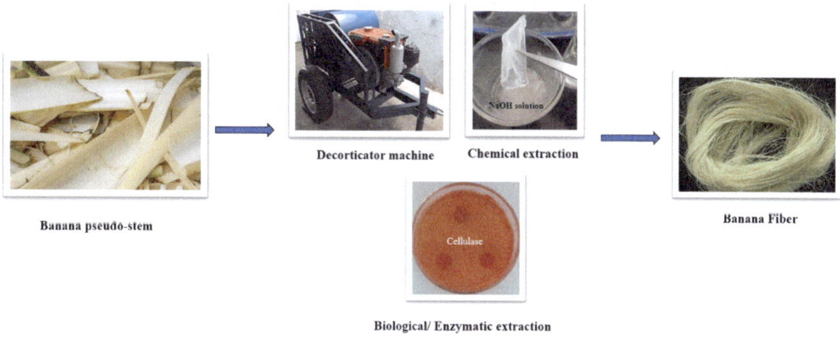

Fig. 6 Banana fiber extraction methods

The effectiveness of retting and mechanical extraction largely depends on the chemical composition of banana fiber [26], particularly the interactions between cellulose, hemicellulose, lignin, and pectin.

Thus, optimizing retting and mechanical processing techniques based on fiber composition enhances quality and efficiency, ensuring sustainable fiber extraction without excessive chemical usage.

Chemical extraction involves treatment with alkalis, acids, or bleaching agents like NaOH, hypochlorite, or hydrogen peroxide to remove lignin and other non-cellulosic substances, producing high-purity fibers. However, this approach raises environmental concerns due to chemical effluents and potential fiber degradation. Biological extraction, leveraging microbial or enzymatic retting processes using species like *Bacillus* or *Fusarium*, offers an eco-friendly alternative that preserves fiber integrity but requires longer treatment durations and controlled conditions.

Given the balance between efficiency and sustainability, mechanical methods are most explored, while chemical methods are often avoided in environmentally focused studies.

Following extraction, the raw banana fibers are still rough and rigid. To make them suitable for spinning into yarn, they undergo softening treatments, which modify the fiber texture and improve flexibility, fineness, and hand feel. These treatments are also categorized into mechanical, chemical, and biological methods (see Fig. 7). Mechanical softening techniques such as beating, carding, and roller pressing physically break down the stiffness and align the fibers, making them more pliable. Chemical softening uses agents such as cationic softeners, silicone emulsions, or mild alkalis to modify the fiber surface. These chemical agents are particularly effective in improving the feel of the fiber but must be handled with care to avoid environmental and health issues. Biological softening, similar in principle to biological extraction, uses specific enzymes like cellulase and pectinase to selectively degrade coarse or rigid components of the fiber surface, resulting in a smoother and more textile-friendly material [25]. This method is gaining popularity for its sustainability and minimal impact on fiber strength.

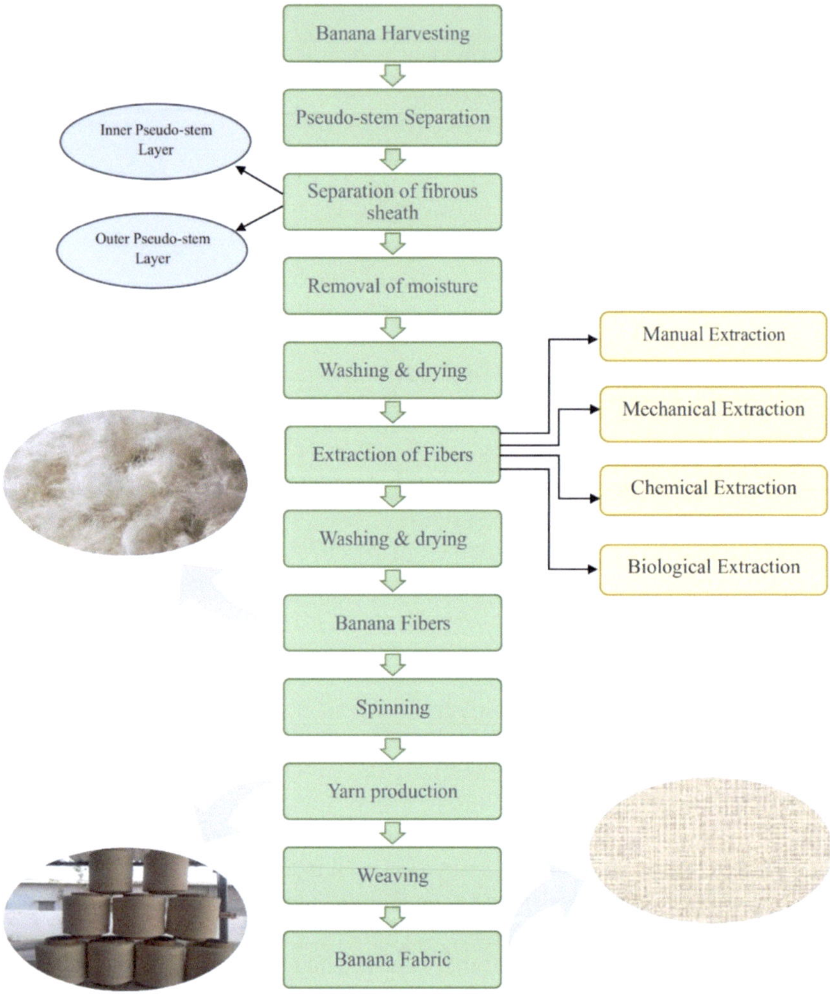

Fig. 7 Processing of banana fibers

Banana yarns have a distinct and appealing hand feel that makes them suitable for a variety of textile applications. They are typically soft and silky, especially when derived from the inner bark (pseudo stem) of the banana plant, offering a comfortable sensation against the skin. The yarns often feel smooth to the touch, with a texture comparable to bamboo or silk.

4 Alkali Treatment of Banana Fabric

The surface modifications and partial removal of lignin, hemicellulose and other impurities from grey banana fabric is realized by alkali treatment under mild conditions. This treatment is used also for enhancing interphase area between fibers and matrix in production of composites. The treatment of fiber with NaOH is creating more accessible amorphous region on crystalline cellulose by breaking and reforming of some hydrogen bonds between cellulose chains. However, increasing alkali concentration causes "delignification" of matrix joining the microfibrils, separation of elementary fibers and great shrinkage of fabric.

Banana fabric from retted fibers are usually treated with 10% NaOH at room temperature for two hours. This treatment is also altering surface roughness, porosity, and crystallinity. The influence of NaOH-treatment under these conditions on properties of banana fabric and their profound impact on fiber properties is discussed in sequel. Key methods, including density measurement, air permeability tests for breathability assessment, and microscopic analysis to investigate fiber structure, provide a comprehensive outline of textile behaviour. Figure 8 shows a more open and porous structure, which can improve adhesion between fibers and matrix in composites.

The microfibril rearrangements can disrupt some hydrogen bonds present in the cellulose structure and reduce fiber strength [15]. The moderate tensile strength suggests that cellulose microfibrils have undergone reorientation [27]. The excessive disruption of hydrogen bonds can reduce overall stiffness of fibers. The modulus being relatively low (refer Table 7) suggests that hemicellulose and pectin removal have softened the fiber structure. Hemicellulose, responsible for binding cellulose

(a) (b)

Fig. 8 SEM images of NaOH-treated banana fabric

fibrils, has been degraded, decreasing fiber rigidity. Pectin, which holds fiber components together, has also been partially dissolved, leading to increased weakness [10].

The structural rearrangement of banana fiber significantly impacts its density, porosity, and surface topology. Raw banana fiber contains natural impurities and hemicellulose, which contribute to higher air permeability. The removal of non-cellulosic materials such as hemicellulose, lignin, and pectin leads to a decrease in fiber density, increasing porosity, making it lighter and more breathable. Additionally, alkali treatment induces fiber swelling, altering microfibril angles, which improves mechanical performance. The naturally high porosity of banana fiber facilitates rapid water absorption and dye penetration [10, 28], making it highly beneficial for fiber processing. Furthermore, NaOH treatment roughens the fabric surface (Fig. 8), enhancing adhesion and mechanical interlocking in composite applications.

Energy-dispersive X-ray spectra of NaOH-treated Banana Fabric indicate the presence of carbon (C) and oxygen (O) elements (see Fig. 9). This indicates a predominantly organic composition, which is expected for natural fibers like banana cellulose. The presence of oxygen suggests oxidized cellulose structures, potentially due to processing treatments like alkali (NaOH) treatment or environmental exposure. High carbon content suggests a dominant cellulose structure.

Lower fabric density indicates a looser weave with more spaces between fibers, allowing greater airflow and fluid penetration [13, 29]. The measured mean air permeability value of 24.55 $cm^3/cm^2/s$ indicates that the NaOH treatment enhances the breathability of the fabric, making it suitable for applications where airflow is desirable. Depending on the concentration and duration of treatment, this modification can lead to improvements in comfort, moisture management, and even potential application in sustainable textiles.

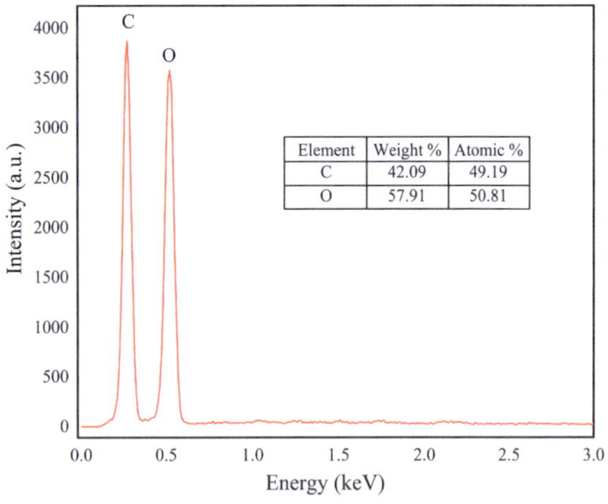

Element	Weight %	Atomic %
C	42.09	49.19
O	57.91	50.81

Fig. 9 Energy-dispersive X-ray spectroscopy of NaOH-treated banana fabric

5 Physical Properties

Banana fiber effectiveness is largely determined by fiber density, lightweight properties, and strength-to-weight ratio, which define their performance and application. Fiber density is a fundamental property that influences the selection and performance of natural fibers in engineering and environmental applications [30]. Table 2 shows the density ranges for typical natural fibers [31].

Among the critical physical properties of natural fibers, density plays a significant role in determining the fibers suitability for various applications, especially in lightweight structural components [31]. The density of the fibers is obviously measured by pycnometers.

Banana fibers have a density ranging from 1.35 to 1.40 g/cm^3, which is relatively lower compared to fibers like ramie (1.38–1.6 g/cm^3) [30, 32] and significantly lower than glass fibers (2.5 g/cm^3). The strength-to-weight ratio determines how much load banana fabric can withstand while remaining lightweight. From the tensile test, NaOH-treated banana fabric has a strength range of 3.09 MPa, which makes them less strong than the cotton and hemp.

Banana fabric is naturally lightweight, making it a preferred choice for breathable, and wearable textiles. With a moisture content of 12.37%, banana fabric remains breathable and comfortable while preventing excessive water retention [33].

Capillary action is the ability of a liquid to flow in narrow spaces without the assistance of external forces. This occurs due to surface tension and adhesion. In banana fiber, surface tension arises because water molecules exhibit strong cohesive forces, making the liquid seek minimal surface area. Adhesion is the attraction between water molecules and the cellulose-based structure of banana fiber. Because banana fiber has micro-capillaries formed by its porous nature, water can travel through these small channels against gravity, making it effective in moisture absorption. Fibers porosity refers to the void spaces within them. Banana fiber is highly porous, meaning it has

Table 2 Fiber density of some natural fibers

Fiber	Density (g/cm^3)
Banana	1.35–1.40
Sisal	0.76–1.58
Cotton	1.5–1.54
Coir	1.25
Jute	1.3–1.46
Kenaf	1.31–1.5
Pineapple	1.32–1.54
Ramie	1.38–1.6
Flax	1.4–1.5
Hemp	1.47–1.48
Abaca	1.5

interconnected spaces that help in water retention. Surface tension occurs because molecules at the surface experience unbalanced cohesive forces, creating a tendency to minimize surface area.

Banana fiber exhibits strong hydrophilic properties due to its high cellulose content and porous structure. The surface tension of water (~ 72.8 mN/m at 25 °C) affects how liquid spreads or interacts with fiber surfaces. Since morphologically NaOH-treated banana fiber has a naturally rough and porous morphology, water molecules experience capillary action, leading to efficient moisture absorption.

Contact angles for water very much below 90° (around 39°) indicate good wettability; the NaOH-treated banana fabric is hydrophilic and strongly interacting with water. Since banana fibers are rich in cellulose and hemicellulose, they contain numerous hydroxyl (–OH) groups, which can form hydrogen bonds with water molecules. The crystalline regions of cellulose do not allow water molecules to penetrate deeply. Only amorphous regions of cellulose show some degree of water interaction, contributing to water retention. This means the high adhesion between water and the fabric surface, promoting capillary action and strong moisture absorption, allowing efficient water uptake in small fiber capillaries due to the chemical composition [33, 34]. Enhanced cohesion within water molecules also enables continuous flow through porous channels. This confirms that banana fabric absorbs water more easily.

The lower contact angle suggests strong adhesion, allowing rapid water absorption via capillary action [35]. This makes banana fiber suitable for applications in moisture-wicking textiles, eco-friendly absorbents, and filtration systems.

6 Mechanical Properties

The mechanical properties of natural fibers, such as tensile strength, initial modulus (stiffness), elongation at break, are pivotal in determining their applicability in industrial settings. Banana fibers are emerging as a promising natural material that is versatile and biodegradable, making them a sustainable option for medium-load applications such as textiles, composites, and reinforcement structures. The mechanical behaviour of banana fiber is largely influenced by its morphology (circular hollow) high cellulose content, significantly contributing to its rigidity and strength. Additionally, compared to cotton, banana fibers have a much higher lignin content, further modifying their structural integrity [36].

NaOH-treated banana fabric was subjected to a tensile test, (ASTM E8) for determining its mechanical properties. When a fiber is subjected to an increasing tensile load, its length changes [37].

Once the fibers align completely, they start resisting the applied force in a linear proportional manner. At this point, the fabric following the equation, $\sigma = E\varepsilon$, where the stress increases in direct proportion to the strain. Modulus E is here characterizing initial stiffness (resistance to deformation). The initial modulus for the unbleached banana fiber was 2.21–4.23 GPa, which suggests that the fiber has relatively low stiffness compared to other fabrics like from cotton. The peak of the stress-strain curve arises around 6 MPa, indicating break. Summary of the mechanical properties of NaOH-treated banana fabric is presented in the Table 3 and for bleached and unbleached banana fabric is given in Table 4.

The alkali treatment increase the aesthetic and tactile properties of banana fabric, they also compromise its mechanical durability, highlighting the need for optimized treatment conditions that balance fiber softness with structural integrity [38].

For comparison are the basic mechanical characteristics of natural cellulose fibers compared in Table 5.

The most of the natural fibers exhibit high tenacity and modulus than banana fibers. To enhance mechanical performance, banana fiber is blended with synthetic or natural reinforcements, forming composites.

The mechanical properties of banana fiber greatly vary depending on the reinforcement material used. Table 6 indicates that pineapple leaf composites demonstrated the high tensile strength (74.5 MPa), making them ideal for strong structural applications. On the other hand, banana-hemp blends exhibited the lowest tensile

Table 3 Summary of the mechanical properties of NaOH-treated banana fabric

Mechanical properties	Banana fabric (unbleached)
Maximum tensile strain (%)	2.6
Tensile stress at maximum tensile strain (MPa)	3.09
Breaking force at maximum tensile strain (N)	70
Tenacity at maximum tensile strain (cN/tex)	112.65

Table 4 Mechanical properties of unbleached and bleached banana fiber

Fibers	Elongation (%)	Tenacity (cN/tex)	Modulus (GPa)	References
Bleached banana fiber	1.76–3.61	13.3–34.7	2.31–3.25	Nkemaja et al. [27]
Unbleached banana fiber	1.63–3.06	11.2–23.5	2.21–4.23	Nkemaja et al. [27]

Table 5 Mechanical properties of main natural fibers

Fibers	Elongation at break (%)	Tenacity (cN/tex)	Modulus (GPa)	References
Cotton	7–8	20–40	5–13	Hsieh [39]
Jute	1.5–3.1	26.5–51.2	13–26.5	Kozlowski et al. [38]
Hemp	1.6–4.5	35.2–48.5	70	Kozlowski et al. [38], Ramesh [8]
Coir	30	10.5–18	4–6	Ramesh [8]
Flax	1.2–3	50.3	27.6	Kozlowski et al. [38], Ramesh [8]

Table 6 Tensile strength of banana blended composites

S. No.	Blend	Tensile strength	References
1.	Banana fiber–glass fiber	39.5	Bhoopathi et al. [40]
2.	Banana fiber–hemp	28	Bhoopathi et al. [40]
3.	Banana fiber–pineapple leaves	74.5	Ramesh et al. [41]
4.	Banana fiber–polyester (composite)	53	Pothan et al. [42]
5.	Banana fiber–epoxy resin (composite)	45.57	Maleque et al. [43]

strength (28 MPa). Each blend has specific advantages, making banana fiber-based composites a sustainable alternative to synthetic reinforcements.

Many industrial applications often require fibers that can withstand higher stress before failure, like hemp and jute [8, 38]. Unlike hemp, which has a significantly higher tensile modulus, banana fiber lacks structural rigidity. Natural banana fiber in its unprocessed form might not withstand extreme mechanical stress, limiting its direct use in high-performance composites. To improve industrial viability, several surface modification and reinforcement techniques are adopted, and many reinforced banana composites are produced.

7 Thermal Properties

Banana fiber as a sustainable material with good insulating properties and low thermal conductivity, making it a viable alternative for thermal insulation applications in construction, textiles, and composites. Morphologically, banana fiber is naturally porous, containing hollow lumens that significantly slow heat transfer. This unique structural composition, coupled with the fibers cellulose and hemicellulose components, makes it a thermal insulator.

Banana fiber has also been studied in various composite formulations to evaluate its thermal insulation efficiency. The composite containing 60% banana fiber, combined with both pineapple and jute fiber, is considered a better insulator, offering maximum thermal resistance and minimum thermal conductivity. It is noted that the thermal resistance increased with fiber thickness, confirming its insulation properties, reinforcing the potential of banana fiber composites for energy-efficient insulation applications [44]. It is observed that alkali-treated banana composite improved fiber adhesion and crystallinity, enhancing its insulating performance [44].

Thermogravimetric analysis (TGA) of banana fabric were conducted under a nitrogen atmosphere to understand its thermal stability and degradation profile. The sample was heated from room temperature to 800 °C at a constant rate of 10 °C/min. The TGA curve (Fig. 10) exhibited distinct stages of weight loss. The first stage below 150 °C showed a 17.86% weight loss, primarily due to desorption of moisture and volatile agents that are common in natural fabrics subjected to mild chemical processing. This was followed by a 7.34% loss in the range of approximately 150–250 °C, attributed to the decomposition of low molecular weight waxes, pectin, and hemicellulose (started to be decomposed at approximately 178 °C).

The major decomposition occurred between 250 and 400 °C with a 63.13% weight loss, corresponding to the thermal degradation of cellulose (at approximately 296 °C). Between 400 and 600 °C, the sample underwent a further 26.25% weight loss due to the progressive breakdown of lignin and thermal carbonization of the previously charred structure. A final 9.20% loss from 600 to 800 °C was observed, likely due to the breakdown of thermally stable aromatic structures, leaving behind a substantial residual mass of 23.78%, indicative of enhanced char-forming ability [45].

The comparison of thermal degradation of banana, jute, flax and cotton fibers is given in Table 7.

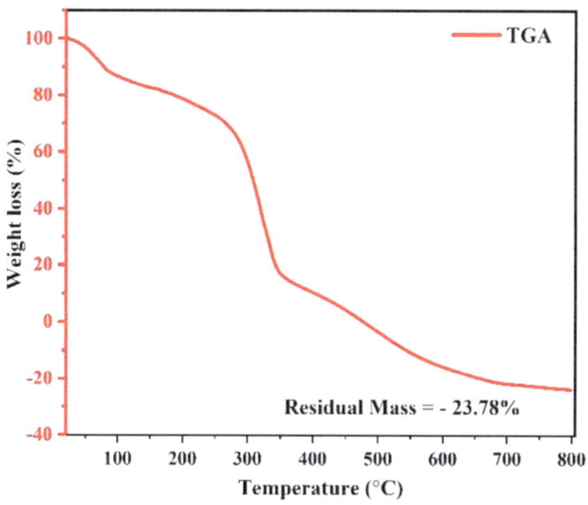

Fig. 10 Thermogravimetric analysis (TG) curves of NaOH-treated banana fabric sample

Table 7 TG Comparison of banana, jute, flax and cotton fibers

Fabric type	Major components	Initial degradation temp (°C)	Major weight loss (%)	Residual mass @ 800 °C (%)	References
Jute	High cellulose, moderate lignin	250–300	61–65%	18–20%	Ovalı et al. [45]
Flax	Rich in cellulose, low lignin	290–320	70%	12–15%	Wang et al. [46]
Cotton	Pure cellulose	300–340	60%	1.3–2%	Yiga et al. [47]
Banana	Cellulose, hemicellulose, lignin	240–270	63.13% (250–400 °C)	23.78	This work

Table 7 findings affirm that banana fabric exhibits moderate thermal stability with char residue, suggesting potential applicability in thermal barrier mats, and composite linings [47].

8 Chemical Composition of Alkali Treated Banana Fibers

It is well known that cellulose serves as the primary structural component of plant fibers [8, 39], followed by hemicellulose and lignin in varying proportions, and pectin, respectively. To investigate crystallinity and molecular arrangement of NaOH treated banana fabric, the X-ray Diffraction (XRD) was performed.

The XRD pattern (Fig. 11), conducted from 10° to 80° (2θ), reveals peaks at 12.24°, 20.01°, and 22.03° 2θ, which indicate key structural attributes of the treated Banana fabric. The peak at 12.24° (2θ) represents amorphous cellulose regions. NaOH treatment causes partial disruption of hydrogen bonding in cellulose, which increases fiber elongation and moisture retention.

The peaks at 20.01° and 22.03° (2θ) correspond to highly crystalline cellulose structures. Generally, these regions in banana fabric restrict movement, making the fabric more rigid. However, NaOH treatment in banana fabric reduces peak intensity or broadens peaks, signalling partial transformation of crystalline cellulose I, which affects rigidity. It is observed that the tensile test results indicate low stiffness (modulus), indicating that alkali treatment on banana fabric has disrupted some crystalline regions and removed portions of hemicellulose and lignin, contributing to structural modifications. The broadening of XRD peaks reflects a loss of structural order, leading to decreased rigidity.

Fourier-Transform Infrared Spectroscopy (FTIR) was conducted on NaOH-treated Banana Fabric to analyse its chemical bonding and structural integrity (see Fig. 10). Table 8 compares the FTIR absorption peaks of NaOH-treated banana fabric

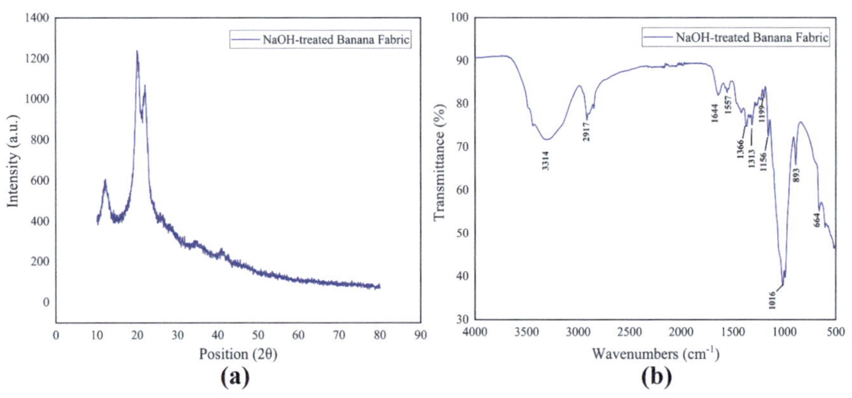

Fig. 11 **a** XRD pattern and **b** FTIR analysis of NaOH-treated banana fabric

with reference wavenumbers commonly associated with cellulose, hemicellulose, pectin, lignin, and other functional groups.

Peaks at 3314 cm^{-1} (O–H stretching), suggesting moderate hydrogen bonding density in cellulose, and 2917 cm^{-1} (C–H stretching) correspond to aliphatic hydrocarbons from cellulose [48], and O–H alcohol (stretching) indicates the presence of pectin and hemicellulose [49]. 1644 cm^{-1} (C=O stretching) aligns with hemicellulose and pectin, indicating their contribution to water retention. 1557 cm^{-1} (Aromatic C=C stretching) confirms lignin presence, responsible for mechanical rigidity. 1366 cm^{-1} (C–H deformation) is linked to methoxy groups in lignin, which affect the hydrophobicity of banana fabric. 1199 and 1156 cm^{-1} (C–O–C stretching) suggest the ether linkages in the cellulose network, affecting fiber durability. 1016 cm^{-1} (C–O stretching) suggests hemicellulose side-group interactions. These side-groups play a role in fiber flexibility and moisture absorption, making the fabric softer and more adaptable. β-glyosidic bond (893 cm^{-1}) confirms the presence of polymeric cellulose units. The lower β-glycosidic bonding affects the better fiber integrity which means the cellulose is not fully crystalline but not amorphous either. 664 cm^{-1} (O–H bending) reflects pectin and lignin [50, 51], possibly influencing fabric hydration behaviour. This strong O–H bending signal implies pectin's role in absorbing and interacting with moisture. While lignin is generally hydrophobic, it still possesses hydroxyl (–OH) groups within its structure.

The structural modifications observed in NaOH-treated banana fabric confirm that alkali treatment significantly impacts crystallinity, tensile properties, and hydration behaviour [39]. The broadening of XRD peaks suggests disruption of highly ordered cellulose regions, making the fiber less rigid. FTIR analysis further supports this by revealing moderate hydrogen bonding density, altered ether linkages (1199, 1156 cm^{-1}), and partial lignin reduction, all of which influence fiber durability and moisture interaction. However, despite these modifications, the NaOH treatment

Table 8 The FTIR Transmittance peaks of NaOH-treated banana fabric

Wavenumbers (cm^{-1})	Functional group	Expected component
3314	O–H stretching (hydroxyl)	Cellulose, pectin
2917	C–H stretching (alkane) O–H stretching (alcohol)	Cellulose, hemicellulose, pectin
1644	C=O stretching	Hemicellulose, pectin
1557	Aromatic C=C stretching	Lignin
1366	C–H deformation (methyl groups)	Lignin, cellulose
1313	C–O stretching	Cellulose, hemicellulose
1199	C–O–C asymmetric stretching	Cellulose
1156	C–O–C asymmetric stretching	Cellulose, hemicellulose
1016	C–O stretching	Cellulose, hemicellulose
893	C–H glycosidic bond	Cellulose, pectin
664	O–H bending	Pectin, lignin

was not fully effective in completely removing hemicellulose and lignin from the fiber. FTIR results still show some stretching related to hemicellulose side-group interactions, which continue to affect fiber flexibility and moisture absorption. Additionally, the aromatic C=C stretching (1557 cm^{-1}) and methyl groups (1366 cm^{-1}) indicate lignin presence, meaning some degree of structural rigidity and hydrophobicity remains. These residual components limit the extent to which the fiber becomes highly crystalline and purely cellulosic.

This incomplete removal affects the mechanical properties of banana fabric, as observed in the tensile test results, which indicate lower stiffness and reduced modulus. The remaining hemicellulose and lignin contribute to structural modifications but prevent the fiber from achieving fully optimized tensile strength. Furthermore, the strong O–H bonding (664 cm^{-1}) in FTIR suggests pectin's continued role in moisture interaction, reinforcing hydration behaviour rather than pure hydrophobicity. This finding is crucial for future optimization of fiber treatment techniques aimed at maximizing mechanical strength, moisture resistance, and industrial usability.

9 Applications of Banana Fabric

Banana fiber has a broad spectrum of global applications due to its properties, sustainability, and versatility, which makes it a perfect choice for eco-conscious consumers (see Table 9). Banana fiber is an eco-friendly alternative that is 100% plant-based and free from harmful chemicals. Its biodegradable nature making it a responsible choice for sustainable fashion.

Table 9 Applications of banana fabric

Application	Products
Fashion and Apparel	Breathable textiles for warm climates Lightweight clothing Casual and summer wear
Industrial	Paper products—tissue paper, parchment paper, wipes Geotextiles—ropes, mats Automobile—seat belts, floor coverings Aerospace—lightweight air craft components Sound-proofing materials Heat-insulated construction products
Packaging	Biodegradable food packaging Plastic-free storage containers Wrapping covers
Decorative	Cushion covers, curtains and table runners Wall hangings Decorative mats Woven baskets and storage boxes

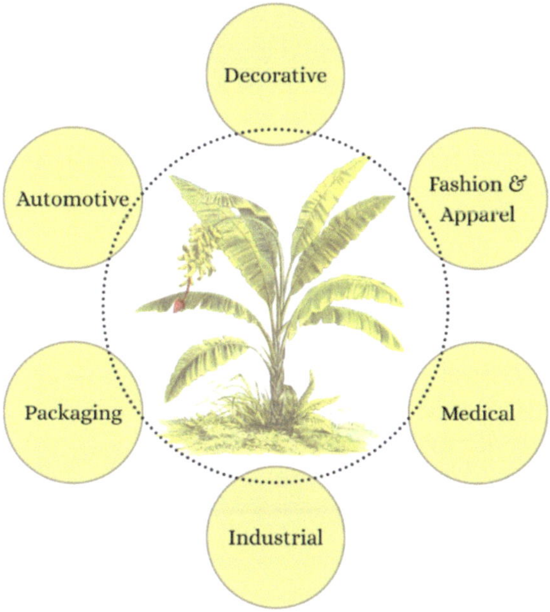

Fig. 12 Applications of banana fiber

Fashion and Apparel

With its moisture-regaining properties, banana fiber fabric offers a cool, breathable experience, ideal for hot and humid climates [30, 32]. Its lightweight texture ensures effortless wear, making it perfect for everyday comfort in textile (see Fig. 12). Biodegradable textiles crafted from banana fiber enhance eco-conscious design choices.

Industrial Textiles

Traditionally, banana fiber composites have been utilized in paper-making, rope, and mat production, offering an eco-friendly material due to its tensile strength, light weight, and moisture-absorbing properties. Many varieties of paper products like tissue paper, wipes, parchment paper, and card sheets are produced by altering the thickness of banana fiber composite. The tensile strength of banana fiber paper is higher than that of traditional wood pulp paper [3, 52]. As an alternate material to fiberglass, banana fiber has also been exposed to composite materials in the production of automobile products like seat backs, door panels, floor coverings, transmission covers, etc. Lightweight yet durable materials are crucial across various industries, from automotive to sports equipment and construction. The interplay between fiber density and strength-to-weight ratio determines a material's efficiency, usability, and performance. Studies indicate banana fiber composites can be used due to their good strength-to-weight ratio. Combined with reinforced components that provide impact resistance while staying lightweight, plays its role in automotive applications. Composite panels offer durability without adding too much weight to construction [36].

Banana fiber has been found to have low thermal conductivity, meaning it does not easily transfer heat. This property makes it highly suitable for use in construction, especially in applications requiring [53]. When incorporated into composites, banana fiber can help regulate temperature by reducing heat absorption and providing insulation. This makes it valuable for building materials such as wall panels, roofing, and flooring that need to withstand extreme climatic temperatures without degrading. Much ongoing research has been conducted on soundproofing material using banana fibers due to its sound-absorbent properties.

Decorative Textiles

Banana fiber unique texture, moisture-absorbing properties, and sustainability make it an excellent option for decorative applications. Woven banana fiber is crafted into wall hangings, decorative mats, and table runners. Banana fabric is also used in cushion covers, curtains, and upholstery due to its natural texture and elegant finish. Many often combine banana fiber with other natural fibers like jute and cotton to create intricate patterns for textured wallpapers and accent fabrics in luxury settings.

Packaging

In Food Packaging, Banana fiber-based biodegradable packaging is being explored to replace plastic-based food containers [3, 52]. In addition, the production of banana fiber packaging does not require harmful chemicals, making it safer for food storage.

Despite its advantages, banana fiber faces scientific challenges and limitations that hinder widespread adoption, including processing difficulties, standardization issues, limited awareness, and durability concerns. However, it aligns with the circular economy model, repurposing agricultural waste into valuable products rather than leaving it unused. By investing in advanced processing techniques, fiber-strengthening treatments, and industry awareness, researchers and innovators are working to overcome these drawbacks.

10 Conclusion

Banana fiber stands as a promising, eco-conscious material with remarkable potential across various industries. Its high cellulose content and biodegradable nature contribute to sustainable product development, reinforcing the global movement toward green materials. From textiles to composites, its natural properties position it as a viable alternative to synthetic fibers, offering reduced environmental impact without compromising functionality. In this chapter, NaOH treatment was used on the banana fabric at room temperature, and its physical, chemical, and mechanical properties were examined. The effects of NaOH treatment on banana fabric have proven better surface roughness and porosity. These modifications improve its compatibility with various matrix materials, making it ideal for applications requiring breathability and moisture management. However, the disruption of hydrogen bonding slightly limits tensile strength, suggesting that optimization techniques could further enhance structural integrity.

Physically, banana fabric exhibits a lightweight nature and strong moisture absorption, enabling its integration into breathable textiles and moisture-retentive applications such as absorbents and wicking fabrics. Although its mechanical properties, such as tensile strength and stiffness, are moderate, blending with other high-performance natural fibers, such as pineapple or hemp, presents opportunities for reinforced composites with superior durability. High char residue formation making banana fiber a strong candidate for eco-insulation solutions. These characteristics expand its utility beyond textiles to construction materials and protective gear. Chemical assessments, including XRD and FTIR studies, confirm the partial breakdown of lignin and hemicellulose during alkali processing. While this transformation improves fiber hydration, residual lignin and incomplete hemicellulose removal slightly hinder full optimization of mechanical performance.

Future research should explore alternative treatment methods to refine these aspects, ensuring that banana fiber reaches its highest potential. Overall, banana fiber showcases an exceptional balance of sustainability, functionality, and adaptability. Continued research and technological advancements in fiber processing and composite integration will further solidify its role as a valuable material in eco-friendly innovations, supporting industries seeking greener alternatives to conventional fibers. While it may not serve as a complete substitute for synthetic fibers, its unique properties offer a sustainable alternative that can effectively reduce dependency on synthetic materials.

References

1. Baloyi, R.B., Gbadeyan, O.J., Sithole, B., Chunilall, V.: Recent advances in recycling technologies for waste textile fabrics: a review. Text. Res. J. **94**(3–4), 508–529 (2024). https://doi.org/10.1177/00405175231210239
2. Brusseau, M.L.: Sustainable development and other solutions to pollution and global change. In: Environmental and Pollution Science, pp. 585–603. Elsevier (2019). https://doi.org/10.1016/b978-0-12-814719-1.00032-x
3. Arafat, K., Nayeem, J., Quadery, A., Quaiyyum, M., Jahan, M.S.: Handmade paper from waste banana fiber. Bangladesh J. Sci. Ind. Res. **53**(2), 83–88 (2018). https://doi.org/10.3329/bjsir.v53i2.36668
4. Zou, F., Tan, C., Zhang, B., Wu, W., Shang, N.: The valorization of banana by-products: nutritional composition, bioactivities, applications, and future development. Foods **11**(20), 3170 (2022). https://doi.org/10.3390/foods11203170
5. Sullivan, A.L., Ball, R.: Thermal decomposition and combustion chemistry of cellulosic biomass. Atmos. Environ. **47**, 133–141 (2012). https://doi.org/10.1016/j.atmosenv.2011.11.022
6. Balda, S., Sharma, A., Capalash, N., Sharma, P.: Banana fiber: a natural and sustainable bioresource for eco-friendly applications. Clean Techn. Environ. Policy **23**(5), 1389–1401 (2021). https://doi.org/10.1007/s10098-021-02041-y
7. Mejía Osorio, J.C., Rodríguez Baracaldo, R., Olaya Florez, J.J.: The influence of alkali treatment on banana fiber's mechanical properties. Ingeniería e Investigación **32**(1), 83–87 (2012). https://doi.org/10.15446/ing.investig.v32n1.28524

8. Ramesh, M.: Hemp, jute, banana, kenaf, ramie, sisal fibers. In: Handbook of Properties of Textile and Technical Fibers, pp. 301–325. Elsevier (2018). https://doi.org/10.1016/B978-0-08-101272-7.00009-2
9. Jústiz-Smith, N.G., Virgo, G.J., Buchanan, V.E.: Potential of Jamaican banana, coconut coir and bagasse fibers as composite materials. Mater. Charact. **59**(9), 1273–1278 (2008). https://doi.org/10.1016/j.matchar.2007.10.011
10. Pothan, L.A., Thomas, S.: Polarity parameters and dynamic mechanical behaviour of chemically modified banana fiber reinforced polyester composites. Compos. Sci. Technol. **63**(9), 1231–1240 (2003). https://doi.org/10.1016/S0266-3538(03)00092-7
11. Sivaranjana, P., Arumugaprabu, V.: A brief review on mechanical and thermal properties of banana fiber based hybrid composites. SN Appl. Sci. **3**(2) (2021). https://doi.org/10.1007/s42452-021-04216-0
12. Sathasivam, K., Haris, M.R.H.M.: Thermal properties of modified banana trunk fibers. J. Therm. Anal. Calorim. **108**(1), 9–17 (2012). https://doi.org/10.1007/s10973-011-1793-1
13. Milani, M.D.Y., Samarawickrama, D.S., Dharmasiri, G.P.C.A., Kottegoda, I.R.M.: Study the structure, morphology, and thermal behavior of banana fiber and its charcoal derivative from selected banana varieties. J. Nat. Fibers **13**(3), 332–342 (2016). https://doi.org/10.1080/15440478.2015.1029195
14. Rukundo, A.: Extraction, treatment and characterization of banana pseudo-stem fibers as potential utility in textile industry. IDOSR J. Biochem. Biotechnol. Allied Fields **9**(1), 23–39 (2024). https://doi.org/10.59298/IDOSR/JBBAF/24/91.2339
15. Li, K., Fu, S., Zhan, H., Zhan, Y., Lucia, L.A.: Analysis of the chemical composition and morphological structure of banana pseudo-stem. Bioresources **5**(2), 576–585 (2010). https://doi.org/10.15376/biores.5.2.576-585
16. Subagyo, A., Chafidz, A.: Banana pseudo-stem fiber preparation, characteristics, and applications. In: Banana Nutrition—Function and Processing Kinetics (2020). https://doi.org/10.5772/intechopen.82204
17. Bello, K., Sarojini, B.K., Narayana, B., Rao, A., Byrappa, K.: A study on adsorption behavior of newly synthesized banana pseudo-stem derived superabsorbent hydrogels for cationic and anionic dye removal from effluents. Carbohydr. Polym. **181**, 605–615 (2018). https://doi.org/10.1016/j.carbpol.2017.11.106
18. Basak, S., Samanta, K.K., Saxena, S., Chattopadhyay, S.K., Narkar, R., Mahangade, R., Hadge, G.B.: Flame resistant cellulosic substrate using banana pseudostem sap. Green Sci. **17**(1), 123–133 (2015). https://doi.org/10.1515/pjct-2015-0018
19. Kavitha, V., Aparna, G.: A review on banana fiber and its properties. Asian J. Pharm. Res. Dev. **9**(3), 118–121 (2021). https://doi.org/10.22270/ajprd.v9i3.956
20. Mostafa, H.S.: Banana plant as a source of valuable antimicrobial compounds and its current applications in the food sector. J. Food Sci. **86**(9), 3778–3797 (2021). https://doi.org/10.1111/1750-3841.15854
21. Thandavamoorthy, R., Devarajan, Y., Kaliappan, N.: Antimicrobial, function, and crystalline analysis on the cellulose fiber extracted from the banana tree trunks. Sci. Rep. **13**(1), 15301 (2023). https://doi.org/10.1038/s41598-023-42160-8
22. Vajpayee, M., Singh, M., Dave, H., Ledwani, L.: Enzymatic surface modification of banana fabric with cellulase: Characterization and antimicrobial finishing with plant extracts. Ind. Crop. Prod. **191**(115), 895 (2023). https://doi.org/10.1016/j.indcrop.2022.115895
23. Xi, L., Qin, D., An, X., Wang, G.: Resistance of natural bamboo fiber to microorganisms, and factors that may affect such resistance. BioRes **8**(4), 6501–6509 (n.d.). https://doi.org/10.15376/biores.8.4.6501-6509
24. Khalid, M.Y., Al Rashid, A., Arif, Z.U., Ahmed, W., Arshad, H., Zaidi, A.A.: Natural fiber reinforced composites: sustainable materials for emerging applications. Results Eng. **11**(100), 263 (2021). https://doi.org/10.1016/j.rineng.2021.100263
25. Badanayak, P., Jose, S., Bose, G.: Banana pseudostem fiber: a critical review on fiber extraction, characterization, and surface modification. J. Nat. Fibers **20**(1) (2023). https://doi.org/10.1080/15440478.2023.2168821

26. Mumthas, A.C.S.I., Wickramasinghe, G.L.D., Gunasekera, U.S.: Effect of physical, chemical and biological extraction methods on the physical behaviour of banana pseudo-stem fibers: based on fibers extracted from five common Sri Lankan cultivars. J. Eng. Fibers Fabr. **14** (2019). https://doi.org/10.1177/1558925019865697
27. Nkemaja, D., Sikametagne, N.R., Anafack, M., Huisken, P.W., Mbout, E., Njeugna, E.: Studies on the tensile properties of Banana stalk fibers from Njombe-Penja-Cameroon. SSRG Int. J. Polym. Text. E Eng. (SSRG-IJPTE) **7**(1) (n.d.). https://doi.org/10.14445/23942592/IJPTE-V7I1P110
28. Richely, E., Zarei, A., Melelli, A., Rajan, D.K., Govilas, J., Gabrion, X., Clévy, C., Legland, D., Perez, J., Guessasma, S., Placet, V., Kallio, P., Beaugrand, J.: Measurement of microfibril angle in plant fibers: comparison between X-ray diffraction, second harmonic generation and transmission ellipsometry microscopies. Compos. Part C: Open Access **11**(100), 355 (2023). https://doi.org/10.1016/j.jcomc.2023.100355
29. Das, D., Ishtiaque, S.M., Das, S.: Influence of fiber cross-sectional shape on air permeability of nonwovens. Fibers Polym. **16**(1), 79–85 (2015). https://doi.org/10.1007/s12221-015-0079-9
30. Yu, C.: Natural textile fibers. In: Textiles and Fashion, pp. 29–56. Elsevier (2015). https://doi.org/10.1016/B978-1-84569-931-4.00002-7
31. Karimah, A., Ridho, M.R., Munawar, S.S., Adi, D.S., Ismadi, Damayanti, R., Subiyanto, B., Fatriasari, W., Fudholi, A.: A review on natural fibers for development of eco-friendly bio-composite: characteristics, and utilizations. J. Mater. Res. Technol. **13**, 2442–2458 (2021). https://doi.org/10.1016/j.jmrt.2021.06.014
32. Elanchezhian, C., Ramnath, B.V., Ramakrishnan, G., Rajendrakumar, M., Naveenkumar, V., Saravanakumar, M.K.: Review on mechanical properties of natural fiber composites. Mater. Today Proc. **5**(1), 1785–1790 (2018). https://doi.org/10.1016/j.matpr.2017.11.276
33. Zouari, R., Gargoubi, S.: Enhancing flame resistance of cellulosic fibers using an ecofriendly coating. Coatings **11**(2), 179 (2021). https://doi.org/10.3390/coatings11020179
34. Oliveira, F.R., Erkens, L., Fangueiro, R., Souto, A.P.: Surface modification of banana fibers by DBD plasma treatment. Plasma Chem. Plasma Process. **32**(2), 259–273 (2012). https://doi.org/10.1007/s11090-012-9354-3
35. Schellbach, S.L., Monteiro, S.N., Drelich, J.W.: A novel method for contact angle measurements on natural fibers. Mater. Lett. **164**, 599–604 (2016). https://doi.org/10.1016/j.matlet.2015.11.039
36. Sheebamercy, D., Annapoorani, S.G., Krithika, S.M.U.: Investigation of natural cellulosic fibers from banana for potential reinforcement in polymer composites. Biomass Convers. Biorefinery (2024). https://doi.org/10.1007/s13399-024-06398-0
37. Schäfer, F., Werner, H.O., Henning, F., Kärger, L.: A hyperelastic material model considering biaxial coupling of tension–compression and shear for the forming simulation of woven fabrics. Compos. A Appl. Sci. Manuf. **165**(107), 323 (2023). https://doi.org/10.1016/j.compositesa.2022.107323
38. Kozlowski, R., Baraniecki, P., Barriga-Bedoya, J.: Bast fibers (flax, hemp, jute, ramie, kenaf, abaca). In: Biodegradable and Sustainable Fibers, pp. 36–88. Elsevier (2005). https://doi.org/10.1533/9781845690991.36
39. Hsieh, Y.L.: Chemical structure and properties of cotton. In: Cotton: Science and Technology, pp. 3–34. Elsevier Inc. (2006). https://doi.org/10.1533/9781845692483.1.3
40. Bhoopathi, R., Ramesh, M., Deepa, C.: Fabrication and property evaluation of banana-hemp-glass fiber reinforced composites. Procedia Eng. **97**, 2032–2041 (2014). https://doi.org/10.1016/j.proeng.2014.12.446
41. Ramesh, M., Dhilip, J.D.J., Akilan, S., Ganeshprasad, B., Shankar, G.H.: Influence of stacking sequence on mechanical and thermal characteristics of banana-pineapple fiber reinforced epoxy composites. J. Nat. Fibers **19**(10), 3879–3892 (2022). https://doi.org/10.1080/15440478.2020.1848734
42. Pothan, L.A., Thomas, S., Neelakantan, N.R.: Short banana fiber reinforced polyester composites: mechanical, failure and aging characteristics. J. Reinf. Plast. Compos. **16**(8), 744–765 (1997). https://doi.org/10.1177/073168449701600806

43. Maleque, M.A., Profile, S., Yousif, B.F., Sapuan, S.M., Belal, F.Y.: Mechanical properties study of pseudo-stem banana fiber reinforced epoxy composite. Arab. J. Sci. Eng. **32** (2007). https://www.researchgate.net/publication/242783313
44. Nensok, M.H., Mydin, M.A.O., Awang, H.: Investigation of thermal, mechanical and transport properties of ultra-lightweight foamed concrete (ULFC) strengthened with alkali treated banana fiber. J. Adv. Res. Fluid Mech. Therm. Sci. **86**(1), 123–139 (2021). https://doi.org/10.37934/arfmts.86.1.123139
45. Ovalı, S., Eryılmaz, O., Uyanık, S.: Exploring the potential of sustainable natural cellulosic fiber from *Sorghum bicolor* (*Sorghum vulgare* var. technicus) stem for textile and composite applications. Cellulose. **31**(5), 3289–3302 (2024). https://doi.org/10.1007/s10570-024-05800-4
46. Wang, P., Dong, Y., Liu, G.: Green and cost-effective carboxylic acid Fe complex functionalized cotton fabrics: sunlight-driven catalytic and antibacterial activities, mechanical and thermal properties. Cellulose **25**(6), 3663–3678 (2018). https://doi.org/10.1007/s10570-018-1812-5
47. Yiga, V.A., Lubwama, M., Olupot, P.W.: Pyrolysis, kinetics and thermodynamic analyses of rice husks/clay fiber-reinforced polylactic acid composites using thermogravimetric analysis. J. Therm. Anal. Calorim. **148**(9), 3457–3477 (2023). https://doi.org/10.1007/s10973-022-11927-y
48. Momzyakova, K.S., Valishina, Z.T., Deberdeev, T.R., Aleksandrov, A.A., Berlin, A.A., Deberdeev, R.Y.: Structural analysis of powder celluloses by FTIR spectroscopy. Polym. Sci., Ser. D. **14**(2), 288–292 (2021). https://doi.org/10.1134/S1995421221020222
49. Collard, F.X., Blin, J.: A review on pyrolysis of biomass constituents: mechanisms and composition of the products obtained from the conversion of cellulose, hemicelluloses and lignin. In: Renewable and Sustainable Energy Reviews, vol. 38, pp. 594–608. Elsevier Ltd. (2014). https://doi.org/10.1016/j.rser.2014.06.013
50. Shi, Z., Xu, G., Deng, J., Dong, M., Murugadoss, V., Liu, C., Shao, Q., Wu, S., Guo, Z.: Structural characterization of lignin from *D. sinicus* by FTIR and NMR techniques. Green Chem. Lett. Rev. **12**(3), 235–243 (2019). https://doi.org/10.1080/17518253.2019.1627428.
51. Zope, G., Goswami, A., Kulkarni, S.: Isolation and characterization of cellulose nanocrystals produced by acid hydrolysis from banana pseudostem. BioNanoScience **12**(2), 463–471 (2022). https://doi.org/10.1007/s12668-022-00960-8
52. Ramdhonee, A., Jeetah, P.: Production of wrapping paper from banana fibers. J. Environ. Chem. Eng. **5**(5), 4298–4306 (2017). https://doi.org/10.1016/j.jece.2017.08.011
53. Mohamed, G.R., Mahmoud, R.K., Fahim, I.S., Shaban, M., Abd El-Salam, H.M., Mahmoud, H.M.: Bio-composite thermal insulation materials based on banana leaves fibers and polystyrene: physical and thermal performance. J. Nat. Fibers **19**(13), 4806–4821 (2022). https://doi.org/10.1080/15440478.2020.1870628
54. Yu, X., Xia,Y., Liang, D., Fu, W., Yin, C.: Effect of warm-water retting pretreatment on the physical properties of banana stem and its fibre. Materials **15**(23) 8462 (2022). https://doi.org/10.3390/ma15238462

South African Wool and Mohair: A Sustainable and Functional Fiber Solution

Adine Gericke, Mohanapriya Venkataraman, and Jiří Militký

1 Introduction

In an era marked by increasing environmental consciousness and a demand for high-performance textiles, natural fibers are experiencing a resurgence. South Africa, renowned for its production of premium merino wool and mohair, offers unique opportunities to explore the sustainable production practices and diverse functional applications of these materials.

Wool and mohair are considered animal fibers produced through sustainable and ethical practices. Current trends highlight a strong focus on ethical animal welfare, evidenced by initiatives like the Sustainable Cape Wool Standards, which emphasise responsible sheep husbandry and minimise environmental impact [1]. Producers are increasingly adopting regenerative agriculture practices, aiming to improve soil health and biodiversity in South Africa [2]. Potential consumers need to make informed choices about the origin of their wool products and ethical production Responsible Mohair Standards were implemented in 2025 [3]. Their responsibility is to provide the industry with the tools to recognise the best practices of farmers and to ensure that mohair comes from farms with an advanced approach to land management, animal welfare, and social responsibility [4].

A. Gericke
University of Stellenbosch, Stellenbosch, South Africa
e-mail: agericke@sun.ac.za

M. Venkataraman (✉) · J. Militký
Department of Material Engineering, Faculty of Textile Engineering, Technical University of Liberec, Liberec, Czech Republic
e-mail: mohanapriya.venkataraman@tul.cz

J. Militký
e-mail: jiri.militky@tul.cz

J. Militký and M. Venkataraman (eds.), *Fibrous Structures for Sustainable Future*, Engineering Materials, https://doi.org/10.1007/978-981-95-1382-6_5

This chapter aims is to demonstrate the significant potential of wool and mohair fibers in meeting the evolving needs of the textile industry and design special socks for diabetics.

2 Wool and Mohair

The term "wool" is generally used to describe the hair of various breeds of domesticated sheep (ovis aries), but is also accepted as the generic name for all animal hair fibers, including the so-called fine fibers of the cashmere and angora (mohair) goats as well as cashgora, camel, vicuna and alpaca, and many others. **Merino wool** is known for its highly valued performance properties, elastic recovery and resilience, fire resistance, and insulation properties [5–7]. The exceptional combination of properties stems from its unique chemical structure and morphology. The chemical structure is composed of keratin polymers. Keratin is made up of at least 18 alpha amino acids joined by amidic groups in a specific sequence. These polypeptide chains contain a wide range of types of side groups, including alkyl and aryl groups, alcoholic and phenolic groups, amino groups, amide groups, carboxyl groups, imido groups (heterocycle proline), and disulfide groups (cystine). The high content of oppositely charged side chains results in cross-links called ionic bonds or salt bridges. The relatively high sulphur content in wool (about 3–4%) is due to the presence of cystine, which forms disulfide bonds covalently joining polypeptide chains. These cystine cross-linkages are a distinctive feature of keratin fibers and form an integral part of the chemistry and physical behavior of wool. Hydrogen bonds act as the stabilizing elements of wool between the amide groups and between various other hydrogen-donating or accepting groups. They play an important role in the shaping of wool fabrics during apparel construction, but are also responsible for the sensitivity of wool to all kinds of hydrogen bond-breaking reagents, and are the reason why absorption of water has a major effect on the physical properties of wool fibers. The terminal groups of polypeptide chains are partly basic (amino groups) and partly acidic (carboxyl groups) as demonstrated by the abbreviated notation: NH_2–wool–COOH. Wool is therefore an ampholyte that ionises NH_3^+–wool–COO^- in water. There are more basic side groups [7]. The isoionic region (isoelectric point) is at pH 4.9 (4.5–6.3). Water is therefore an alkali for wool (as dangerous as 9% H_2SO_4). Wool fiber also contains about two percent of external/internal lipids.

The basic structural unit of wool is alpha-keratin, formed by a spatial spiral with intramolecular hydrogen bonds. The planar elongated structure (roughly twice as long) of beta-keratin is created by 120% stretching of alpha-keratin. Under the influence of supercontraction in an aqueous environment, e.g., during felting and crabbing, meander-shaped gamma-keratin is created, even more contracted than alpha-keratin [8].

Wool can provide groups for various types of reactions and bind several compounds, making it capable of irreversibly absorbing different types of gaseous substances such as formaldehyde (40–100 g per kg of wool), toluene (around 20 ng per kg of wool), sulfur dioxide, nitrogen oxides and odor components [7].

Submicroscopically, the innermost part of the wool fiber is the medulla. This segment carries the pigment of colour of the fiber, but may be absent (as in course or dead wools) [8, 9]. The cortex forms the main part of the wool fiber and makes up 70–90% of the fiber mass. The cortex is covered with a protective layer of scales composed of epi, exo and endo cuticle. Scale number is about 900 3500 per 1 mm^2. Fine wools have one scale on the fiber circumference and coarserwools have more scales on the fiber circumference. Scales are opened from root to tip Smooth scales are from para cortex (deeper overlap) and rippled scales are from ortho cortex. The scales are nearly rectangular sheet (width 20 μm, a length of 30 μm, and a thickness of 0.5–0.8 μm) and slightly bent [7].

The cortex consists of two components, namely the orthocortex and paracortex. Orthocortex has better defined fibrils (more tyrosine by 42%) and glycine (by 18%), better dyed and hydrolyzed (it is mainly in Mohair). The paracortex contains more cystine (by 20%) and transverse S–S bridges, causing more stiffness. It is stiffer and contains less amorphous matrix. This bilateral structure of wool (presence of ortho and para cortex) is the cause of its natural crimping, because ortho and paracortex wrap the fiber in a helix, where the more deformable ortho cortex is always outside. This bilateral internal structure of wool is the cause of its natural crimping. Especially in coarse and dead wool fibers, the medulla, i.e. tissue of strongly pigmented square cells, is also present The three-dimensional crimp structure of wool increases the springiness and bulk of wool and makes it resilient and easy to spin into yarns [8, 9].

The outer layer of the cortex consists of a fiber layer of small, overlapping scales. Cuticle and cortical cells are separated by the cell membrane complex composed from internal lipids and proteinsFine wools may have as few as one scale on the fiber circumference, and coarser wools, the scale number can be as high as 900–3500 per mm^2. Scales open from root to tip. Scales from the para cortex are smooth (with a deeper overlap), and those on the ortho cortex appear rippled. The scales form a nearly rectangular (slightly bent) sheet with a width of 20 μm, a length of 30 μm, and a thickness of 0.5–0.8 μm [7]. The scales on the fibers contribute to the dry feel of wool, even when it has absorbed a lot of water, but are also the cause of excessive felting and shrinking during washing [9].

The thickness of wool fibers varies from extrafine wool (diameter 14–17 μm) and fine wool (diameter 18–21 μm) to crossbred wool (diameter 30–38 μm). Smaller fiber diameters are an indication of more comfortable fibers. The fine fibers are used to manufacture high-quality textiles for luxury clothing, and the coarser fibers for applications like carpets, blankets, and other interior textiles. Recently, wool has become a popular component in technical textiles, and with popular applications ranging from sports and leisure wear to acoustic and thermal materials used in the construction industry for insulation of walls, roofs and ceilings. Commercial applications include various wool fabric structures for oil sorbents and geo-textiles. The significance of wool geo-textiles lies in the fact that, as opposed to synthetic fibers produced from nondegradable polymers, wool is a naturally biodegradable fiber. The use of wool for the production of geotextiles is normally reasonably priced and therefore economically justified, especially if made from poor quality wool that would have been regarded as waste or a troublesome byproduct of sheep breeding. Wool is completely biodegradable in the right conditions, with the degradation of fibers depending on the wool characteristics, product parameters, and several environmental conditions [10, 11].

The basic process affecting the behavior of wool is its stabilization (decatizing), which is conditioned by the glass transition temperature or the transformation of disulfide bridges into thiol groups, supported by a reducing environment. In cohesive stabilization (usually only in water at a temperature above Tg), the stabilization effect is canceled by immersion in water at 20 °C. When using appropriate conditions and chemical reducing agents (usually sodium dithionite), permanent stabilization can be achieved, which is not broken even in hot water. This is the principle of wool stretching in a reducing environment (OPTIM technology).

Wool is one of the materials with unique properties that are largely dependent on its fineness. It has been verified that wool fibers with a thickness of max. 19 μm have sufficient drapeability, softness and are not "scratchy", i.e. their ends do not cause prickly sensations due to their low stiffness. Wool absorbs UV radiation, especially of shorter wavelengths (transmittance in the dangerous UV-B range is only about 9%) and significantly blocks its penetration to the skin.

Mohair (*Capra hircus ancryrensis*) (the fibers from Angora goats) originated in Ankara, Turkey, but today it is widespread in many countries, and many breeds have derived from it. In South Africa, Angora goats thrive in the Eastern Cape and Karoo regions. These areas have become a significant economic contributor to goat production through the development of the market demand for mohair fibers. Mohair fibers are relatively strong and not considered to be crimped, but rather described as having a "wavy" appearance. Fiber diameters vary between 24 and 40 μm, and a fiber length of 300 mm can be achieved for a full year's growth [6, 12]. Mohair's lustre, smoothness, low friction and low felting properties are related to its surface scale structure. The mohair scales differ from those on wool and have a much lower scale frequency (one half of wool fibers) and wider distance between the cuticle scale margins. They are thin and flat and hardly overlap. They are attached much more closely to the body of the fiber, which is the reason for their lustrous appearance, smoothness, and low felting tendency. The main part of the fiber is formed by the ortho cortex only. Medullated fibers in mohair can be problematic, especially the more extreme form known as kemp. Kemp fibers feature a central medulla with cell residues and air pockets, making them appear chalky white and dye lighter. They also negatively affect the mohair's handle, stiffness, and prickliness [6, 12]. Images of mohair and wool fibers are depicted in Fig. 1a, b. The reduced prominence and overlap of the sheet-like scales on mohair fibers (compared to wool) are observed.

For centuries, mohair has been regarded as one of the most luxurious and superior fibers known to man. The luxurious appearance, bulky structure and resilience of mohair yarns make them popular in products like clothing items (suiting, coats, jerseys, scarves, socks, etc.), blankets, furnishings, upholstery, curtains, carpets, and

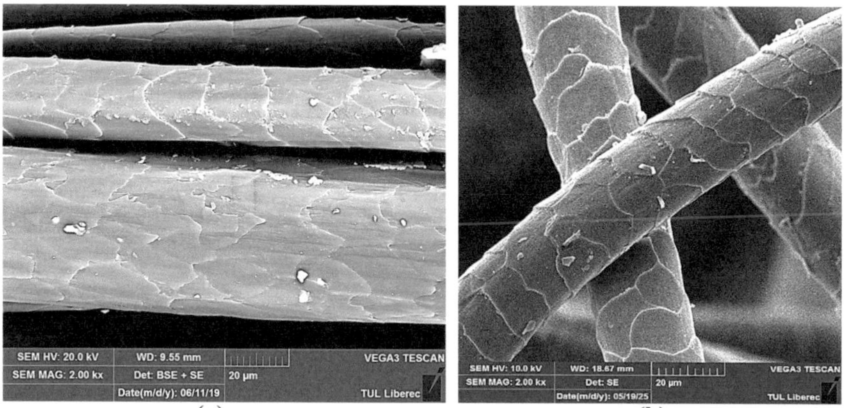

(a) (b)

Fig. 1 SEM images of **a** mohair fibers and **b** wool fibers

more. Emerging research has highlighted mohair's unique performance properties (often combined with those of wool), driving a shift towards functional applications and innovative blends. In many of these applications, the thermal resistance and moisture management properties of the fabrics play a pivotal role in ensuring their functionality. Current end-use applications include those related to sportswear, wellness, and healthcare products. In these applications, the role of clothing is taken beyond the traditional requirements associated with clothing, towards enhancing the performance of the wearer by providing ultimate thermo-physiological comfort, even under extreme conditions of temperature and activity. Socks for diabetics is one example of an application where mohair, wool and nylon are combined with bamboo in purpose-made socks that meet specific patient requirements.

3 Comfort of Fabrics

Recent studies confirm that, of all the factors that play a role in consumer decision making, clothing comfort is consistently rated among the highest. It is strongly recommended that designers and manufacturers consider comfort as a quality aspect that contributes to the total product performance and user satisfaction [13, 14]. The global functional textile market is driven by the increasing consumer expectation, on the one hand, and technological developments in the textile and apparel industry, on the other [15]. An increased consumer awareness, together with access to information, leads to increased market competitiveness, which encourages research and innovation [16, 17]. Fan and Hunter [18] refer to the creation of apparel fabrics and functional textiles as "engineering", which implies that textiles are designed to meet the performance requirements and specifications appropriate to their specific end-uses.

The design and development of apparel fabrics and garments present a much greater challenge compared to other products. These difficulties are attributed to the length and complexity of the textile manufacturing pipeline and the narrow tolerances allowed, which include variables that are often difficult to control or measure [18]. These challenges increase in the design and development of functional textiles. Functional textiles are "textiles with integrated functions of controlling and adjusting according to application" and are meant to keep a wearer cool or warm, have anti-bacterial properties, or provide protection against specific hazards, to name but a few. Functional clothing is used in sports and leisure wear, healthcare, personal protection, military applications, geo-textiles, and other emerging markets. "Athleisure" wear, for example, is an interesting new emerging trend in the fashion market that embraces the principles used in athletic wear in everyday life [18, 19]. Utilising textiles that provide protection against specific hazards in a way that does not compromise the protective clothing's acceptable levels of human comfort poses a tremendous challenge to textile engineering and has become a strong focus in research [17, 20].

Thermo-physiological comfort is considered an indispensable design requirement in functional clothing such as high-performance garments for sportswear, workwear, and protective wear [21]. Textile scientists describe "comfort" as a state of being unaware of your clothing, both on a physiological and psychological level. As such, comfort is also referred to as a pleasant state of psychological, physiological, and physical harmony between a human being and their environment, which, for this study, includes clothing [22]. Comfort is also defined as the absence of perceived pain and discomfort [5]. Adds that comfort is a complex mix of highly subjective sensations.

The human body maintains a comfortable state by exchanging heat and moisture with the environment [17, 22, 23]. During normal activity and in regular atmospheric conditions, the human body liberates metabolised heat through a combination of conduction, convection, and radiation of dry heat [23, 24]. The textiles that surround the body form a barrier between the skin and the environment and, thereby, create a micro-climate between the layers of clothing and the skin. Proper regulation of the micro-climate is crucial in the design of functional textiles and clothing [22]. Thermal resistance and moisture transfer are considered the main parameters that determine clothing comfort, and specifically thermo-physiological wear comfort [13]. Thermal resistance refers to the resistance of a textile or clothing system (which could include multiple layers of textiles) to the transfer of heat and has a considerable influence on a wearer's comfort or the performance of a technical product. Thermal resistance R_t (m^2 K/W) is inversely proportional to thermal conductivity $R_t = h/\lambda$ (where R_t is thermal resistance, h is the textile thickness, and λ (W/m K) is the thermal conductivity) [25]. Thermal resistance of clothing is commonly expressed in clo, where 1 clo equals 0.155 m^2 K/W [26].

The complex mechanism of heat transfer through textiles may involve conduction through the fibrous phase, radiation through fabric interstices, and convection of air within the structure. At low wind speeds, convection within a fabric is negligible. Heat loss through conduction is, thus, determined by the thermal conductivity of its fibrous matter and the thickness of the fabric [27–31]. Other factors that play a role in heat transfer by conduction are fabric planar weight called as gsm (g/m^2) closely connected to fabric volume porosity. These characteristics are closely related to the amount of air trapped inside the fabric. The fact that air has a thermal conductivity $\lambda_a = 0.024$ W/(m K) leads to the surmise that air trapped within the fabric may have an overbearing effect on specific fiber properties [32].

Moisture management refers to the controlled movement of water vapour and liquid water (sensible and insensible perspiration) through a fabric. In clothing, this refers to moisture moving from the surface of the skin to the environment. If this process is not inhibited in any way by the clothes that form a barrier between the skin and the environment, an acceptable level of comfort can be maintained [22].

The maintenance of thermal insulation depends upon effective moisture transmission in a textile fabric. The body loses up to 60 ml of water per hour through insensible perspiration, during which perspiration is liberated in vapour form. This amount can increase to 450 ml per hour during moderate exertion. At higher activity levels, or if the environmental temperature rises, the heat production increases further, and the sweat glands start producing liquid perspiration, known as sensible perspiration. If the liquid moisture is not dissipated effectively, the fabric will become "wet". As little as 3–5% of moisture content in a garment can lead to a sensation of wetness and the development of discomfort [22, 23]. Moisture accumulating in the inner layer of the fabric system reduces the thermal insulation of the textiles [5].

Water (in liquid form) can accumulate in the fabric interstices between fibers, on fiber surfaces, or it can be absorbed into the internal structure of the fibers, depending on the fiber properties. The water absorption capacity of the fibers will, in turn, affect the rate of moisture loss from the fabric structure (drying). This postulation emphasises the importance of recognising the complex phenomena of moisture management in fabrics, both in warm and cold weather, when designing high-performance applications.

4 Thermal Comfort Properties of Fabrics Containing Wool and Mohair

Recent developments that focus on expanding the use of mohair (often used in combination with wool) in technical and functional textile products necessitate the need for an improved understanding of how the presence of mohair in fabrics will impact its thermal properties. Despite its popularity, limited research has been reported on the quantification of the thermo-physiological comfort properties of mohair fabrics. Gericke et al. [33] investigated the effect of fibrous matter composition and fabric structural parameters on conductive and convective heat transfers to study the thermal comfort properties of mohair fabrics.

A special set of fabrics, woven from filament polyester (PES) warp and with the weft yarns either from mohair, wool, acrylic, cotton, polyester, or bamboo-based viscose, was woven under similar conditions for the investigation. The average porosity of fabrics was 71%. The thermal resistance, conductivity, and fabric thickness were measured using the Alambeta device (Sensora, Czech Republic) according to ISO 11092. The instrument measures heat flow through the fabric between hot and cold plates at 200 Pa, simultaneously determining fabric thickness (h).

It is known that the thermal conductivity of wool and mohair fibers differs significantly from those reported for the other fibers in this set of fabrics. When these fibers are spun and woven into fabric form, however, the fabric structure comprises a combination of fiber mass and trapped air, both contributing factors that should be considered when the performance of these fibers in a fabric is evaluated.

The thermal resistance of the samples containing mohair and wool fibers in the weft yarns (about 0.019 m^2 K/W) was higher than that of those containing other fibers (in the range 0.013–0.016 m^2 K/W). Samples featuring mohair and wool fibers in their weft yarns exhibited lower thermal conductivity (in the range of 0.032–0.037 W/m K) compared to samples with other fibers (0.039–0.041 W/m K). This finding aligns with the thermal resistance measurements, which would show an inverse relationship. It is also possible that the polyester warp yarns in the test fabrics "masked" the effect of the different weft yarns, leading to a reduced magnitude in the observed effect. The effect was, however, statistically significant. This outcome is attributed to the dominant role of trapped air within the fabric, whose low thermal conductivity substantially increased the overall thermal resistance. The contribution of fiber composition to fabric properties can nevertheless not be overlooked.

5 The Effect of Mohair and Wool Fiber on the Microclimate Inside Socks for Diabetics

Diabetes, a worldwide health concern, is estimated to cause a limb loss every 30 s globally, often resulting from diabetic foot ulcers [34]. These ulcers can arise from seemingly minor skin injuries, such as abrasions or dryness, and progress to severe conditions requiring amputation. The prevention of these complications involves mitigating factors that contribute to skin damage and microbial infections [34, 35].

Specialised socks can play a crucial role in reducing diabetic foot ulcer incidence. Beyond providing a cushioning effect, these socks should create and maintain a stable microclimate between the skin and sock, regardless of environmental changes or activity levels [36]. Critically, they must also prevent excessive dryness, which can lead to skin cracking and infection [37].

The stability of this microclimate depends on the sock's thermo-physiological properties, primarily heat transfer resistance (thermal resistance) and moisture management [38, 39]. The skin's natural moisture production in the form of both insensible and sensible perspiration necessitates effective dissipation of both moisture vapour and liquid water by the sock. Knitted fabrics facilitate moisture vapour transfer through interstitial spaces, but liquid moisture management requires effective wicking, or alternative mechanisms of moisture management, to prevent wetness accumulating on the fabric surface that lies against the skin [22, 23]. The presence of shoes, with varying moisture and air permeability, further complicates this process. This factor must be considered when selecting test methods for evaluating sock fabrics.

Synthetic fibers like polyester offer good liquid moisture transmission due to low moisture regain, allowing for rapid drying [5, 40]. However, when footwear restricts vapor escape, the inability of polyester fibers to absorb water will cause liquid moisture to accumulate either between the sock and the shoe or against the skin inside the sock. Mohair and wool, known for their superior moisture management, possess hygroscopic properties that enable them to absorb a significant amount of moisture without feeling wet. This helps regulate microclimate humidity, preventing excessive dryness and mimicking natural skin hydration [40].

Heat transfer resistance is generally influenced by fiber thermal conductivity, fabric thickness, and porosity [27, 29]. Cellulose and polyester fibers have a much higher thermal conductivity than mohair and wool. However, entrapped air in fabric structures, with its low thermal conductivity, has a significant effect on its thermal resistance [24, 28]. Mohair's inherent three-dimensional crimp and stiffness result in yarns and fabrics with many "air-trapping" spaces. Up to two-thirds of the volume of these structures can be occupied by air spaces, kept intact by the resiliency of the fibers [28].

To understand how the fiber content of diabetic socks (specifically the inclusion of mohair and wool fibers) impacts the complicated process of maintaining the required moisture and temperature conditions in their microclimate, Gericke and Venkataraman [41] employed a selection of recognised test instruments and methods to evaluate the thermal and moisture management attributes of selected sock samples. Thermal and moisture transfer were considered the two key parameters for the objective evaluation of thermo-physiological comfort in textiles to maintain a stable microclimate between the skin and surrounding textiles (the main requirements for diabetic socks) [38, 39]. To enhance the reliability of the prediction of in-wear sock performance, a variety of established test methods or instruments for this specific end-use test were performed, and both fabric specimens and full socks were tested. From this, a novel method that quantifies fabric moisture sorption and its influence on the sock's microclimate was developed.

For investigation of fiber composition and design of knitted socks on their thermal properties and microclimate, the three yarn types (MW (mohair/wool), BV (bamboo viscose), and PETR (recycled polyester)) were compared (see Fig. 2).

The first yarn (MW) was used in two of the sock samples and was spun of predominantly mohair and wool fibers, blended with a small amount of nylon staple fibers to improve the spinnability and tensile strength. The other two yarns were spun from bamboo viscose staple fibers (BV) and recycled polyester staple fibers (PETR). A small percentage of nylon-elastane covered filament yarn (NE) (< 10%) was added

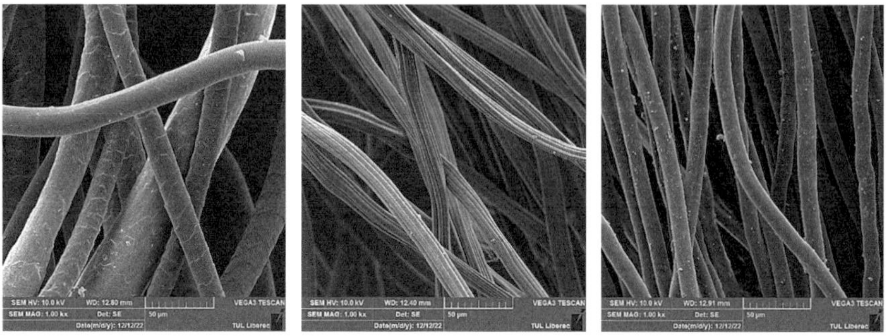

MW BV PETR

Fig. 2 SEM images of yarns referenced MW (mohair/wool), BV (bamboo viscose), and PETR (recycled polyester). SEM magnification 1000×

during knitting with an extra feeder to provide elasticity. The yarns are described in Table 1.

The MH yarn reveals the characteristic scaled surface of wool and mohair fibers, where overlapping, hardened cuticle cells are oriented towards the fiber tip. Notably, mohair scales are less pronounced and exhibit less overlap compared to wool scales [28]. In contrast, the PETR yarn displays the smooth surface typical of synthetic polyester fibers, while the BV yarn distinctly shows the striated texture associated with viscose yarns [40].

Three distinct sock designs, referenced as S-MW, S-PETR, and S-MWV (visualised in Fig. 3), were selected. Socks S-MW and S-PETR were constructed using, respectively, MW and PETR yarns. In sock S-MWV, MW and BV yarns were strategically alternated, resulting in a 30% MW and 70% BV yarn distribution across the sock's surface. The rationale behind this design was to leverage the BV yarn's higher thermal conductivity in the upper foot and ankle regions, aiming to reduce overall sock thermal resistance and mitigate potential foot overheating. Conversely, the placement of MW yarn in areas crucial for diabetic foot care was intended to replicate the thermal and moisture management properties observed in sock S-MW. All the socks were produced on 4-in. cylinder, 95-needle automatic knitting machines, using a standard pile knit construction with a 3 cm ribbed cuff, to provide a secure fit around the leg without inducing undue pressure.

The ANOVA (focused on data where thickness fell within the predefined range) is depicted in Fig. 4, clearly indicating a statistically significant difference in thermal resistance R_t between the MW, BV, and PETR fabrics. The MW fabric exhibited the highest Rt, while the BV fabric displayed the lowest R_t. This is in accordance with the known thermal conductivity of wool, polyester, and viscose fibers [28, 42].

Further testing on the Thermal Foot Manikin (TFM) enabled the measurement of the Rt of the socks as a whole during wear (Fig. 5) [39].

Here, thermal resistance was evaluated in conditions that more closely resemble real-world use (in contrast to assessments of individual fabric pieces). The TFM is

Table 1 Description of yarns and socks

Description of yarns							Description of sock fabrics			
Yarn Ref.	Mohair (%)	Wool (%)	Nylon (%)	Bamboo-viscose (%)	Recycled polyester (%)	Yarn (Tex)	Sock Ref.	Aerial weight (g/m^2)	Thickness (mm)	Porosity (%)
MW	40	40	20			40	S-MW	510	4.2	90.4
PETR					100	33	S-PETR	815	4.6	87.5
BV				100		36	S-MWV	MW:615 BV: 588	4.4	90.5

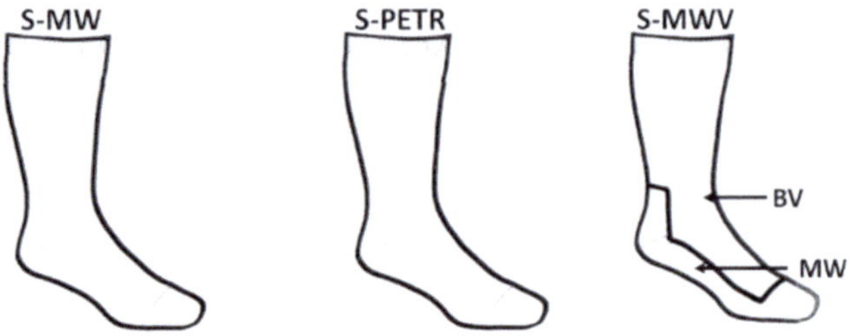

Fig. 3 Sock designs for socks referenced S-MW, S-PETR, and S-MWV

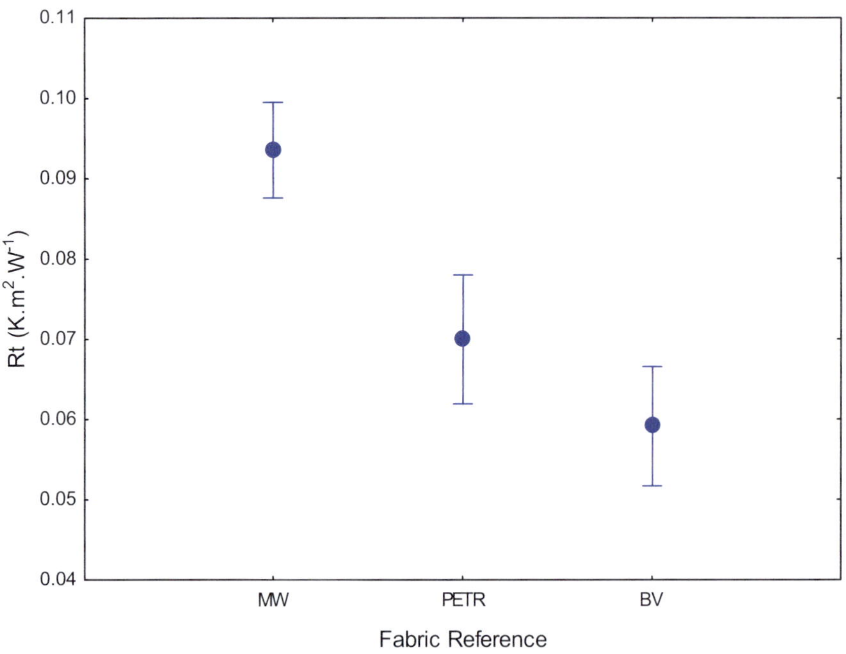

Fig. 4 Effect of fiber content on thermal resistance (R_t) as measured on the Alambeta (vertical bars denote the 95% confidence intervals)

designed to assess the thermal resistance of socks and comprises 13 independently heated segments, each maintained at 35 °C.

A statistical comparison was conducted on the R_t of the three sock variants to determine the influence of their fiber content, but the findings (illustrated in Fig. 6) showed no significant difference between the socks made of MW and PETR. However, the S-MWV sock, which incorporated panels of MW and bamboo viscose, exhibited a significantly lower thermal resistance. This analysis supports the intended design of the S-MWV sock, where strategically placed BV areas aimed to create an overall cooling effect, and this objective was achieved. (The difference in Rt

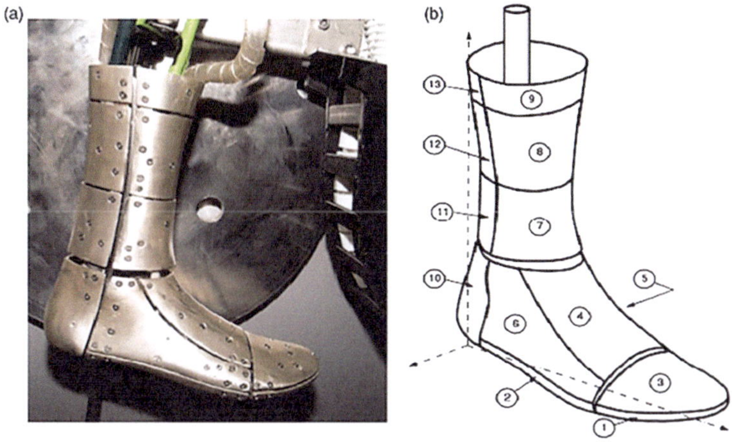

Fig. 5 The mechanical leg of the TFM divided into 13 units

between MW and BV fibers is confirmed in Fig. 6.) A key observation was the strong interdependence between sock Rt (TFM) and variations in sock fabric thickness.

The effects of variations in fabric thickness, as well as sock design, on the thermal resistance (Rt) of the socks (as measured on the TFM) are illustrated in the 3D

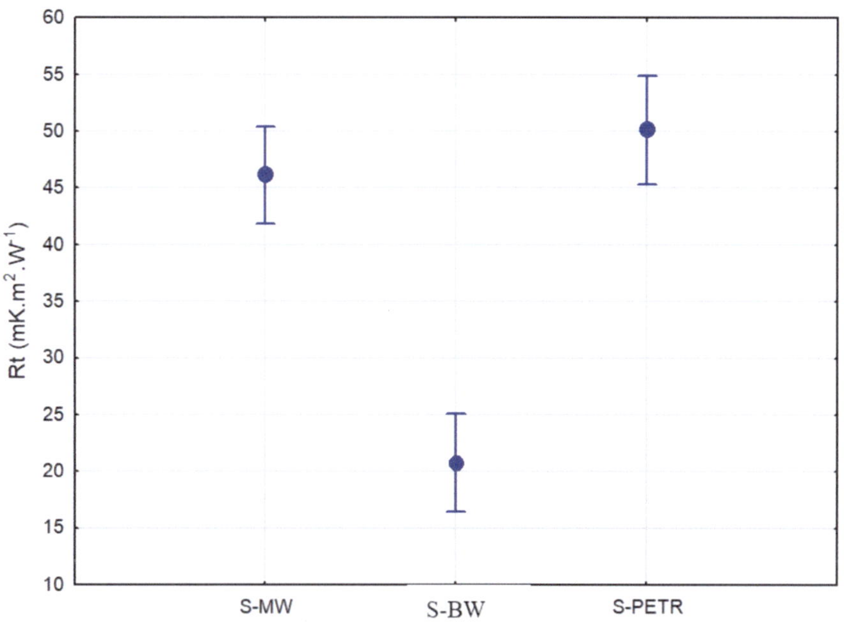

Fig. 6 ANOVA of thermal resistance (Rt) of socks as measured on TFM, indicating the differences between the three sock variants (vertical bars denote the 95% confidence intervals)

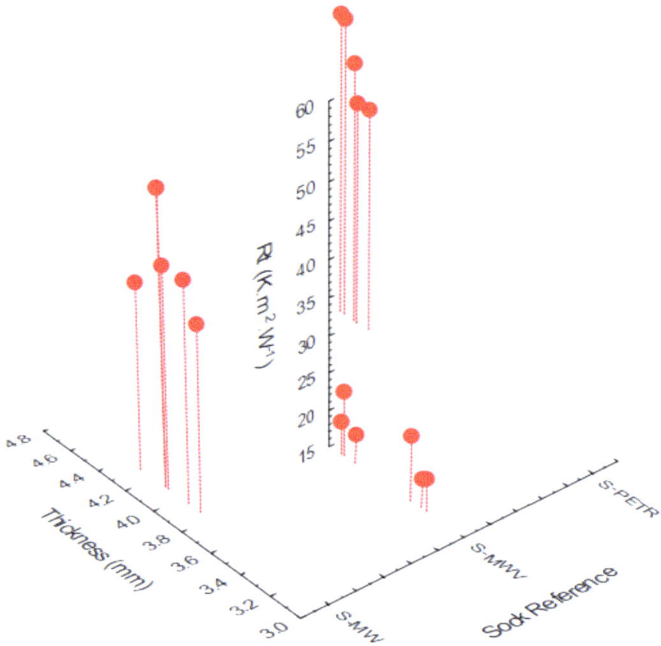

Fig. 7 Scatterplot of fabric thickness (h) and thermal resistance (Rt) (measured on TFM) plotted against sock reference

scatterplot of thermal resistance (Rt) plotted against fabric thickness (h) and sock reference in Fig. 7.

Comparing the R_t results obtained from Alambeta data (cut samples) and TFM data (whole socks) underscores the limitations of using a single method to evaluate a complex property like R_t in a garment.

The Thermal Foot Manikin (TFM) measures thermal resistance (R_t) solely under dry conditions, presenting a limitation to the reflection of actual conditions in the microclimate of socks during wear. More comprehensive testing prioritised methods that assessed both thermal and moisture management within the microclimate during use. In vivo wear trials were conducted, during which small thermochrom data loggers called iButtons were used to monitor temperature and relative humidity inside the socks. The iButtons were carefully attached to the skin inside the socks and recorded data every 60 s over two time segments [41]. Wear trials simulated typical office or home activities, with the participant remaining largely inactive for the first time segment (TS1) that lasted 40 min, with brief movements every 10 min to ensure normal blood flow in the legs and feet. This was followed by 20 min of gentle cycling on a stationary bike without significant effort (TS2). Leather shoes were worn over the socks throughout the trials. The iButtons utilised in the study are depicted in Fig. 8.

Fig. 8 Thermochron iButton used to log temperature and humidity in the microclimate during wear trials (reused from Gericke et al. [28] under CC-BY Permission)

The temperature patterns observed within the microclimate of both TS1 and TS2 align with the Rt measurements obtained from the TFM (Fig. 9). It is important to distinguish that the iButton readings represent actual measured temperatures, whereas the TFM quantifies the thermal resistance (Rt) of the sock under static conditions. Notably, the temperatures recorded when the S-MWV sock was worn were consistently lower than those measured inside the other two variants, particularly during the TS2 (more active) phase, thus corroborating the TFM findings. Despite a trend emerging (where MW values appeared slightly lower than PETR later in the cycle), the accurate measurement of humidity proved difficult due to the small, enclosed space within a sock, especially in TS2, where sweat production was more likely. It's worth noting that direct contact of the iButton with any liquid moisture during the test immediately impacts the measurement, as evidenced by the sharp peaks in the graph.

Building on the insights from the wear trials, the CJ test method was developed to address the potential concern about the reliability of the wear trials due to the confined microclimate within which the iButtons operated. A laboratory setup was designed to simulate in-use microclimate conditions and allow more reliable measurements. Based on the traditional "upright cup method" for moisture vapor permeability (ref), the customized covered jar (CJ) method utilized a laboratory setup with at least four glass jars, each containing a fixed volume of water. Conditioned specimens were securely mounted over these jars, creating a sealed system where the only path for water vapour inside the jar to escape into the surrounding environment was by passing through the test specimen. The test setup allows for the mounting of multiple specimens simultaneously. This method comprises two distinct parts.

In first part, the average weight change of the specimens after 60 min provided data on the percentage moisture sorption in the fabric. Note that for this test, the term "% moisture sorption" refers specifically to the ability of the fibers in the specimen to absorb moisture vapour from air. This differentiates the test from others that measure the ability of a fabric structure or fibers to absorb liquid moisture [23, 43]. The results

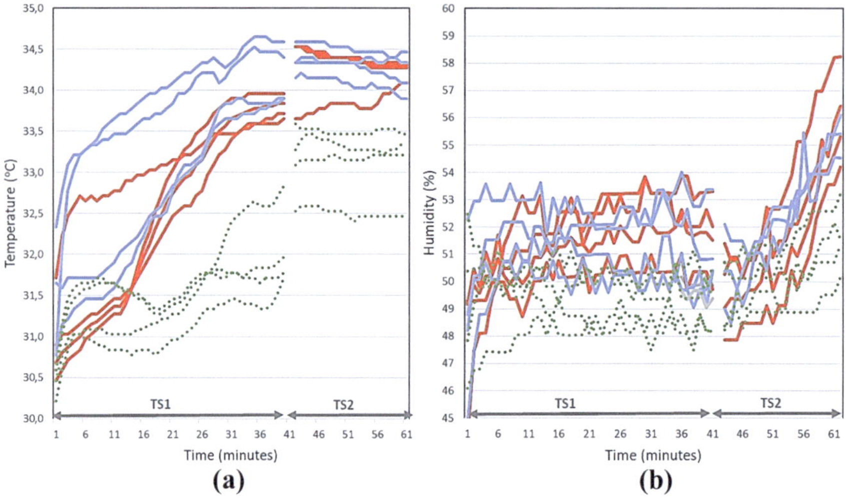

Fig. 9 The temperature (**a**) and humidity (**b**) measurements in the microclimate logged with thermochron iButtons during wear trials during which the S-PETR, S-MW, and S-MWV socks were worn

of this test reflect differences in moisture sorption of fabrics with comparable water vapour permeability.

For the test procedure glass jars with a diameter of 55 mm were used. The jars have screw-on lids that mount samples tightly over a standardized central opening. This allows secure mounting of specimens in either single or multiple layers. The only way for evaporated water inside the jar to escape to the environment was through the specimen(s). Test parameters like the air gap underneath the specimen, water temperature, ambient temperature and time of exposure were carefully controlled. To establish a standardized air gap of 40 mm between the sample and the water surface, the same volume of water was used in all the jars. The water bath was used to keep the water inside the jars at a constant temperature of 34 °C. The ambient conditions in the laboratory were 20 ± 2 °C and 60–65% relative humidity. To determine the sorption, specimens were weighed before and after a 60-min running time to record W_0 and W_{60}. The laboratory setup is illustrated in Fig. 10. The % sorption is calculated according to equation $\% \text{ Sorption} = ((W_{60} - W_0)/W_0) \times 100$, where % sorption refers to the amount of water vapour absorbed by the fabric specimen during the test and W_0 and W_{60} are the specimen weights recorded before the start of the test and after 60 min.

In the second part of the test, changes in humidity and temperature inside the jars were recorded with thermocron iButtons. To mount the iButtons inside the jar, a "scaffold" was created immediately below the sample using a nylon mesh fabric. Care was taken to make sure that the button did not touch the sample surface or the water. Data was logged and recorded every 60 s for a total of 60 min.

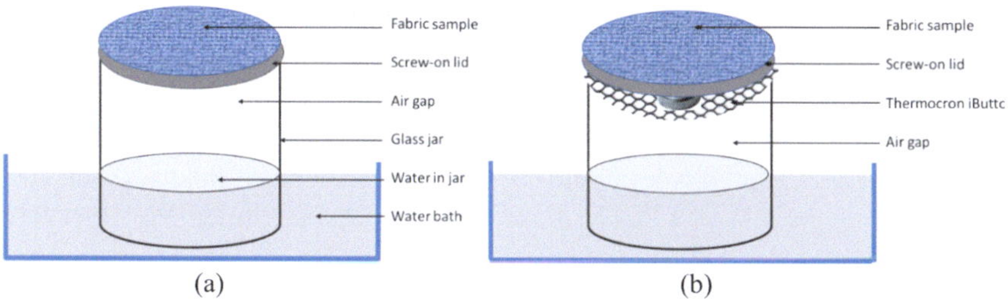

Fig. 10 The laboratory set-up of the CJ method, where **a** indicates the measurement of the % moisture sorption (part1) and **b** the temperature and humidity inside the microclimate (part 2)

Tests were conducted on conditioned single-layer specimens as well as on double layer assemblies (where a specimen was covered with a second fabric "cover layer" resembling that used in shoes). Two cover layers were used, namely CP1, a knitted polyester that resembles the fabric in training shoes in terms of thickness and water vapour permeability, and CP2, a coated woven polyester with a low water vapour permeability commonly used in shoes for leisure wear. Only PETR and MW samples (cut from socks S-MW and S-PETR) were evaluated, as these represent the fiber compositions of the areas around the foot, which are critical for meeting the requirements of diabetic socks. The results are shown in Fig. 11.

The CJ method supported and clarified the performance differences between S-MW and S-PETR socks that were observed in the wear trial outcomes. For the uncovered (single-layer) fabrics, the initial temperature and humidity in the jar's

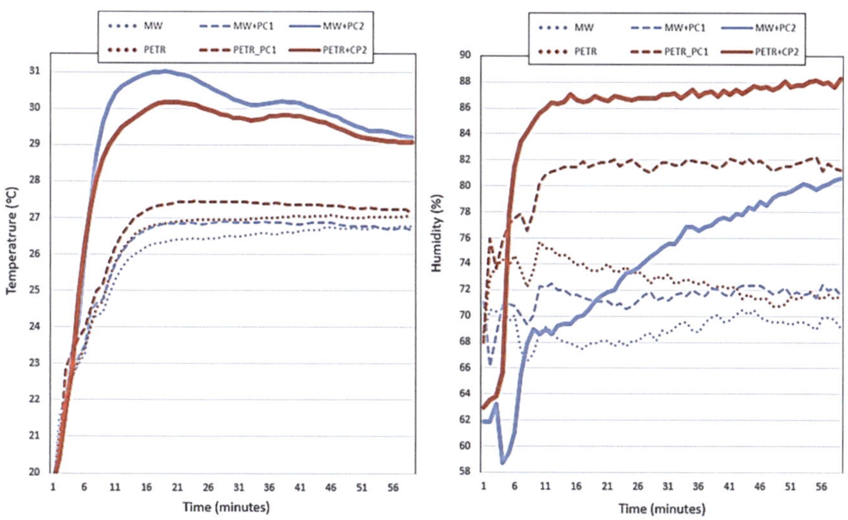

Fig. 11 The temperature and humidity measurements in the microclimate logged over 60 min with thermochron iButtons during the CJ method (part 2) tests on the PETR and MW fabrics

microclimate (beneath the PETR specimens) increased more than with the MW specimens. This stabilized after approximately 13 min and remained higher for the remainder of the test. When the test specimens were covered with CP1, a similar trend was observed for the microclimate temperature, but the difference between MW and PETR was greater. Humidity levels did not change significantly. With the denser CP2 fabric covering the MW and PETR, the temperature profiles shifted. PETR showed an immediate rise above 30 °C, but the overall increase for the MW specimens was larger. The values converged as they stabilized, but remained considerably higher than those for the more permeable cover CP1 and the uncovered samples. These findings are consistent with the R_t measurements from both the Alambeta and TFM instruments. Notably, the higher R_t of the MW fabric became more evident as the cover fabric's permeability decreased. These results are visually depicted in Fig. 11. The average temperatures recorded in the CJ tests were lower than those during the wear trials, which can be attributed to the different sizes of the respective microclimates. Future improvements to the CJ test could consider reducing the air gap within the jar between the water and the specimen. The contrasting moisture management characteristics of the MW and PETR fabrics are also illustrated in Fig. 11. For uncovered samples, and those covered with the highly moisture vapor-permeable CP1, water vapor readily dissipated into the surrounding environment, limiting microclimate humidity. Under these conditions, the hygroscopic MW fibers exhibited superior performance in maintaining lower microclimate humidity compared to the synthetic PETR fibers. Mean microclimate relative humidity values for the single-layer, sample + CP1, and sample + CP2 configurations were 69%, 71.7%, and 75.4% for MW, and 72.7%, 81.5%, and 87.5% for PETR, respectively (Table 2).

CP2 layer with the low moisture vapor permeability, resulted in a significant elevation of microclimate humidity, more pronounced in the PETR fabric system. The inclusion of the Consistent with existing literature, the hygroscopic nature of MW, enabling the absorption of moisture vapour without causing fabric wetness, results in slower changes in the microclimate conditions over time. While humidity did increase as the fabric approached saturation, it did not reach the same levels as observed with PETR. In the initial phase of the CJ method, specimen moisture sorption was quantified after 60 min of exposure to permeating water vapor. Analysis of variance (ANOVA; Fig. 12) revealed a statistically significant difference in moisture absorption between MW and PETR specimens.

This difference is likely attributable to the enhanced capacity of MW fibers to absorb and retain moisture, particularly under conditions where the CP2 cover restricts water vapor diffusion. By absorbing moisture, MW fibers mitigate its accumulation within the microclimate, supporting the hypothesized self-regulating mechanism. In contrast, PETR fibers exhibited minimal moisture sorption, resulting in higher relative humidity due to the presence of unabsorbed water vapor.

Table 2 Summary of results of R_t (Alambeta), MVP (Permetest) R_t (TFM), microclimate temperature and humidity (wear trials and CJ method) and moisture sorption (CJ method)

Sock Ref.	Fabric Ref.	Rt (Alambeta) (mK m^2 W^{-1})	MVP (Permetest) (%)	Rt (TFM) (mK m^2 W^{-1})	Temperature (microclimate) (wear trial) (°C)		Humidity (microclimate) (wear trial) (%)		Sorption (CJ method) (%)
					TS1	TS2	TS1	TS2	
S-MW	MW	93.5	23.6	46.1	33.5	34.3	51.4	52,9	2.20
S-PETR	PETR	70.0	25.3	50.2	32.9	34.2	51.4	52.3	0.53
S-MWV	BV	59.1	21.5	27.4	31.6	33.1	49.3	50.3	
	PC1		62.7						
	PC2		8.6						
	PC1 + MW		23.1						1.70
	PC2 + MW		17.3						7.38
	PC1 + MW		19.7						0.55
	PC2 + BW		9.5						2.87

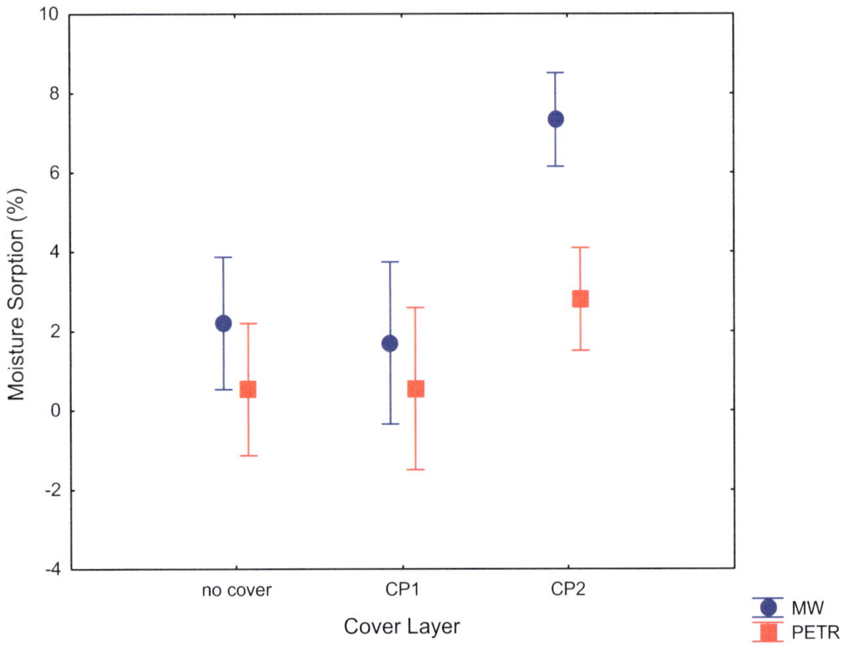

Fig. 12 Analysis of variance depicting the % sorption measured during the CJ method (part 1) tests on the PETR and MW fabrics, with no cover and covered with CP1 and CP2

6 Conclusion

The exceptional qualities of mohair fibers are due to thin and flat cuticle scales with relatively low scale frequency, wide distance between the cuticle scale margins, and close attachment to the body of the fiber. This is the main reason for their visible luster, smoothness, and low felting tendency. The main part of the fibers is from the ortho cortex only, so the fibers are not crimped.

The effect of fiber type on conductive thermal resistance was not consistently statistically significant. This outcome is attributed to the dominant role of trapped air within the fabric, whose low thermal conductivity substantially increased the overall thermal resistance. However, the contribution of fiber composition to fabric properties cannot be overlooked. Notably, fabrics containing mohair or mohair/wool exhibited consistently greater thickness compared to fabrics from other fibers, even under controlled weaving conditions. This phenomenon can be attributed to the inherent physical characteristics of mohair fibers, which facilitate superior air entrapment within yarns and the overall fabric architecture. These thicker structures not only provide enhanced thermal insulation by maximising trapped air but also impede convective heat flow. This research emphasises the distinctive properties of mohair and its promising applications in functional and technical textiles.

The thermal and moisture management properties of knitted socks for diabetic patients, the impact of fabric structure, especially fabric thickness, on thermal resistance, was presented. This is in accordance with previous findings on woven fabrics by Gericke et al. [33]. This implies that in the design of the socks, individual preferences can be taken into account for the design of socks with different levels of thermal resistance. By varying the fabric structure, a product that suits the thermal requirements of the wearer can be designed without changing fiber content.

Concerning moisture management, mohair-wool socks exhibited enhanced moisture management compared to their polyester counterparts. A novel test procedure demonstrated that humidity within the microclimate was consistently lower in mohair-wool socks, irrespective of the simulated shoe lining. With the new test method developed for this purpose, results illustrated that in the mohair-wool sock, the humidity inside the microclimate stayed significantly lower than in the polyester sock, irrespective of which cover was used to simulate the shoe. The improved performance is linked to the consistently higher moisture vapour sorption of mohair-wool, which is vital for maintaining appropriate skin hydration and preventing dry conditions. These results suggest that mohair-wool socks can aid in maintaining a healthy microclimate around the skin, making it suitable for diabetic patients, where foot care is critical.

References

1. Africa, C.W.S.: Sustainable Cape Wool Standard good practice principles for sustainable wool production in South Africa. Sustainable Cape Wool Standard (2020). https://www.capewools.co.za/content/sustainable-cape-wool-standard
2. Department of Agriculture Land Reform and Rural Development: Dept Agriculture Annual Report 2023–2024 (2024). https://doi.org/10.51227/ojafr.2024.40
3. Mohair, S.A.: Mohair South Africa—luxury mohair production in South Africa (2025). https://www.mohair.co.za/
4. Mtenjwa, B., Ikusika, O., Mpendulo, C.T., Gajana, S.C.: Angora goats and Mohair production in South Africa: a review. Online J. Anim. Feed. Res. **14**(6), 347–357 (2024). https://doi.org/10.51227/OJAFR.2024.40
5. Hatch, K.L.: Textile Science. West Group (1993)
6. Hunter, L.: Mohair: A Review of Its Properties, Processing and Applications. CSIR Division of Textile Technology (1993)
7. Simpson, W., Crawshaw, G.H.: Wool science and technology. In: Simpson, W., Crawshaw, G.H. (eds.) Wool Science and Technology (2002). https://doi.org/10.1533/9781855737648
8. Kuffner, H.: Wool fibers. In: Kozlowski, R.M. (ed.) Handbook of Natural Fibres, Volume 1: Types, Properties and Factors Affecting Breeding and Cultivation, vol. 1, pp. 171–194 (2012)
9. Collier, B.J.: Understanding Textiles, 7th edn. Pearson Education, Inc. (2007)
10. Broda, J.: Biodegradation of sheep wool geotextiles designed for erosion control. In: Environmental Chemistry and Recent Pollution Control Approaches, pp. 1–19. IntechOpen (2019). https://doi.org/10.5772/intechopen.80247
11. Broda, J., Przybyło, S., Kobiela-Mendrek, K., Biniaś, D., Rom, M., Grzybowska-Pietras, J., Laszczak, R.: Biodegradation of sheep wool geotextiles. Int. Biodeterior. Biodegrad. **115**, 31–38 (2016). https://doi.org/10.1016/j.ibiod.2016.07.012

12. Hunter, L., Hunter, E.L.: Mohair. In: Frank, R. (ed.) Silk, Mohair, Cashmere and Other Luxury Fibers, pp. 60–117. Woodhead Publishing Limited (2001). https://doi.org/10.1533/9781855737594.68
13. Kaplan, S., Okur, A.: The meaning and importance of clothing comfort: a case study for Turkey. J. Sens. Stud. **23**(5), 688–706 (2008). https://doi.org/10.1111/j.1745-459X.2008.00180.x
14. Sanad, R.A.: Consumer attitude and purchase decision towards textiles and apparel products. World J. Text. Eng. Technol. **2**, 16–30 (2016)
15. Persistence Marketing Research: Functional Textiles Market: Global Industry Analysis and Forecast 2016–2024 (2021). https://www.persistencemarketresearch.com/market-research/functional-textiles-market.asp
16. Global Market Insights: Functional textile finishing agents market 2021–2027. Industry Report (2021). https://www.gminsights.com/industry-analysis/functional-textile-finishing-agents-market
17. Hes, L.: Laboratory measurement of thermo-physiological comfort. In: Song, G. (ed.) Improving Comfort in Clothing, pp. 114–137. Woodhead Publishing Ltd. (2011)
18. Fan, L.J., Hunter, L.: Engineering Apparel Fabrics and Garments. Woodhead Publishing Ltd. (2009)
19. Persistence Market Research: Global Market Study on Wool: Industry Focusing on Research and Development as Demand Continues to Wane (2021). https://www.persistencemarketresearch.com/market-research/wool-market.asp
20. Song, G.: Improving Comfort in Clothing. Woodhead Publishing Ltd. (2011)
21. Memiş, N.K., Kaplan, S.: Wool fabric having thermal comfort management function via shape memory polyurethane finishing. J. Text. Inst. **111**(5), 734–744 (2020). https://doi.org/10.1080/00405000.2019.1661936
22. Das, B., Das, A., Kothari, V.K., Fanguiero, R., de Araújo, M.: Moisture transmission through textiles: Part I: Processes involved in moisture transmission and the factors at play. Autex Res. J. **7**(2), 100–110 (2007)
23. Bhatia, D., Malhotra, U.: Thermophysiological wear comfort of clothing: an overview. J. Text. E Sci. Eng. **06**(02) (2016). https://doi.org/10.4172/2165-8064.1000250
24. Holcombe, B.V., Hoschke, B.N.: Part IV: Prediction of tensile properties. Textile Res. J. **53**, 368–374 (1983)
25. Özgen, B., Altaş, S.: The investigation of thermal comfort, moisture management and handle properties of knitted fabrics made of various fibers. Tekstil ve Konfeksiyon **24**(3), 266–271 (2014)
26. Roy Choudhury, A.K., Majumdar, P.K., Data, C.: Factors affecting comfort: human physiology and the role of clothing. In: Song, G. (ed.) Improving Comfort in Clothing, 1st edn, pp. 3–60 (2011)
27. Afzal, A., Ahmad, S., Rasheed, A., Ahmad, F., Iftikhar, F., Nawab, Y.: Influence of fabric parameters on thermal comfort performance of double-layer knitted interlock fabrics. Autex Res. J. **17**(1), 20–26 (2017). https://doi.org/10.1515/aut-2015-0037
28. Gericke, A., Militký, J., Venkataraman, M., Steyn, H., Vermaas, J.: The effect of mask style and fabric selection on the comfort properties of fabric masks. Materials **15**(7), 2559 (2022). https://doi.org/10.3390/ma15072559
29. Kaplan, S., Karaman, C.: Thermal comfort performances of cellulosic socks evaluated by a foot manikin system and moisture management tester. Int. J. Cloth. Sci. Technol. **1**, 1–12 (2019). https://doi.org/10.1108/IJCST-06-2018-0080
30. Majumdar, A., Mukhopadhyay, S., Yadav, R.: Thermal properties of knitted fabrics made from cotton and regenerated bamboo cellulosic fibers. Int. J. Therm. Sci. **49**(10), 2042–2048 (2010). https://doi.org/10.1016/j.ijthermalsci.2010.05.017
31. Mao, N., Russell, S.J.: The thermal insulation properties of spacer fabrics with a mechanically integrated wool fiber surface. Text. Res. J. **77**(12), 914–922 (2007). https://doi.org/10.1177/0040517507083524
32. Rengasamy, R.S.: Improving moisture management in apparel. In: Song, G. (ed.) Improving Comfort in Clothing, 1st edn, pp. 182–188. Woodhead Publishing Ltd. (2011)

33. Gericke, A., Militký, J., Venkataraman, M., Steyn, H.J., Vermaas, J.: Investigation of thermal comfort properties of fabrics containing mohair. J. Text. Inst. **113**(4), 616–627 (2022). https://doi.org/10.1080/00405000.2021.1896158
34. Wu, S.C., Driver, V.R., Wrobel, J.S., Armstrong, D.G.: Foot ulcers in the diabetic patient, prevention and treatment. Vasc. Health Risk Manag. **3**(1), 65–76 (2007)
35. Iraj, B., Khorvash, F., Ebneshahidi, A., Askari, G.: Prevention of diabetic foot ulcer. Int. J. Prev. Med. **4**(3), 373–376 (2013)
36. Elsayed, E.M., Elsaman, S.I.L.: Synthesis of smart medical socks for diabetic foot ulcers patients. Fibers Polym. **18**(4), 811–815 (2017). https://doi.org/10.1007/s12221-017-1256-9
37. Borkow, G., Zatcoff, R.C., Gabbay, J.: Reducing the risk of skin pathologies in diabetics by using copper-impregnated socks. Med. Hypotheses **73**(6), 883–886 (2009). https://doi.org/10.1016/j.mehy.2009.02.050
38. Rogina-Car, B., Skenderi, Z., Vrljičak, Z.: Thermal resistance of viscose socks. Koža & Obuća **68**(3), 18–21 (2019). https://doi.org/10.34187/ko.68.3.6
39. Rogina-Car, B., Skenderi, Z., Vrljičak, Z.: Thermophysiological wear comfort of viscose and tencel socks. Koža & Obuća **68**(4), 22–29 (2020). https://doi.org/10.34187/ko.68.4.3
40. Kadolph, S., Marcketti, S.: Textiles, 12th edn. Pearson Education, US (2016)
41. Gericke, A., Venkataraman, M.: Thermal and moisture management in the microclimate of socks for diabetic foot care: the role of mohair-wool content. Fibers **12**(53), 1–18 (2024). https://doi.org/10.3390/fib12070053
42. Militký, J., Křemenáková, D.: Thermal conductivity of wool/pet weaves. In: 6th International Conference on Heat Transfer, Fluid Mechanics and Thermodynamics, July 2 (2008)
43. Marolleau, A., Salaun, F., Dupont, D., Gidik, H., Ducept, S.: Influence of textile properties on thermal comfort. IOP Conf. Ser. Mater. Sci. Eng. **254**(18) (2017). https://doi.org/10.1088/1757-899X/254/18/182007

Physical Properties and Dyeing Performance of Sustainable Paper Yarn

Guocheng Zhu, Lingya Wu, Yan Wang, Jiří Militký, and Mohanapriya Venkataraman

1 Introduction

In recent years, with the growing global emphasis on environmental protection and sustainable development, paper-based yarns and fabrics, as emerging eco-friendly materials, have garnered significant attention in textile industry, especially in the fields of clothing and garments and packaging [25].

Paper yarn is made by cutting and twisting base paper, and its development is inseparable from the progress of papermaking technology [4, 23]. Since Cai Lun invented papermaking in the Eastern Han Dynasty, papermaking technology has become increasingly mature, with abundant sources of paper and relatively low prices. In the case of lacking clothing and quilts, people used paper to make paper clothes, paper jackets, and paper quilts to keep warm in the Tang Dynasty [22].

G. Zhu · L. Wu · Y. Wang
College of Textile Science and Engineering, Zhejiang Sci-Tech University, Hangzhou, China
e-mail: gchengzhu@zstu.edu.cn

Y. Wang
e-mail: yan.wang@zstu.edu.cn

G. Zhu
Zhejiang-Czech Joint Laboratory of Advanced Fiber Materials, Zhejiang Sci-Tech University, Hangzhou, China

Y. Wang
Xiangshan Knitting Research Institute, Zhejiang Sci-Tech University, Xiangshan, China

J. Militký · M. Venkataraman (✉)
Department of Material Engineering, Faculty of Textile Engineering, Technical University of Liberec, Liberec, Czech Republic
e-mail: mohanapriya.venkataraman@tul.cz

J. Militký
e-mail: jiri.militky@tul.cz

J. Militký and M. Venkataraman (eds.), *Fibrous Structures for Sustainable Future*, Engineering Materials, https://doi.org/10.1007/978-981-95-1382-6_6

Alongside China, Japan and South Korea shared a profound papermaking tradition that has fostered unique paper cultures. In 606 AD, the monk Tanjo introduced papermaking to Japan, where the craft gradually evolved over centuries. The history of Japanese washi yarn dates back to the Nara Period (710–794), when Buddhism was established as the state religion, prompting extensive use of washi for copying Buddhist scriptures [5]. Japanese artisans crafted washi yarn by integrating traditional papermaking techniques with high-fiber plants like Manila hemp, kozo (paper mulberry), and ramie, yielding a material renowned for its toughness and breathability. The production of washi yarn involves an intricate process, demanding highly refined skills and generations of accumulated expertise.

In 1638, in Volume V of Kefukigusa, there is a fabric called "Shifu" mainly made from paper mulberry, a tree species that grows rapidly and has bark rich in highly tough fibers, making it ideal for papermaking [8]. During the production process, craftsmen first soak, peel, and wash the bark to extract high-purity fibers, then make pulp through steps such as pounding, filtering, and blending. Finally, they pour the pulp onto a screen and form washi yarn through processes like extrusion and drying. This technique ensures that washi yarn not only retains the light texture of traditional paper but also has good durability and toughness, becoming an intangible culture protected and inherited in Japan [6].

In South Korea, paper yarn is called Hanji Yarn, with a structural characteristic similar to filament and a unique texture and touch [14]. Hanji Yarn is a yarn made from cellulose fibers of the inner bark of mulberry trees, whose manufacturing process is adapted from traditional Korean papermaking methods. The unique raw materials and novel production technology enhance the aesthetics and functionality of textile products [9].

Before the invention of paper yarn spinning machines, paper yarn was mainly twisted by hand or spun using wooden spinning wheels. Due to low early production efficiency and the fact that paper yarn had not yet reached the level of natural fibers in terms of abrasion resistance and softness, it failed to occupy a mainstream position in large-scale textile production. Its main applications included daily necessities, simple packaging materials, and use as a substitute during periods of scarcity of other natural fiber resources.

In the nineteenth century, with the advancement of the Industrial Revolution and the development of materials science, the manufacturing process of paper yarn ushered in breakthrough progress. In 1898, Emil Clavie first applied for a patent on a paper yarn spinning machine [3]. In 1959, Marks further applied for a patent for the "Continuous Paper Yarn Manufacturing Process and Equipment" [12], which proposed a continuous production process capable of completing wetting, slitting, and twisting on a single device. This technological breakthrough significantly improved the production efficiency and quality of paper yarn, making it more competitive in the market.

Early studies found that pure paper yarn had defects in softness, tensile properties, processability, and knittability [15, 17]. Researchers improved the strength, durability, and elasticity of paper yarn by adding different natural fibers or synthetic materials.

Sun Li applied carboxymethyl pea starch onto paper yarn by sizing technology, and found that the tensile properties and abrasion resistance of CMPS-sized paper yarn were significantly improved [18]. Hu Diefei knitted three types of blended yarns of paper yarn/regenerated polyester (100/0, 70/30, 60/40) into weft plain knitted fabrics, and found that with the increase in the content of regenerated polyester, the breaking strength and elongation at break of the blended yarns increased. The wearing performance of the paper yarn and regenerated polyester fabric was optimal when the proportion was 60/40 [7]. Liang Yanteng analyzed the characteristics of washi/polyester blended yarns, and found out that the blended yarns exhibited a spiral arrangement, uniform yarn evenness, fewer hairiness, and high breaking strength and elongation at break. Under tensile force, the yarn showed certain elongation, and the cross-section shrank simultaneously, in which the washi yarn broke before the polyester filament [11]. Memon evaluated the structure and properties of paper yarns made by cutting and twisting base papers of 12 g/m^2 and 18 g/m^2, found out that the paper yarns had good linear density, appropriate twist, and hairiness ranging from 1.3 to 1.33. Their cross-sections showed a layered structure, and the longitudinal surface had groove characteristics. The breaking strength was 0.77–1.05 cN/dtex, and the wash resistance was excellent, maintaining more than 89% of the strength after 15 washes [13].

Sun Shiyuan used paper yarn and wool as raw materials to produce paper yarn/wool worsted fabrics, and found that the paper yarn/wool worsted fabrics had good softness, drapability, breathability, antistatic properties [19]. Park dyed warps with chemical stock solutions and natural dyes, used paper yarn and core-spun yarn as wefts through blending and 1:1 insertion respectively, and developed denim fabrics with stable morphology and good mechanical properties. And found out that paper yarn had a fluff-free surface, lower breaking stress, high breaking strain, and stable structure. Compared with existing denim fabrics, the developed denim fabrics showed more stable dimensional changes during washing and drying, better color fastness to washing, and faster drying speed [16].

At present, the research on dyeing of paper yarn is still in its initial stage. Especially, the in-depth discussion on paper yarn dyeing technology and process is relatively less. Therefore, the development of dyeing technology for paper yarn has become an urgent issue to be solved. Sun et al. [18] used natural plant dyes to dye paper yarn, and found that the color fastness to wet friction only reached to level 2–3. This phenomenon occurs mainly because the binding force between direct dyes and paper yarn is weak. The dyes are combined with the hydroxyl (–OH) groups in the fiber through electrostatic effects and hydrogen bonds. The stability of this binding method is not enough to meet the color fastness requirements of the clothing field. In comparison, reactive dyes can form stronger covalent bonds with paper yarn fibers during the dyeing process, which can significantly improve the stability of the dyeing effect [24].

For tuning the mechanical and sorption characteristics of yarns, the hybrid yarns created from mixture of paper strips with other fibres as polyesters should be investigated and used.

The high requirements for color fastness of paper yarn in daily wear, lead to use reactive dyes to dyeing. By optimizing the parameters of the dyeing process, including dyeing temperature, pH value, immersion time and dye concentration, it is possible to provide a theoretical basis and practical guidance for the paper yarn dyeing process to achieve an ideal dyeing effect. Optimizing the dyeing process helps to improve the market competitiveness of paper yarn products.

2 Paper Yarns

Manufacturing of paper yarn is comprising from three main steps.

1. Preparation of paper sheets by paper making or selection of suitable wastes containing them.
2. Cutting a paper sheet into a strips (tapes) with rectangular cross section and width in few millimeters.
3. Preparation of paper yarn by twisting the paper tapes or their mixture with other fibres. Twist is usually about 400–800, and z-twist is used.

The base paper can be obtained through specific processing conditions of paper-making technology or just selected from paper products. Thickness of papers sheets vary in huge range (e.g. copy paper is between 0.07 and 0.10 mm) and depends on aim of their use. Then, the base paper is cut into strips with width of few millimeters, which are then twisted to form paper yarn. In addition, paper strips can be spun with other fibers, or single-strand paper yarn can be twisted to form double-strand or multi-strand paper yarn to produce textile product [4, 23].

Base paper is commonly prepared from wood pulp. The plant wood pulp is put into a pulper, with dyes and water added for beating. Through the mechanical action during beating, the degree of fiber fibrillation is improved, the interweaving probability and bonding strength between fibers are increased, and more hydrogen bond combinations on the fiber surface are promoted. After processing, the cell walls of fibers deform and displace, the outer layers of the primary and secondary walls rupture. While the fibers absorb water and swell, they undergo fibrillation, and partial fiber fragments appear during fiber cutting, so that the pulp has further papermaking properties and paper-forming properties to meet the requirements of paper formation. After beating for a certain time, it is sent to the stock mixing tank, and auxiliary materials such as color fixatives, wet strength agents, softeners, etc. are added in sequence and fully mixed to obtain the finished stock. The stock is evenly interwoven into paper by a paper machine to obtain wet paper sheets, which are put into a wet paper sheet press for dehydration, and then dried in an electric hot air-drying oven at 60 °C for 24 h to prepare the base paper.

The base paper is often cut by a paper cutting machine to strips with a width of 1–4 mm and the edges of the obtained strips are flattened. The paper strips are twisted into slivers by a spinning machine, and then several slivers are rotated and twisted

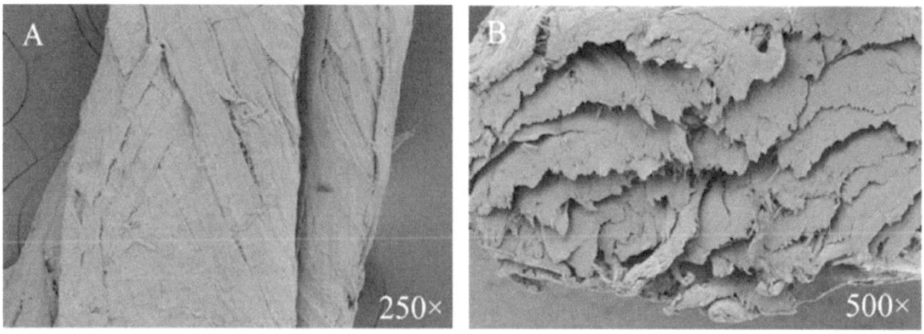

Fig. 1 Microscopic views of paper yarn: **a** surface (250×), **b** cross section (500×) (reused from Memon et al. [13] under CC-BY Permission)

into yarn along a pre-determined direction. The yarn is subjected to heat treatment at 90 °C to fix its shape, resulting in paper yarn.

The paper yarn has a special surface with visible rectangular strips (tapes) and small voids (see Fig. 1a). Cross-section of the paper yarn showed a layered structure created by rotation of strips during yarn spinning (see Fig. 1b).

Typical feature of paper yarns is absence of hairiness because individual papers strips are no be split and are sufficiently long.

Properties of paper yarns are dependent on basic paper thickness and areal weight and on the width of cut strips mainly. In recent work [13] the paper yarns from 3 mm paper strips with the areal weights of 12 g/m^2 (S1) and 18 g/m^2 (S2) were compared. The basic characteristics of both paper yarns are given in Table 1.

The tensile properties of paper yarn have a great influence on the quality of yarn and on weaving and knitting processing.

The tensile properties of paper yarn (see Table 1) were tested according to GB/T 3916-2013 standard. It was found that paper yarn, S1 has smaller tensile strength and higher elongation till break and (0.77 $\pm$ 0.07 cN/dtex, 3.67 $\pm$ 0.38%,) in comparison with paper yarn S2 (1.05 $\pm$ 0.10 cN/dtex and 2.29 $\pm$ 0.29%). The higher tensile strength of S2 is due to to the higher areal weight.

The tensile strength and elongation of hybrid paper/polyester yarns with the same width has a significant decrease as the increase of mass portion of paper yarn. Even though all the hybrid yarns have enough tensile strength for further process such as knitting according to the standard of FZ/T 71005-2014 "Natural cotton yarn for knitting", the elongation of blended paper yarn decrease to a very low value, especially

Table 1 Characteristics of different paper yarns [13]

Paper yarn	Paper areal weight (g/m^2)	Linear density (Tex)	Diameter (mm)	Twist (Twist/10 cm)
S1	12	42.2 $\pm$ 0.10	0.11 $\pm$ 0.01	46.4 $\pm$ 2.03
S2	18	59.0 $\pm$ 0.19	0.12 $\pm$ 0.01	47.5 $\pm$ 2.22

for paper yarn which decreased to 3.67%. The low elongation may cause breakage during knitting or weaving process (Fig. 2).

The mechanical properties are frequently changed due to repeated washing. For this reason, the paper yarn was washed in a fully automatic washing shrinkage tester (Model: Y089D, Wenzhou Fangyuan Instrument Co., Ltd., Wenzhou, China) according to GB/T 3921-2008 standard for 40 min at temperature 30 °C After repeated washing the tensile strength and elongation of paper yarn decreased gradually (see Fig. 3) but both paper yarn (see Table 1) exhibited relatively acceptable stability. The decrease of tensile strength and elongation of paper yarn is mainly caused by structure loosening due to breaking of some hydrogen between cellulose chains caused by penetration of water molecules.

The yarn used for weaving or knitting should have certain hygroscopicity. Under the same relative humidity conditions, yarn with good hygroscopicity is more

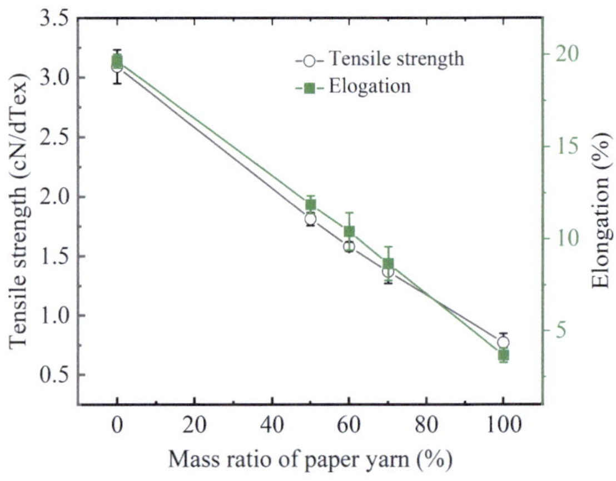

Fig. 2 Decrease in the tensile strength of hybrid paper/polyester yarns

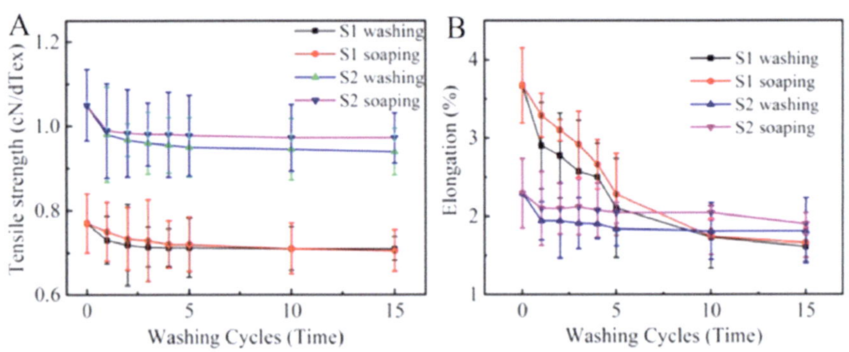

Fig. 3 Ultimate tensile properties of paper yarn after cyclic washing. **a** Tensile strength decrease, **b** break elongation decrease (reused from Memon et al. [13] under CC-BY Permission)

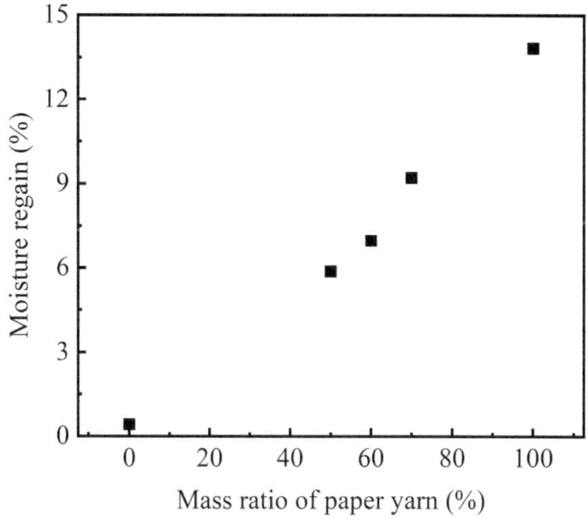

Fig. 4 Moisture regain of blended paper/polyester yarns

conducive to the stability of yarn twist and the improvement of extensibility, so that the yarn has good spinnability.

The moisture regains of hybrid paper/polyester yarns increased when the portion of paper yarn increased (see Fig. 4).

The paper yarn is composed of plant wood pulp fibers, which contain a large number of hydrophilic groups. Especially in pure paper yarn, the content of hydrophilic groups such as –OH is high, endowing it with good moisture absorption and water absorption properties. Polyester, lacking hydrophilic groups and with high crystallinity, has extremely poor moisture absorption capacity. Combined with the analysis of morphology of paper yarn, it can be seen that there are many gaps inside the paper yarn, and the fiber structure is loose and stacked, which is conducive to the passage of moisture. Moreover, the longitudinal surface of the fibers has voids making it easier to adsorb water molecules, therefore, exhibiting good moisture absorption performance.

The wicking characteristics of pure paper yarns can be evaluated by a capillary effect tester (YG(B)871, Wenzhou Darong Textile Instrument Co., Ltd., Wenzhou, China) according to FZT 01071-2008 standard. The wicking height of paper yarns is generally concave increasing function of time with decreasing rate (see Fig. 5).

The wicking height of paper yarns S1 is significantly higher than of paper yarns S2 due to smaller packing (areal mass).

The wicking height of hybrid paper yarn increases with increase of paper yarn mass portion, that is, the water conductivity of the paper yarn is proportional to the mass ratio of plant wood pulp fibers (see Fig. 6).

The reason is that the plant wood pulp fibers contain a large number of hydrophilic hydroxyl groups, and the fibers form many grooves on the longitudinal surface of the paper yarn, with a layered stacking cross-section. The surface tension between the paper yarn and water molecules is small, allowing water molecules to enter the

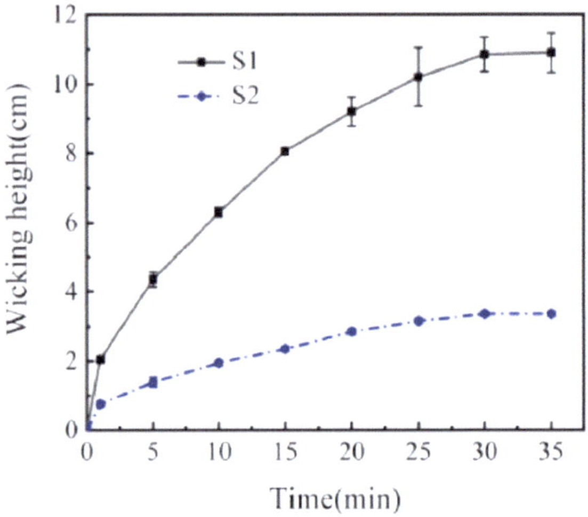

Fig. 5 Wicking kinetic of paper yarns (reused from Memon et al. [13] under CC-BY Permission)

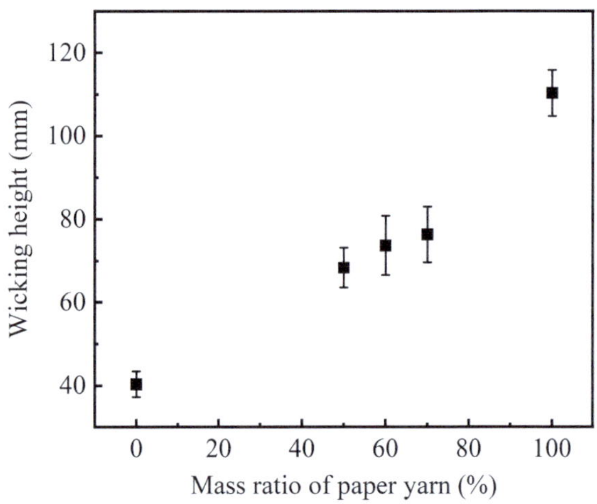

Fig. 6 Wicking of hybrid paper yarn

interior of the yarn more quickly and be absorbed, thus leading to good wicking performance. However, due to the low content of hydrophilic groups in polyester and the small gaps between fibers, which are not conductive to water molecule transport. Wicking height of hybrid polyester/paper yarn increase with increasing of mass portion of paper yarns.

3 Reactive Dyeing of Paper Yarns

Process of paper yarns reactive dyeing is schematically shown in Fig. 7.

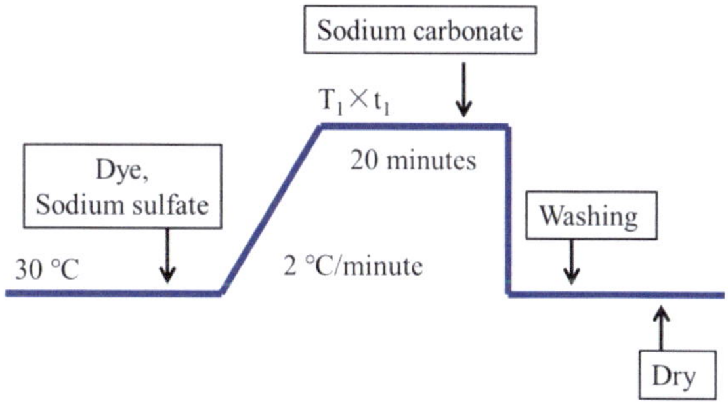

Fig. 7 Dyeing process of paper yarn

Paper yarns are immersed in the dye liquor when the temperature is 30 °C, and kept for a certain dyeing time after heating to the maximum temperature. The holding time is 20 min. After dyeing, the paper yarns are soaped in 95 °C hot water for 10 min (the soap liquor composition is 2 g/L soap flakes and 1 g/L sodium carbonate, and the bath ratio is 1:50). After soaping, the paper yarns are dried at 30 °C for 1 h.

Typical dyeing formula is: bath ratio is 1:20, dye concentration is 2% (owf), dyeing temperature range: 30–65 °C, dyeing time range: 25–60 min, sodium sulfate concentration range: 5–40 g/L, sodium carbonate concentration range: 5–40 g/L.

The dyed paper yarn is tested with Datacolor 1050 colorimeter under D65 light source and 10° observation angle. The characteristics of color space L, a, b, and K/S values of the dyed paper yarn are evaluated.

For color fastness evaluation, the methods specified in the standard GB/T 3920-2008 "Textiles - Tests for color fastness - Color fastness to rubbing", the color fastness to dry abrasion, and the color fastness to wet abrasion were used.

Washing fastness of the samples was measured according to the experimental method specified in the standard GB/ T3921-2008 "Textiles - Tests for Color Fastness - Color Fastness to Washing: Test 1", and the discoloration and staining grade was evaluated.

3.1 *Effect of Dye Concentration on Paper Yarn Dyeing*

The dye concentrations were varied from 0.5 to 6%, and the other parameters were kept constant, i.e., dyeing temperature was 60 °C, dyeing time was 40 min, sodium sulfate concentration was 30 g/L, and sodium carbonate concentration was 15 g/L.

The dye uptake rate and K/S values first increase and then gradually stabilize with the increase of dye concentration. In the low dye concentration range (0.5–2%), the dye uptake and K/S of the three kinds of dyes showed a significant increase. At higher dye concentrations (2–6%), the increase of dye uptake rate gradually slowed down and tend to a peak. When the dye concentration was 6%, some curves showed a slight decline. The reason could be that water molecules diffused into the yarn in the dye solution, breaking up the hydrogen bonds between the fiber macromolecule chains, causing the fiber to swell and form dye diffusion channels [21]. At the same time, in the low concentration range, the concentration of dye molecules in the dye solution is low, while the adsorption sites on the paper yarn fibers are not yet saturated. At this time, the dye molecules have more opportunities to react with the fibers, which is manifested in a rapid increase in dye uptake and an increase in K/S value. As the dye concentration continues to increase, the adsorption sites of the paper yarn are gradually occupied, and the dyeing gradually reaches equilibrium. At this time, the adsorption rate of the dye molecules decreases. Even if the concentration continues to increase, the proportion of dye adsorbed by the fiber per unit time will decrease, so the increase in dye uptake slows down, and the K/S value tends to be stable. However, too high a dye concentration will cause competition and aggregation effects between dye molecules in the dye solution. Aggregated dye molecules are more difficult to diffuse into the fiber, resulting in the inability of some dyes to be effectively utilized, which is manifested in a decrease in dye uptake. The presence of excessive dye may lead to uneven distribution of dye on the fiber surface. At the same time, dark areas may have a thick dyeing effect, resulting in uneven color and increased dyeing costs. It may also lead to a decrease in the mechanical properties and wet friction color fastness of paper yarn [1, 20].

According to experiments, when the dye concentration exceeds 4%, the dyeing rate and K/S value are not significantly improved, and high concentration may cause waste, so the concentration should not exceed 4%. Effect of dyeing concentration on dyeing uptake rate and K/S value for three typical reactive dyestuffs are shown in Fig. 8.

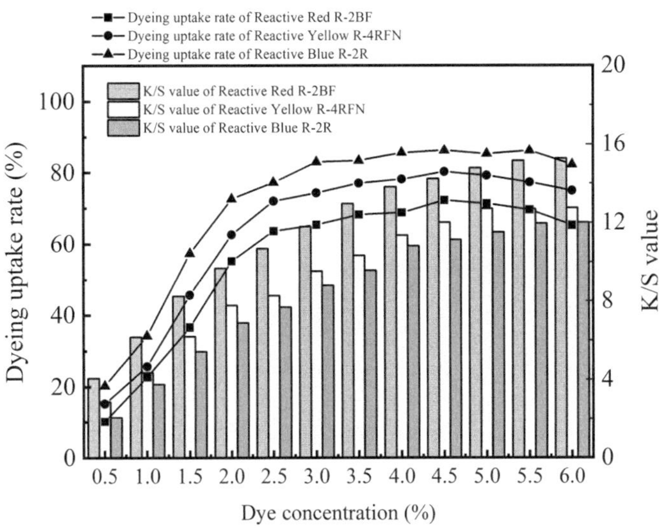

Fig. 8 Effect of dyeing concentration on dyeing uptake rate and K/S value

3.2 *Effect of Dyeing Temperature on Paper Yarn Dyeing*

In order to explore the dyeing effect of dyeing temperature on paper yarn, dyeing solutions of different colors and the same concentration were prepared. The dyeing temperature was adjusted to 30–65 °C, and other experimental conditions were fixed, including: dyeing time was 40 min, sodium sulfate concentration was 30 g/L, and sodium carbonate concentration was 15 g/L. The effect of dyeing temperature on paper yarn for three typical dyestuffs is shown in Fig. 9.

As shown in Fig. 9, for reactive red R-2BF and reactive yellow R-4RFN, the dyeing rate gradually increases with increasing temperature, reaches a peak value between 55 and 60 °C, and then shows a downward trend. The trend of the K/S value is consistent with the dye uptake. The adsorption and fixation efficiency of the dye is optimal within this temperature range, and a higher color depth can be presented. This shows that appropriately increasing the temperature can accelerate the diffusion rate of the dye and enhance the reactivity with the fiber [10, 11], thereby increasing the dye uptake and K/S value. However, too high a temperature may cause the dye to hydrolyze in the solution, reduce the binding of effective dye molecules to the fiber, and thus reduce the dye uptake and color depth. For reactive blue R-2R, the dyeing rate continues to increase with increasing temperature, but the amplitude gradually slows down, and its K/S value is generally low and the increase amplitude is not obvious with increasing temperature. The optimization of dyeing temperature should be to achieve the ideal dyeing rate and K/S value without causing excessive hydrolysis, and the temperature is selected as 60 °C.

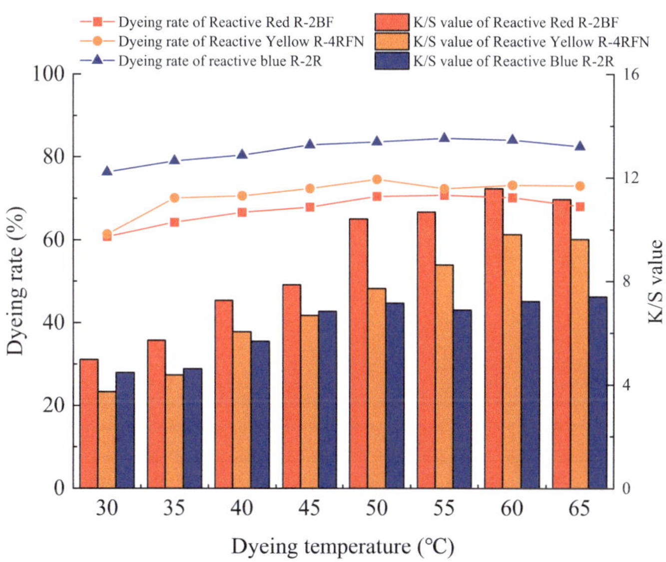

Fig. 9 Effect of dyeing temperature on paper yarn dyeing

3.3 *Effect of Dyeing Time on Paper Yarn Dyeing*

In order to explore the effect of dyeing time on paper yarn, dye solutions of different colors and the same concentration were prepared. The dyeing time was adjusted to 25–60 min, and other experimental conditions are fixed, including: dyeing temperature is 60 °C, sodium sulfate concentration is 30 g/L, and sodium carbonate concentration is 15 g/L. The dyeing effect of dyeing time on paper yarn is shown in Fig. 10.

In Fig. 10 for Reactive Red R-2BF and Reactive Yellow R-4RFN, the dyeing rate increased rapidly between 25 and 40 min, then stabilized and decreased after 50 min, while the K/S value reached a peak between 35 and 40 min and then gradually decreased, which is consistent with the trend of the dyeing rate. This result shows that in the early stage of dyeing, the dye molecules diffused rapidly to the surface of the paper yarn and entered the fiber interior, so the dyeing rate and K/S value increased rapidly. As time goes by, the combination of dye and fiber tends to balance. After a certain period of time, the dye may decolorize or hydrolyze, resulting in a decrease in dyeing rate and affecting the dyeing depth [24]. Reactive Blue R-2R showed a slower dyeing speed and higher time stability. Its dyeing rate increased slowly between 25 and 45 min, then stabilized, and the K/S value reached the highest value around 45 min. This shows that the dyeing process of the dye is relatively gentle, and its diffusion and reaction processes are relatively stable. For the faster-reacting Reactive Red R-2BF and Reactive Yellow R-4RFN, the dyeing time can be controlled to about

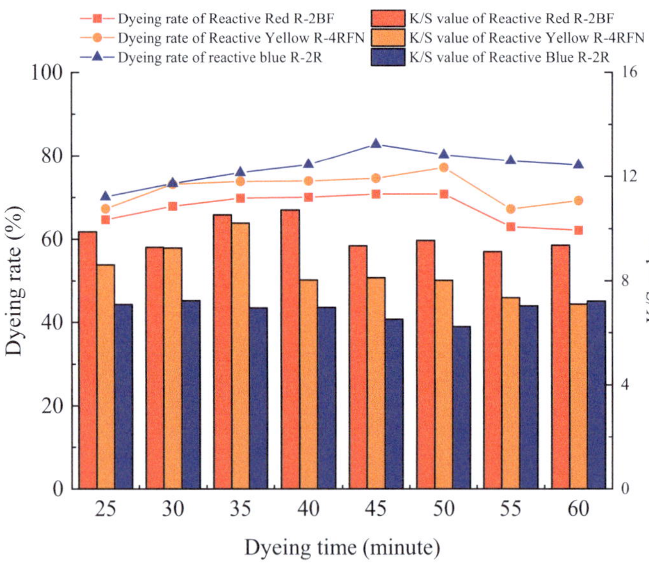

Fig. 10 Effect of dyeing time on paper yarn dyeing

40 min to achieve the best effect; for the slower-reacting Reactive Blue R-2R, 45 min may be the ideal dyeing time.

3.4 Dyeing Effect of Sodium Sulfate on Paper Yarn

In order to explore the dyeing effect of sodium sulfate concentration on paper yarn, dye solutions of different colors and the same concentration were prepared. The sodium sulfate concentration was adjusted to 5–40 g/L, and other experimental conditions were fixed, including: dyeing temperature of 60 °C, dyeing time of 40 min, and sodium carbonate concentration of 15 g/L. The dyeing effect of sodium sulfate on paper yarn is shown in Fig. 11.

As shown in Fig. 11 the dyeing rate and K/S values of the three dyes increased with the increase of sodium sulfate concentration, and then tended to stabilize or slightly decreased. However, the increase of Reactive Blue R-2R was limited at higher concentrations (30–40 g/L), indicating that the dyeing depth of the dye was close to saturation at lower concentrations, indicating that its dyeing process was less dependent on the sodium sulfate concentration. Sodium sulfate can reduce the electrostatic repulsion around the dye molecules, making it easier for the dye molecules to diffuse to the surface of the paper yarn and be adsorbed, thereby increasing the dyeing rate. An appropriate amount of sodium sulfate can effectively promote the dyeing of the dye, enhance the combination of the dye and the fiber, and increase the color depth, but excessive sodium sulfate may cause excessive precipitation or

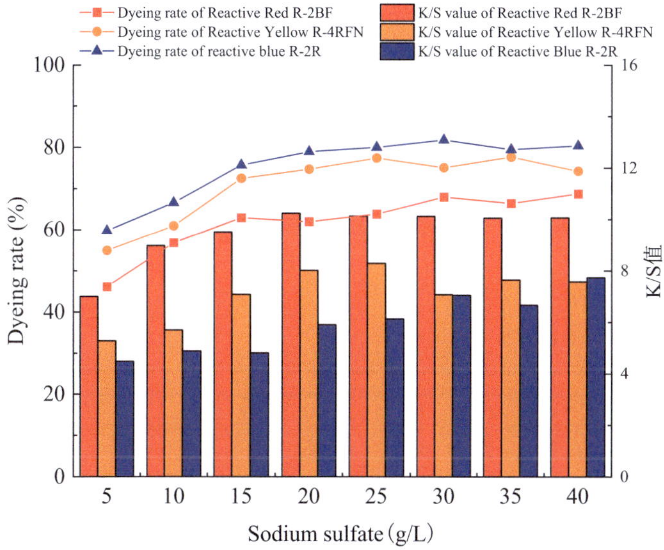

Fig. 11 Effect of sodium sulfate concentration on paper yarn dyeing

agglomeration of the dye molecules, reducing the effective dye concentration [2]. For Reactive Red R-2BF and Reactive Yellow R-4RFN, the sodium sulfate concentration can be controlled at 30 and 25 g/L; while the sodium sulfate concentration of Reactive Blue R-2R can be controlled at 35 g/L.

3.5 Dyeing Effect of Sodium Carbonate on Paper Yarn

In order to explore the dyeing effect of sodium sulfate on paper yarn, dye solutions of different colors and the same concentration were prepared. The sodium carbonate concentration was adjusted to 5–40 g/L, and other experimental conditions were fixed, including: dyeing temperature of 60 °C, dyeing time of 40 min, and sodium sulfate concentration of 30 g/L. The dyeing effect of sodium carbonate on paper yarn is shown in Fig. 12.

As shown in Fig. 12, the dyeing rate and K/S value of Reactive Red R-2BF and Reactive Yellow R-4RFN increased significantly with the increase of sodium carbonate concentration, reaching a maximum value near 25 g/L, and then appeared slightly decline. The dyeing rate of Reactive Blue R-2R also continued to increase with the increase of sodium carbonate concentration, but the growth leveled off after 30 g/L. The K/S value of Reactive Red R-2BF reaches a peak at 25 g/L, while the K/S values of Reactive Yellow R-4RFN and Reactive Blue R-2R show a slowly rising trend as the sodium carbonate concentration increases. Sodium carbonate can provide an alkaline environment during the dyeing process, which can promote the covalent

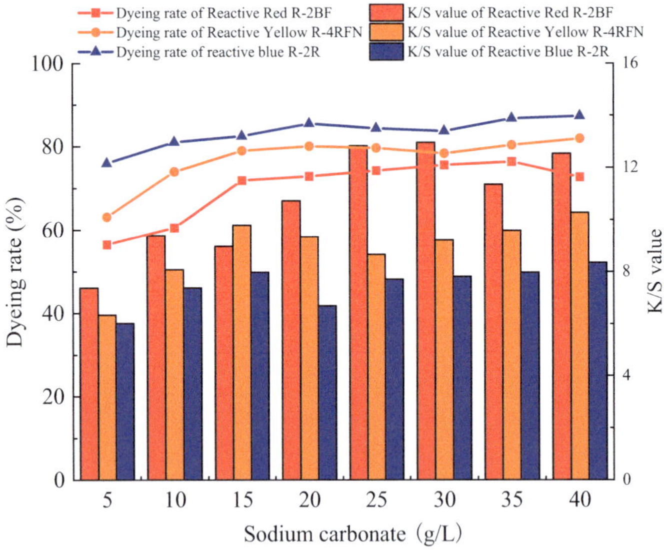

Fig. 12 Effect of sodium carbonate concentration on paper yarn dyeing

bonding of dye molecules and paper yarns, improving the dyeing rate and dyeing depth. Within an appropriate amount, alkalinity can enhance the reactivity of the dye, but excessive sodium carbonate may cause hydrolysis of the dye, precipitation on the fiber surface, or cause certain damage to the fiber, thereby reducing dyeing uniformity and K/S value. Reactive Red R-2BF is more sensitive to alkaline environment, while Reactive Yellow R-4RFN and Reactive Blue R-2R have a relatively wider range of applicable concentrations. However, too high alkali concentration has side effects. In order to avoid excessive use of sodium carbonate, which will affect the dyeing quality Decrease, control the sodium carbonate concentration of Reactive Red R-2BF to 30 g/L, and control the sodium carbonate concentration of Reactive Yellow R-4RFN and Reactive Blue R-2R to 35 g/L.

3.6 Analysis of Color Parameters of Paper Yarn under Optimized Process

According to the influence of different factors of reactive dyes on dyeing effect in the above experiments, the optimized dyeing process of paper yarn with three dyes can be obtained. The paper yarn is further dyed according to the optimized process, and its corresponding color characteristic value is analyzed. The color characteristic values of reactive red R-2BF, Reactive Yellow R-4RFN and Reactive Blue R-2R after dyeing paper yarn at different concentrations are shown in Tables 2, 3 and 4.

As the dye concentration increases, the amount of active dye adsorbed on the paper yarn increases, the L value decreases, and the K/S increases. It can be seen from Table 2 that the measured a-value increases significantly with increasing concentration, indicating that the red hue becomes stronger and stronger. At the same time, both the b value and the c value increased, but the increase in the b value was smaller. From Table 3, it can be seen that the decrease in the L value of Reactive Yellow R-4RFN is not as significant as that of Reactive Red R-2BF and Reactive Blue R-2R. At the same time, it can be seen that its c value increases significantly with the increase

Table 2 Color characteristic values of reactive dye red R-2BF at different concentrations

Dye concentration/%	L	a	b	c	K/S value
0.5	54.51	48.61	− 7.46	49.18	4.756
1	46.28	51.17	− 3.80	51.32	6.236
1.5	44.42	53.45	− 2.65	53.52	8.996
2	43.74	55.60	− 0.21	55.60	10.969
2.5	43.17	56.22	− 0.30	56.22	12.036
3	41.46	55.18	− 0.24	55.19	12.912
3.5	40.74	54.24	1.36	54.24	13.256
4	38.35	52.61	3.37	52.72	14.059

Table 3 Color characteristic values of reactive yellow R-4RFN at different concentrations

Dye concentration/%	L	a	b	c	K/S value
0.5	74.91	17.58	61.02	63.51	3.256
1	74.43	25.11	69.41	73.82	5.023
1.5	70.18	26.64	69.01	73.98	6.925
2	71.13	30.88	73.78	79.99	8.663
2.5	69.64	32.31	73.46	80.26	9.986
3	68.46	34.48	74.3	81.92	11.453
3.5	67.72	32.07	73.59	80.28	12.429
4	66.94	32.84	73.21	80.24	13.069

Table 4 Color characteristic values of reactive blue R-2R at different concentrations

Dye concentration/%	L	a	b	c	K/S value
0.5	54.61	− 4.64	− 24.65	25.09	2.9638
1	47.04	− 2.43	− 26.28	26.40	4.2979
1.5	42.41	− 1.52	− 29.31	29.36	6.4995
2	38.60	− 0.53	− 28.66	28.67	7.5696
2.5	36.79	− 0.04	− 29.29	29.29	8.4469
3	33.75	0.86	− 27.19	27.21	9.8902
3.5	32.99	1.05	− 28.13	28.16	10.638
4	30.76	1.59	− 27.62	27.68	11.815

in concentration, indicating that the color vividness of the dye is greatly affected by its concentration, which has a better visual effect.

Table 4 shows that the a and b values of Reactive Blue R-2R do not change much with increasing concentration, and its c value gradually increases with increasing concentration, but the increase is smaller than that of Reactive Red R-2BF and Reactive Yellow R-4RFN.

3.7 Color Fastness of Paper Yarn

The color fastness to rubbing and color fastness to soaping of the paper yarn under the optimized process were tested, and the results are given in Table 5.

According to Table 5, after the paper yarn was dyed with the reactive dye, its color fastness to rubbing and soap washing was good, both above level 4, indicating that the color fixation performance of the reactive dye and the dyeing adaptability of the yarn were at a high level. This shows that the dyed yarn can not only meet the high color fastness requirements under laboratory conditions, but also has good practical

Table 5 Evaluation of color fastness grades of reactive dyes

	Color fastness to rubbing/grade		Color fastness to soap washing/grade		
	Dry	Wet	Cotton	Hair	Discoloration
Reactive red R-2BF	4–5	4	4–5	4–5	4–5
Reactive yellow R-4RFN	4–5	4	4–5	4–5	4–5
Reactive blue R-2R	4–5	4	4–5	4–5	4–5

application prospects, and is suitable for industrial production and a variety of textile uses.

4 Conclusions

The structural characteristics of paper yarn, composed from twisted flat strips of paper are shown. The yarn has a smooth surface practically without hairiness. The lower linear density paper yarns have higher wicking abilities. Sufficient water-washing resistance was demonstrated. The hybrid paper/polyester yarns obey possibility to tune mechanical robustness and hydrophilicity, required for various textile applications.

Three reactive dyes (Reactive Red R-2BF, Reactive Yellow R-4RFN and Reactive Blue R-2R) were used to dye paper yarn, and the effects of different dyeing factors (dye concentration, dyeing temperature, dyeing time, sodium sulfate concentration and sodium carbonate concentration) on the dyeing effect of paper yarn were explored. The following conclusions were drawn:

When the dye concentration is less than or equal to 4%, the dyeing rate and K/S value of the paper yarn increase significantly with the increase of dye concentration; when the dye concentration exceeds 4%, the dyeing rate and K/S value of the paper yarn do not increase significantly. Under the condition of dye concentration of 4%, the optimized process is:

Reactive Red R-2BF: The dyeing temperature is 60 °C, the dyeing time is 40 min, and the concentrations of sodium sulfate and sodium carbonate are 30 g/L and 25 g/L, respectively.
Reactive Yellow R-4RFN: dyeing temperature is 60 °C, dyeing time is 40 min, and the concentration of sodium sulfate and sodium carbonate is 25 g/L.
Reactive Blue R-2R: Dyeing temperature is 60 °C, dyeing time is 45 min, and the concentration of sodium sulfate and sodium carbonate is 35 g/L.

The paper yarn after the optimized process was dyed and its related color fastness was tested. The yarn's color fastness to rubbing and color fastness to soap washing were both above level 4, which is suitable for textile use.

References

1. Bian, P.: Effect of gardenia extract on dyeing process and color fastness of silk fabric. Print. Dye. Aux. Agents **41**(07), 51–56 (2024)
2. Cai, Z.: Physical Chemistry of Dyeing, pp. 211–212. China Textile Press, Beijing (2023)
3. Claviez, E.: Process for the manufacture of yarn or thread from paper and apparatus therefor: CA, 58620D (1898)
4. Dang, Y., Wang, Z., Li, H.: Research on the base paper of textile paper thread. Trans. China Pulp Pap. **4**, 56–61 (2006). https://doi.org/10.3321/j.issn:1000-6842.2006.04.014
5. Feng, T.: Interpretation of the Production Process and Symbolic Culture of "Washi", pp. 18–22. Minzu University of China, Beijing (2008)
6. Feng, T.: The inheritance of the technique of "washi" under the intangible cultural property protection system of Japan. Chin. Cult. Herit. **2008**(2), 84–89 (2008)
7. Hu, D., Wang, Y., Yao, J.: Properties of paper yarn/recycled polyester composite yarn and its knitted fabrics. Mod. Text. Technol. **31**(2), 152–158 (2023)
8. Karuno, H.: Shifu: a traditional paper textile of Japan (2016)
9. Lee, K.E., Sanders, E.A.: Hanji, the mulberry paper yarn, rejuvenates nature and the sustainable fashion industry of Korea. Green Fash. **1**, 159–184 (2016)
10. Li, X., Cui, Y., Lv, L.: Analysis of dyeing performance of mangosteen shell pigment on cotton knitted fabrics. Knitt. Ind. **3**, 47–50 (2015)
11. Liang, Y., Zhu, R., Nie, S.: Structure and properties of paper/polyester composite yarn. Shanghai Text. Sci. Technol. **52**(12), 27–29 (2024)
12. Marks, R.H.: Continuous process and apparatus for making paper yarn: US, 63448057 (1959)
13. Memon, H., Hu, D., Wu, L.: Structure, properties, and fabric applicability of sustainable paper yarn with high washing stability. Heliyon **10**(5), e27467 (2024)
14. Park, T.Y., Joo, C.W.: Evaluation of tensile property and surface structure of hanji paper yarn with different twist numbers. Text. Sci. Eng. **49**(3), 147–152 (2012)
15. Park, T.Y., Lee, S.G.: A study on coarse Hanji yarn manufacturing and properties of the hanji fabric. Fibers Polym. **14**(2), 311–315 (2013)
16. Park, T.Y., Kim, M.O.: Manufacture and physical properties of the denim fabrics using Hanji paper yarn as weft yarn. Fash. Text. **5**, 1–13 (2018)
17. Peterson, J., Eckard, A., Hjelm, J.: Mechanical-property-based comparison of paper yarn with cotton, viscose, and polyester yarns. J. Nat. Fibers. **18**(4), 492–501 (2021)
18. Sun, L., Yang, M., Xia, J.: Research on the dyeing performance of natural plant dyes on paper yarns. Knitt. Ind. **5**, 43–47 (2022)
19. Sun, S., Dou, H., Jiang, X.: Development of new natural and environmentally friendly materials, paper yarn and their products. Shanghai Text. Sci. Technol. **4**, 51–52 (2006)
20. Wang, B., Chen, J., Yang, Y.: Reactive brittleness: damage of reactive dyes to dyed fabrics—a review of some dyeing and finishing technologies. Print. Dye. **39**(17), 44–48 (2013)
21. Wang, J., Liu, X., Fang, K.: Improvement of anti-wrinkle performance of cotton fabrics by dyeing with divinylsulfone reactive dyes. Acta Text. Sin. **44**(2), 207–213 (2023)
22. You, X.: Paper clothes and paper quilts. Anc. Mod. Agric. **1**, 2 (1996)
23. Zhang, B.: Research and development of paper yarn base paper, pp. 2–4. Tianjin University of Science and Technology, Tianjin (2021)
24. Zhang, S., Yang, J.: The current situation and prospect of reactive dyes. Dyes Dye. **1**, 1–7 (2008)
25. Zhang, X., Zhang, H., Cheng, Y.: Research progress, application status and prospect of paper-based packaging materials. China Pulp Paper **39**(11), 53–69 (2020)

Sustainable Textiles from Recycling Materials for Thermal Insulation Applications

Divan Coetzee, Mohanapriya Venkataraman, Jakub Wiener, and Jiří Militký

1 Introduction

The textile and garment industry is one of the largest polluting industries in the world by contributing to approximately 10% of global greenhouse gas emissions, 20% of clean water pollution, and 35% of microplastic generation [5]. For decades, the textile industry has operated on a "take-make-dispose" model driven by consumer consumption for financial gain that contradicts green development concepts. More recently, sustainability initiatives have taken the forefront, and there is a drive to push for the creation of a circular economy [1, 24]. This change in attitude and policies has driven the push for sustainable garment production. Sustainable garment design thinking has emerged as a crucial pathway to harmonize industry operations with environmental and societal needs. Three principal methods of sustainable garment design have been identified by analyzing practices that include promoting sustainability through versatile, durable, and upgradable design approaches; ensuring raw material sustainability through environmentally friendly, renewable, or recycled materials; and strengthening production sustainability by adopting energy-efficient, low-emission processes [39]. Textile waste is challenging to recycle due to its typical multi-component complexity. Current waste treatment practices do not allow for this kind of waste processing and therefore limit the circular economy potential. It is thus

D. Coetzee · M. Venkataraman (✉) · J. Wiener · J. Militký
Department of Material Engineering, Faculty of Textile Engineering, Technical University of Liberec, Liberec, Czech Republic
e-mail: mohanapriya.venkataraman@tul.cz

D. Coetzee
e-mail: divan.coetzee@tul.cz

J. Wiener
e-mail: jakub.wiener@tul.cz

J. Militký
e-mail: jiri.militky@tul.cz

J. Militký and M. Venkataraman (eds.), *Fibrous Structures for Sustainable Future*, Engineering Materials, https://doi.org/10.1007/978-981-95-1382-6_7

essential that the end of use be incorporated into the initial design of the product [39, 100].

Perhaps the most crucial element determining a garment's sustainability is its fiber selection. It is difficult to find a single material that is both cost-effective and environmentally friendly [73]. It has been shown that fiber selection dictates 83–92% of a garment's environmental impact, with water and chemical use being the most significant factors [102, 112]. Conventional cotton production can use up to 2700 L of water per kilogram of produced fiber, compared to industrial hemp with 300–500 L and recycled cotton with 150 L [9, 17]. Synthetic fibers generally create less water and land stress, but they are considered more long-term pollutants due to the use of fossil fuels and the generation of microplastics. Polyester fiber production generates about 3.8 kg of CO_2 equivalent due to fossil fuel use and polymerization. There are more environmentally conscious alternatives, such as Lyocell, which generates 1.2 kg due to closed-loop solvent recovery systems.

One of the most common methods of utilizing textile waste is exploiting its functional property of thermal insulation to create thermally insulating materials in a process illustrated in Fig. 1. Experiments have shown that recycled textile waste, which has been repurposed as insulation material, is comparable to conventional insulation materials, with thermal conductivity coefficients of approximately 0.039 W/m K [86]. Mixed fiber garments are notoriously difficult to recycle due to their different physical properties and compositions. This is typical for garments such as certain denim or polycotton products that blend cotton, polyester, and elastane, which cannot be efficiently recycled through mechanical methods because the intertwined fibers resist simple separation. This requires advanced techniques such as selective dissolution, acid-catalyzed hydrolysis, or enzymatic processes to isolate each material [20, 38]. This complexity is not only limited to clothing, upholstery textiles also exhibit these challenges, as they often incorporate additional materials such as foams, adhesives, and chemical finishes that hinder conventional recycling approaches [109]. This difficulty in separating and isolating these materials of proper grade for reuse not only complicates recycling but also contributes to the downcycling of valuable fibers, impeding efforts to establish a sustainable circular economy in the textile sector [47]. These processes are often costly and pose their own environmental challenges. Creating thermal insulation from textile waste, especially its blends, is a cost-effective recycling method. The recycled product may be of lower value than the original garment, but its true value lies in the cost savings by not employing expensive and environmentally damaging recycling techniques [84, 85]. The produced thermal insulation materials can also be improved by adding flame retardants such as sulfonated kraft lignin, kraft lignin, and nano clays to increase their value [113]. The properties of these materials are closely linked to those of the fibers used. Therefore, selecting which fibers to be recycled is essential since this can also impact the recycled material's environmental footprint. This chapter will look at various natural and synthetic fibers used in the garment industry to investigate their sustainability from a fiber perspective and their thermal properties to be used in thermal insulation applications.

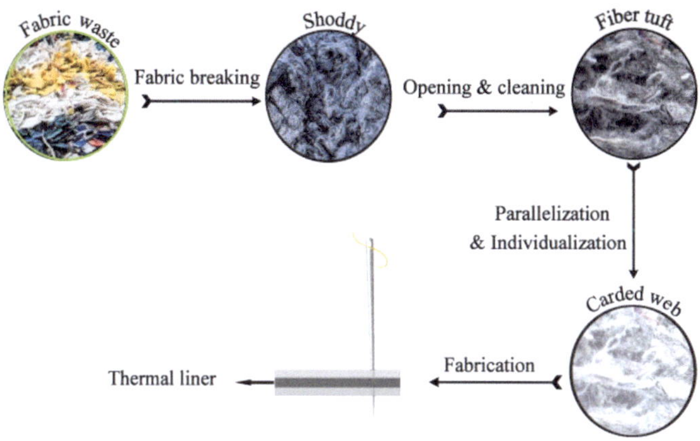

Fig. 1 Process for recycling of fabric waste to produce thermal insulation (reused from Bhuiyan et al. [11] under CC-BY Permission)

2 Textile Structures and Their Thermal Insulation

Textile structures are defined by a hierarchical arrangement featuring a small volumetric proportion of fibrous material and high volume porosity (Φ). Key physical properties of textile structures (woven, knitted, or nonwoven) encompass planar mass (σ) (g/m^2) and thickness (h) (m), although thickness is usually measured in centimeters (cm). Fabric density (ρ_T) is described as:

$$\rho_T = \frac{w}{h} = \rho_F v_F + \rho_a(1 - v_F) = \rho_a + v_F(\rho_F - \rho_a) \tag{1}$$

where v_F is the volume portion of the fibrous phase, which is equal to 1 − Po, where Po is the textile structure volume porosity. The ρ_F is fiber density, and ρ_a is air density. The air density at 21 °C, 65% relative humidity, and an elevation above sea level of 1 300 m is 1.15 kg/m^3. This is a very low value in comparison with typical fiber density (usually from 900 to 1600 kg/m^3) and therefore, Eq. 1 can be simplified to

$$\rho_T = \frac{w}{h} \approx \rho_F v_F \quad \text{and} \quad P_o \approx 1 - \frac{\rho_T}{\rho_F} = 1 - \frac{w}{\rho_F h} \tag{2}$$

A common problem is selecting the appropriate fiber density, which can be associated with fiber crystallinity, morphology (such as the presence of voids), and the presence of other compounds like water or impurities. This is particularly true for natural cellulosic fibers, where true fiber density is often utilized instead of the approximate density of the cell wall, which is 1520 kg/m^3 [22].

Figure 2 illustrates the relationship between the apparent density of polyester fibers (F = 1360 kg/m^3) and fabric thickness for various planar masses, while Fig. 3

depicts the relationship between the volume portion of polyester fibers and fabric planar mass for different fabric thicknesses [64].

Figure 2 indicates that the volume portion of the fibrous phase for heavy fabrics with low thickness is nearly 15%. This corresponds to a total porosity of 85%. Low planar mass combined with high thickness results in a fiber volume fraction below 5%, meaning porosity exceeds 95%. The thermal conductivity of textile fibers

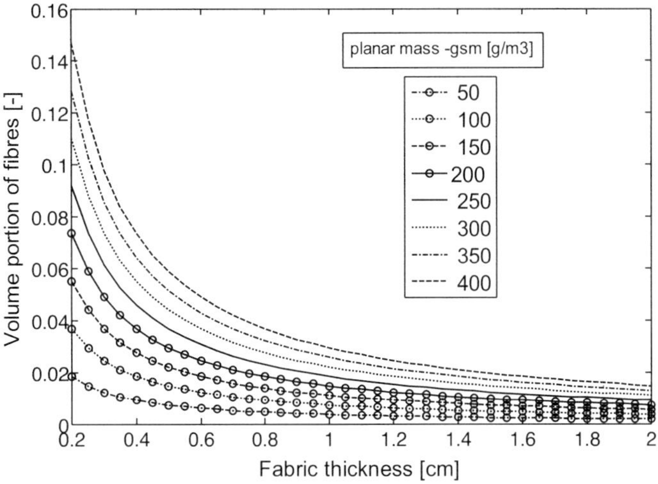

Fig. 2 Dependence of fabric thickness on volume portion of fibers for various planar mass (reused from Militký et al. [64] under CC-BY Permission)

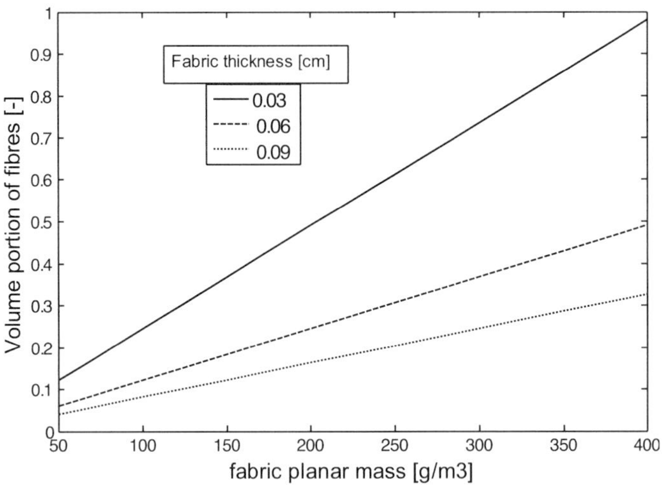

Fig. 3 Dependence of fabric planar mass on volume portion of fibers for various low fabric thicknesses (reused from Militký et al. [64] under CC-BY Permission)

depends on their chemical composition, structure (e.g., crystallinity), morphology (e.g., voids), and moisture content. It is generally directionally dependent (see Table 1).

Thermal conductivity within the fiber mass arises from the specific alignment of polymeric chains. Due to the oriented and closely packed arrangement of these chains, axial thermal conductivity (parallel to the fiber axis) is significantly higher than radial thermal conductivity (perpendicular to the axis). Measured thermal conductivity values also vary depending on experimental conditions and the instrumentation used. For instance, Rengasamy and Kawabata reported a thermal conductivity of 0.352 W/(m K) for dry cotton fibers, derived from measurements on cotton yarns [77].

The small portion of fibrous phase in fibrous structures (high porosity) is the main reason why the total thermal conductivity is governed by air volume fraction (conductivity $\lambda_a = 0.024$ W/(m K)) but not by the thermal conductivity of fibers. This is shown in Fig. 4, where the thermal conductivity of the fabric is computed from the equation based on a combination of serial and parallel arrangement of fibrous and air phases.

$$\lambda_T = 0.5\left((1 - v_F)\lambda_a + v_F\lambda_f + \frac{\lambda_a\lambda_f}{(1 - v_F)\,\lambda_f + v_F\,\lambda_a} \right) \tag{3}$$

Dependence of polyester fabric thermal conductivity (for polyester is $\lambda_f = 0.304$ W/(m K)) on thickness for various planar masses is in Fig. 4 [105].

Total heat resistance R_T is from the total conductivity of the textile structure and its thickness h.

$$R_T = \frac{h}{\lambda_T} \tag{4}$$

Selecting textile structures of sufficient thickness is necessary for effective thermal insulation, especially at low temperatures.

Table 1 Thermal conductivity along the fiber axis and perpendicular to the fiber axis

Fiber type	Thermal conductivity along the fiber axis (W/m K)	Thermal conductivity perpendicular to the fiber axis (W/m K)
Cotton	1.13	0.16
Ramie	1.66	0.21
Mulberry silk	0.83	0.16
Mohair	0.35	0.17
Polyester	0.97	0.19
Polyamide	0.59	0.27
Acrylic	0.74	0.22
Viscose	0.72	0.19

Reused from Militký [62] under CC-BY Permission

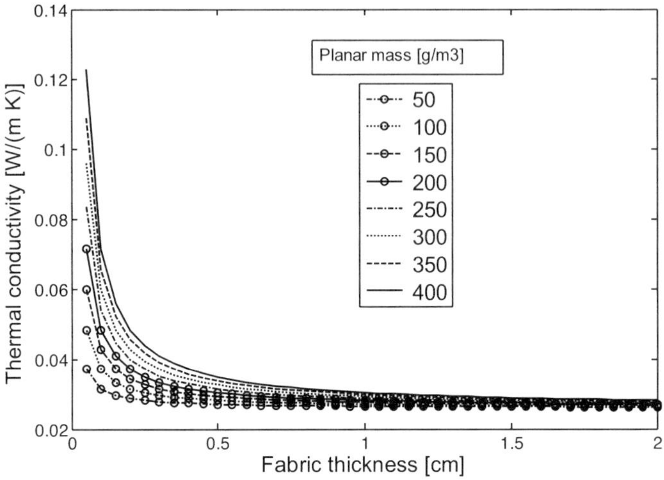

Fig. 4 Dependence of the thermal conductivity of polyester fabric on thickness for various planar masses (reused from Militký et al. [64] under CC-BY Permission)

3 Natural Fibers

Natural fibers, especially cellulose-based ones, although less harmful upon degradation, can be very resource-intensive to grow. Approximately 2000 species of plants are used as a source of natural fibers, but only a few of them are industrially viable [81].

Not all natural fibers are the same and have varying degrees of resource requirements. Table 2 gives the chemical composition of the main cellulosic natural fibers.

Cellulose is a polyalcohol with one primary and two secondary OH groups (see Fig. 5). Its backbone contains ether bonds (glycoside link) –C–O–C–. The surface electrical charge of cellulosic fibers in water is negative and increases with the bath's pH.

Conventional cotton can use up to 2700 L of water per kilogram of produced fiber, compared to industrial hemp's 300–500 L and recycled cotton's 150 L [9, 17].

Table 2 Composition of main natural cellulosic fibers

Fiber	Type	Cellulose (%)	Hemicellulose (%)	Lignin (%)	Pectin (%)
Cotton	Seed	85–92	5–8	0	0–1
Flax	Bast	65–80	14–20	1–4	5–12
Hemp	Bast	60–77	18–24	3–10	5–14

Reused from Militký [62] under CC-BY Permission

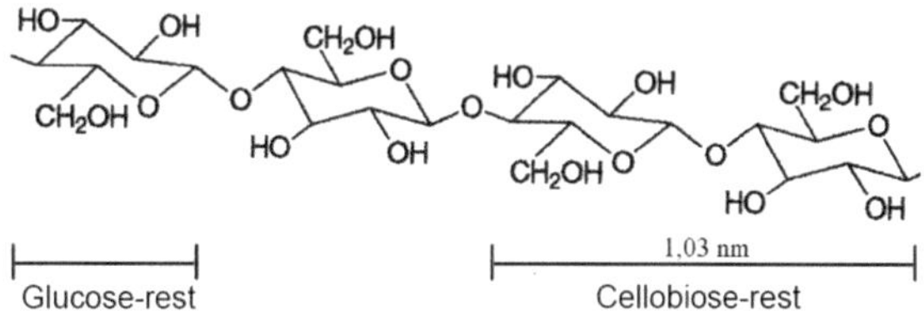

Fig. 5 Structure of cellulose (reused from Militký et al. [64] under CC-BY Permission)

Natural fibers generally have a lower carbon footprint than synthetic fibers due to lower fossil fuel consumption for production [57].

The natural fibers of plant origin (lignocellulosic) contain elementary fibers composed of several cells. These cells are formed by cellulose-based microfibrils, which are connected by amorphous lignin and hemicellulose. Several layers of cellulose, lignin, and hemicellulose in one primary and three secondary cell walls combine to form a multilayer composite that forms the fiber. The cell walls differ in their composition (ratio between cellulose and lignin/hemicellulose). Hemicellulose is formed by complex polysaccharides, such as xyloglucans, xylans, mannans, glucomannans, and beta-glucans, and constitutes the matrix with lignin for the cellulose microfibrils [81].

The main disadvantages of using natural fibers for thermal insulation purposes are that they are more sensitive to moisture and are often flammable [48]. This can also cause biological growth on the fibers if moisture exposure persists. In addition, cellulose fibers are able to regulate the humidity level in the immediate environment and ensure thermal stability due to their high heat capacity [50]. Therefore, it is important to study thermal insulation based on natural fibers, not only in terms of heat but also in terms of humidity.

3.1 Cotton

Cotton is one of the most used fibers in garment production due to its comfort properties, such as breathability, thermal and moisture management, good mechanical properties, and biodegradability [16]. It accounted for about 25% of fiber use in 2018 [7]. However, its use in the garment industry has come under pressure due to environmental concerns, such as intensive water and insecticide use by the agricultural sector. This has led to the recycling of cotton fibers to reduce their ecological impact through life extension [32, 46].

Cotton fibers exhibit thermal insulation properties due to a relatively high degree of porosity, primarily from their unique microstructure and physical arrangement. The internal structure of cotton fibers consists of a hierarchy of microfibrils and fibrils. The average degree of polymerization (DP) of cotton is very high (9000–15,000), and the average crystallinity is around 73%, which also hinders heat transfer through the closely packed cellulose structure. These values are very high compared to other lignocellulosic materials such as wood pulp (DP 600–1500 and crystallinity 35%) and viscose (DP 250–450 and crystallinity 60%) [71].

The ribbon-like shape of the cotton fiber, which twists on its fiber axis, also creates inter-fiber pores that entrap air, reducing the thermal conductivity of cotton fibrous assemblies [6, 111]. Due to more vibrational freedom, heat transfer primarily occurs through the amorphous parts, which is also where moisture is absorbed [67, 68, 76]. Due to its relatively high moisture content, cotton fibers are susceptible to mold damage therefore a dry operating environment is required [107].

The surface of cotton fibers features a cuticle composed of waxes, pectin, and proteins. Beneath this lies the primary wall, which is 0.1–0.2 μm thick and consists of fibrillar bundles organized into transient lamellae. The secondary wall, accounting for 95% of the fiber's mass, includes three layers:

- S1: The outermost secondary layer, 0.1 μm thick, contains fibrillar bundles arranged in 30 nm spirals.
- S2: The central layer (4 μm thick) comprises 25–30 concentric fibrillar sheets, each 0.2 μm thick. These sheets consist of spiraling fibrillar bundles whose direction alternates at specific weak points (20–30 points/cm). Approximately 45% of fiber breaks originate between these points.
- S3: The innermost layer (0.1 μm thick) borders the lumen.

Cotton fibers are single cells primarily composed of alpha-cellulose (with a polymerization degree exceeding 200). Their structure forms concentric layers, beginning with the dense surface cuticle, followed by the primary and secondary walls, and culminating in the central lumen (see Fig. 6) [71].

Cotton fibers have a width ranging from 12 to 20 μm [91]. Their structure consists of a secondary wall (3–6 μm thick) enveloped by a primary wall approximately 200 nm thick [37]. Mature cotton fibers exhibit a total thickness of 25 μm, with the central lumen occupying 19–22 μm of this dimension [90].

The theoretical density of a perfect alpha-cellulose crystal is 1582 kg/m^3. However, cotton's crystallinity falls below 100%, resulting in a lower true density. For

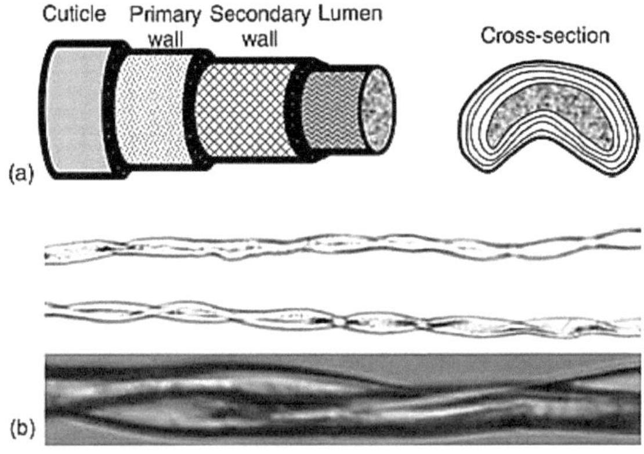

Fig. 6 **a** Schematic structure and **b** longitudinal section image of the cotton fiber (reused from Wang and Wang [106] under CC-BY Permission)

cellulose containing 4% moisture, the measured density is approximately 1460 kg/m^3 [95].

The thermal conductivity of fibrous structures is usually much smaller due to the presence of air (porosity). The thermal conductivity of cross-laid and perpendicularly laid nonwovens from dry cotton fibers was reported to be between 0.045 and 0.058 W/m K, depending on factors such as density and fibrous assembly structure [45]. Recycled cotton for use in thermal insulation structures (bats) was found to have a thermal conductivity ranging between 0.0381 W/m K (ρ = 30 kg/m^3, T = 10 °C, RH = 17%) and 0.0546 W/m K (ρ = 50 kg/m^3, T = 30 °C, RH = 80%). The authors also found that fiber orientation had a negligible effect on the thermal conductivity; however, a higher moisture content led to an increase [7]. Recycling cotton fibers could cause fiber damage due to the mechanical processes required to shred and reform a textile structure, resulting in shortened fibers and compromising their natural loft and alignment, thereby reducing the overall ability to trap air efficiently [14]. Other approaches to reducing cotton's environmental footprint are to use the cotton stalk fibers for insulation materials, which show thermal conductivity as low as 0.055 W/m K [48, 114]. Cotton is often blended with polyester fibers in garments; however, its separation can be costly. Research has shown that textile structures from mechanically recycled cotton and polyester fibers have proven to be an effective thermal insulator, whilst lowering the environmental footprint of cotton by exhibiting a comparable thermal conductivity coefficient of around 0.039 W/m K [86]. The most common technique for nonwovens (bat) formation is needle punching. However, environmentally friendly binders such as sodium alginate have also been employed [7, 55].

3.2 Hemp

Historically, hemp was widely used to make textile products; however, its commercial use declined due to regulatory pressure in the 1900s. Due to more recent regulatory changes and the search for more sustainable fibers, hemp has regained interest [49]. Hemp is seen as a very sustainable fiber due to its relatively low water consumption of around 300–500 L per kilogram of fiber produced [9]. The main chemical components of hemp fibers are cellulose, hemicellulose, lignin, and pectin [49].

Hemp fiber bundles are distributed across multiple layers within the stalk. Each bundle comprises multiple single-cell fibers bound by a pectin-rich middle lamella. Bundles in the inner layers are typically shorter and finer than those in the outer layers. Individual fibers exhibit a cross-section ranging from triangular to heptagonal with rounded edges and a prominent central lumen. Processing hemp loosens and dissolves the lamella. Fiber dimensions include a wall thickness of 5–10 μm, a diameter of 5–50 μm, and an average length of 10–40 mm (though lengths vary between 5 and 100 mm). The fibrils form Z-shaped spirals, and fiber bundles measure 1500–2500 mm in length [41].

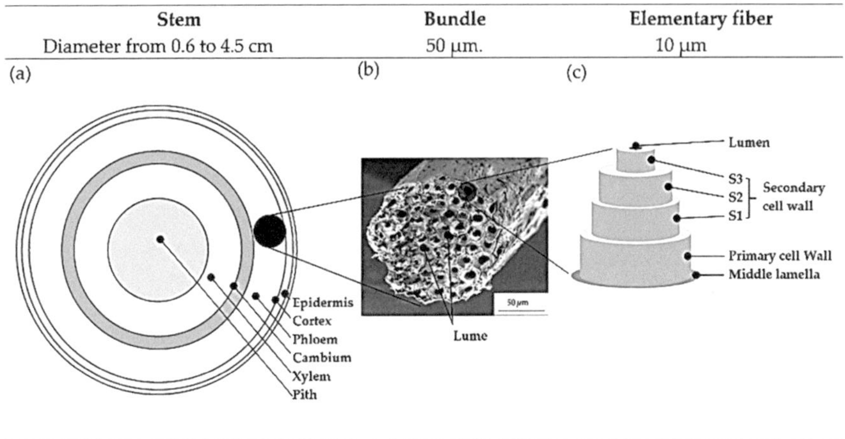

Fig. 7 Cross-section of hemp (reused from Manaia et al. [61] under CC-BY Permission)

Hemp's thermal conductivity stems from its hierarchical cellulose microfibril arrangement. Alkali pretreatment enhances crystallinity, creating ordered regions that facilitate phonon-mediated heat transfer [83]. Secondary cell walls contain amorphous domains that scatter thermal energy, while lumen size variations reduce transverse thermal conductivity by up to 40% compared to solid regions due to air entrapment [60]. Although moisture typically increases thermal conductivity, natural waxes on hemp fibers limit moisture absorption, preserving their insulating properties [50]. A cross-sectional view of hemp fiber is shown in Fig. 7. The density of hemp is approximately 1470 kg/m^3.

Hemp textile structures make very good insulation materials since bulk hemp fibers have been shown to have a thermal conductivity lower than 0.045 W/(m K) for a fabric density around 28 kg/m^3 [50, 93]. Research shows that hemp-based insulation panels produced using wet-laid technology with polylactide (PLA) and viscose binders achieved thermal conductivity values as low as 0.027 W/m K when composed of 80% hemp waste, thus outperforming traditional materials like mineral wool [36]. Increased batt densities ranging 40–80 kg/m^3 using PLA binders with wool fibers exhibited slightly higher thermal conductivity values due to reduced air pockets [104]. Sair et al. found that increasing the amount of hemp fibers when combined with a polyurethane matrix in a composite structure led to an increase in thermal insulation, primarily due to more air trapped in the matrix structure [82]. Similar observations were made by another group using pure hemp fibers, where an increase in batt density led to an increase in thermal conductivity, with the authors obtaining thermal conductivity values between 0.055 and 0.065 W/m K [30].

3.3 *Flax*

Linen (flax fiber) is derived from the stem of the flax plant (*Linum usitatissimum* L.), which consists of bark, fiber bundles, and a woody core containing xylem, shives, and elementary fibers. These elementary fibers feature a primary wall and a secondary wall surrounding the lumen, bound together in bundles by a pectin-rich lamella (see Fig. 7). The primary wall comprises cellulose microfibrils embedded in a matrix of hemicellulose, pectin, and glycoproteins. The secondary wall has a three-layer structure (S1–S3), with the middle layer (S2) constituting 80% of the fiber's mass. The cellulose fibrils in this layer spiral in an S-shaped pattern [81].

Individual flax fibers (single cells) exhibit a polygonal cross-section, with a wall thickness of 5–10 μm, a diameter of 10–20 μm, and an average length of 20–40 mm. Fiber bundles, however, measure 1500–2500 mm in length [51].

Flax fibers are renowned for their breathability, strength, and thermal insulation—properties largely attributed to their hollow core (see Fig. 8). Composed of long cellulose chains, these fibers exhibit low moisture retention. Linen products are considered luxury items in the garment industry due to their higher production costs [12]. Linen's thermal conductivity ranges from 0.029 to 0.325 W/(m K), depending on processing, a result of its hierarchical structure. This includes high cellulose content (up to 77.5% crystallinity), aligned microfibrils, and intrinsic air pockets within the fiber matrix [42].

The density of a flax fiber is about 1430 kg/m^3. Flax fibers exhibit inherent flammability risks due to their lignocellulosic composition [15]. Flax fibers are susceptible to mold and insects, which can be improved with fluorination treatments [96]. Needle punching remains a common form of bonding flax fiber nonwoven batts, although bio-based binders such as Arabic natural glue have also been successfully used [27]. Flax noils, a byproduct of flax processing, were integrated into ventilated insulation systems and exhibited a heat transfer resistance similar to that of regular flax

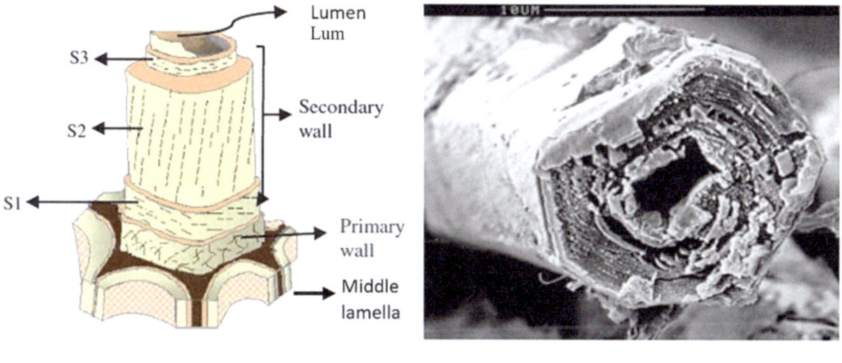

Fig. 8 Flax fiber **a** structure and cross section, **b** SEM images of the fiber cross section (reused from Sadrmanesh and Chen [81] under CC-BY Permission)

fibers [78]. Modified flax fiber slabs exhibited a thermal conductivity range of 0.042–0.047 W/m K with little variation, even with density ranging from 45 to 90 kg/m^3 [79].

3.4 Wool

Wool is a keratin fiber, chemically similar to alpaca, human hair, and mohair, as it shares comparable elemental compositions: approximately 50 wt% carbon, 7 wt% hydrogen, 22 wt% oxygen, 16 wt% nitrogen, and 5 wt% sulfur. The sulfur content arises from the presence of cystine. Wool is a natural polymer composed of α-amino acids (H_2N–CHR–COOH), where the R group varies and can include alkyl, aryl, hydrogen, alcoholic, phenolic, amino, carboxylic, heterocyclic, or sulfur-containing groups. Its isoelectric point—the pH at which wool exhibits maximum chemical stability—is approximately pH 4.9. The primary structural form of wool is α-keratin, organized into a helical configuration with a repeat distance of 0.52 nm and stabilized by intramolecular hydrogen bonds. When stretched to 120% of its original length, α-keratin transitions into β-keratin, which adopts a planar structure with a repeat distance of 0.98 nm. A third, highly contracted form called γ-keratin exists, characterized by a disordered, meandering arrangement. Wool fibers consist of two distinct cortical regions: the ortho-cortex and para-cortex. The ortho-cortex contains well-defined fibrils, higher tyrosine (~ 42%) and glycine (~ 18%) content, enhanced dye affinity, and greater susceptibility to hydrolysis. In contrast, the para-cortex is richer in cystine (~ 20%) and disulfide (S–S) crosslinks, making it harder and less amorphous. This bilateral structure drives wool's natural curling: the ortho- and para-cortices twist into opposing spirals due to differential shrinkage, with the more deformable ortho-cortex always positioned on the fiber's exterior [52].

Wool fibers are characterized by a scaly surface known as the cuticle (see Fig. 9). The cuticle contains between 900 and 3500 scales per mm^2, with approximately one-third of each scale protruding above the fiber surface. Finer wool typically features a single scale per circumference, while coarser wool has more. These scales overlap like shingles and are oriented from the root to the tip of the fiber. Smooth scales originate from the paracortex (with a deeper overlap), whereas rippled scales are derived from the orthocortex.

Each scale consists of three distinct layers. The outermost layer is the hydrophobic and mechanically weak **epicuticle**, which is about 5–10 nm thick and contains lysine. This layer is chemically inert and gives rise to the Alwörden reaction: when undamaged wool is immersed in sodium hypochlorite, bubbles form on the surface. This phenomenon highlights the paradox of keratin fibers, which have a hydrophobic surface surrounding a hydrophilic core, thereby combining both properties.

The middle layer, called the **exocuticle**, is approximately 0.15 μm thick (note: likely a correction from "0.15 nm" to "0.15 μm" for biological plausibility) and is rich in cystine, containing one disulfide (S–S) bond per five amino acid residues, or about 35% sulfur. The innermost layer, the **endocuticle**, contains less cystine (around

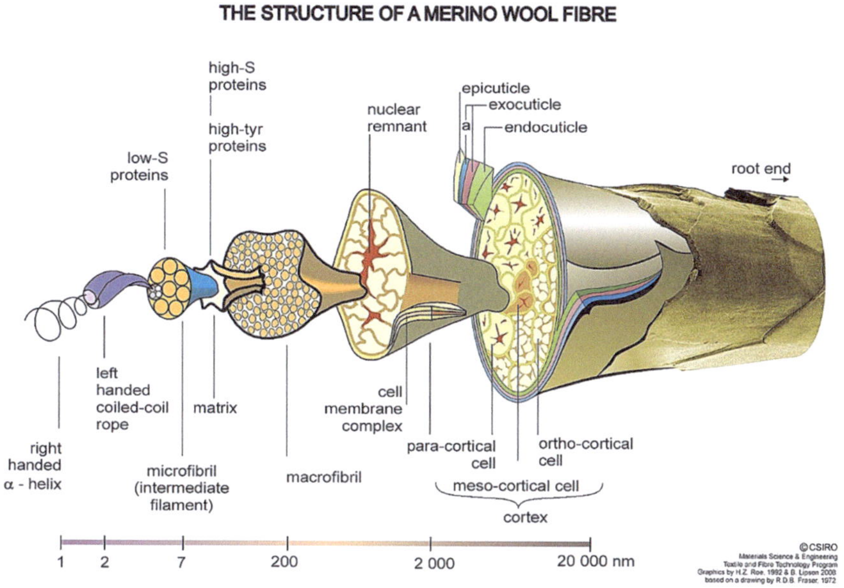

Fig. 9 Structure of the wool fiber. Reused from CSIRO under CC-BY Permission

20% sulfur), includes germ cortical cells, and accounts for about 8% of the scale's mass. The endocuticle is both chemically and mechanically resistant. Cuticle cells are attached at the root end and point toward the fiber tip [52].

Beneath the scales lies the cellular cortex, which constitutes approximately 90% of the fiber's weight. It is composed of tapered cortical cells (4 μm in diameter and 100 μm in length) bonded by a matrix of globular chains rich in sulfur. The cortex features a fibrillar structure originating from a protofibril—a pair of protofilaments forming an alpha helix (8 chains, or 4 dimers). A bundle of protofibrils forms a microfibril measuring 20 nm in thickness and 10 μm in length. These microfibrils, grouped into bundles of 500–800 and surrounded by the matrix, create macrofibrils with diameters of 100–300 nm. Each cortical cell contains 5–20 macrofibrils. Wool exhibits a total crystallinity of about 20%, a thermal conductivity of approximately 0.039 W/(m K), and a density of 1320 kg/m^3.

Wool's natural crimp arises from its bilateral structure, which traps air between fibers in fibrous assemblies, reducing the thermal conductivity of textile structures. In coarse wool fibers, a loosely packed porous region called the medulla resides at the center. This medulla enhances insulation by trapping air [106]. It forms a central canal—a hollow or tubular network—composed of strongly pigmented angular cells and air pockets.

The sustainability aspect regarding wool fibers has different viewpoints. Wool production does require significant resources in terms of water and soil since it comes from an animal that needs to be kept alive. However, wool is generally a biodegradable by-product of the animal in the meat industry and is therefore considered more ecological. Farming practices for sheep and wool processing generate around 11.25–24.9 kg CO_2/kg of fibers produced [13, 26, 66].

Wool fibers exhibit a natural flame retardancy due to their ability to self-extinguish by char formation. This is primarily attributed to their relatively high nitrogen content of around 15–16% and moisture levels of 10–14% under standard conditions [19, 72, 88]. Wool fibers, due to their high moisture content and keratin nature, can be susceptible to mold and insect damage; therefore, protection is needed [70, 107]. Experiments have shown recycled wool fiber bats to be comparable to virgin wool, with thermal conductivity ranging between 0.03 and 0.06 W/m K, with fabric densities ranging between 40 and 80 kg/m^3 for both straight and corrugated batts. Wanza et al. used recycled wool textiles to produce needle-punched batts with fabric densities of 8.135 and 10.583 kg/m^3 and obtained thermal conductivities of 0.03476 and 0.03827 W/m K, respectively [108]. It was found that the thermal conductivity of sheep wool fiber batts decreased with the decrease of temperature, moisture, and density, and with the increase of thickness [43]. Common binding methods include needle punching since wool fibers are excellent at entangling, and also using chemical binders such as low-melt bicomponent polylactide (PLA), which upon melting at around 60–80 °C forms an adhesive bond that does not negatively influence the hygroscopic properties of the wool fibers. Chitosan has also been shown to be an effective binder. However, this is generally more expensive than PLA [80, 103, 104].

3.5 Silk

The *Bombyx mori* silkworm produces the silk used for garments. The silkworm spins its cocoon from raw silk produced in its salivary glands. The fibers in cocoons are laid in the shape of an 8. First, a net is created on leaves and branches, followed by the cocoon from outer to inner layers. The cocoon contains 3–4 km of continuous fiber [10].

Silk fibers are comprised of two fibroin filaments (70–80%) forming the core enveloped partially by non-fibrous sericin (20–30%), also called the bave, as shown in Fig. 10.

Sericin is composed mainly of glycine, alanine, tyrosine, and leucine, and fibrous fibroin is from planar β keratin with a typical amino acid sequence: serine-glycine-alanine-glycine-alanine-glycine. Silk is composed of simple amino acids with a lack of amides, few polar groups (basic and acidic), a very low portion of S–S bonds, and a lot of hydrogen bonds.

The sericin is removed by placing the cocoons in hot water, which frees two triangular-shaped fibroin silk filaments and readies them for reeling [74]. These fibroin filaments have a complex hierarchical structure characterized by tightly packed antiparallel β-sheet structures interspersed with random coil and α-helix conformations. These β-sheet structures, formed by polyalanine sequences, contribute significantly to silk's exceptional tensile strength while creating a protein matrix with excellent thermal insulation capabilities. The density of the silk fibroin

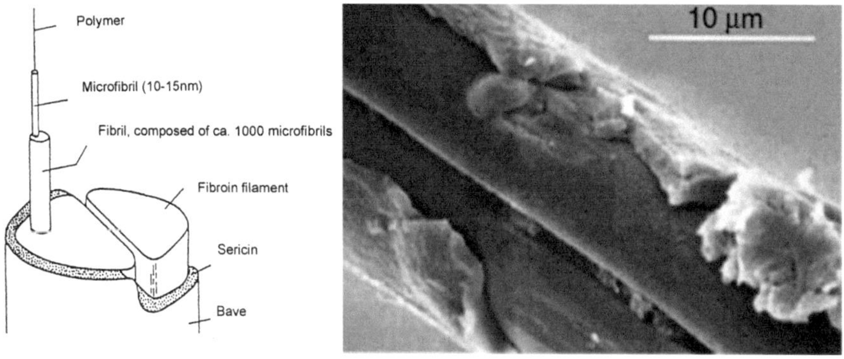

Fig. 10 Silk fiber structure and longitudinal view (reused from Babu [10], Ude et al. [99] under CC-BY Permission)

filament is 1370 kg/m^3, and its thickness is 9–15 μm [8, 35]. The unique molecular arrangement of silk fibers, combined with its ability to create microscopic air pockets within its protein structure, enables it to maintain low thermal conductivity, about 0.775 W/m K at room temperature, providing remarkable temperature regulation in textiles [97]. The recycling rate of silk is very low (< 10%) compared to other natural fibers such as wool and cotton [18]. Silk production has a very high carbon footprint, with 25.4 kg/kg CO_2 equivalent from production to finished fiber, and it's also very agriculturally intensive [74]. It is for this reason that silk is so expensive.

Studies have shown that silk fibers are prone to mold damage and therefore require a dry storage or operating environment [107]. Like wool, silk fibers are less flammable, because they also self-extinguish once the flame source has been removed [34]. Silk fibers are very expensive and, therefore, traditionally not the first choice for producing a low-cost thermal insulation material, since there are less expensive fibers with better thermal insulation properties, however, for specific thermal conductive applications, value can be added to the silk waste to produce a material with unique properties such as enhanced thermal conductivity. An example is with boron nitride nanosheets/silk fibroin (BNNS/SF) nonporous films, which exhibit exceptional thermal conductivity values of 27.58 W/m K in-plane and 1.77 W/m K out-of-plane, with high radiative cooling efficiency (Rsolar = 0.89 and εLWIR = 0.84) [110]. As part of production sustainability, the silkworm cocoon is typically underutilized and therefore can be repurposed into insulation material. Silk cocoon waste has been transformed into needle-punched nonwoven fabrics, achieving thermal insulation values of 0.201 m^{2}K/W for 100% silk cocoon waste and 0.212 m^{2}K/W for blends with comber noil (50/50) [53]. For shorter fibers, the waste is transformed into silk fibroin solutions that can be electrospun into nanofibrous mats (50–300 nm diameter) using volatile solvents like formic acid, achieving surface areas up to 40 m^2/g for biomedical scaffolds [89].

3.6 Regenerated Cellulose Fibers

Viscose was the first regenerated cellulose fiber produced from cellulose or wood pulp, and it is widely used to improve the comfort properties of garments. Because of its composition and physical structure, viscose shares similarities with other natural cellulosic fibers and possesses a silk-like feel due to its surface characteristics. The term "viscose" refers to the viscous liquid state of the sodium cellulose xanthate spinning solution. The primary material source for viscose is typically spruce or beech wood, which contains about 88% alpha cellulose. During coagulation, CS_2 gas is generated within the fibers and subsequently diffuses out. The local pressure created by this process results in depression and overpressing, which together produce the multilobal cross-section seen in Fig. 11.

Viscose fibers develop a radially variable structure due to the conditions of coagulation and simultaneous drawing. The surface skin (cuticle) consists of a highly oriented, dense, and well-organized structure (with high crystallinity) containing small pores measuring 5–20 nm. The core, which constitutes about 35% of the fiber mass, comprises small, well-oriented crystallites and features a porous macrostructure with pores ranging from 25 to 150 nm. These fibers exhibit a fibrillar structure, where fibrils contain alternating crystalline K layers (8 nm thick) and amorphous A layers (12 nm thick) connected by intrafibrillar chains. The base structure follows the cellulose II monoclinic system, which is approximately 15–20% less oriented than cotton and exhibits a crystallinity of 27–35%.

Regular viscose fibers remain widely produced, primarily due to being the most affordable synthetic fiber on the market. However, they are also the weakest and most degradable regenerated cellulose fiber, particularly in wet conditions. Their wet strength drops to half of their dry strength, often reaching critically low levels. Additionally, viscose fibers are highly sensitive to chemicals, necessitating the use of only weak alkalis, shorter reaction times, and moderate temperatures during processing.

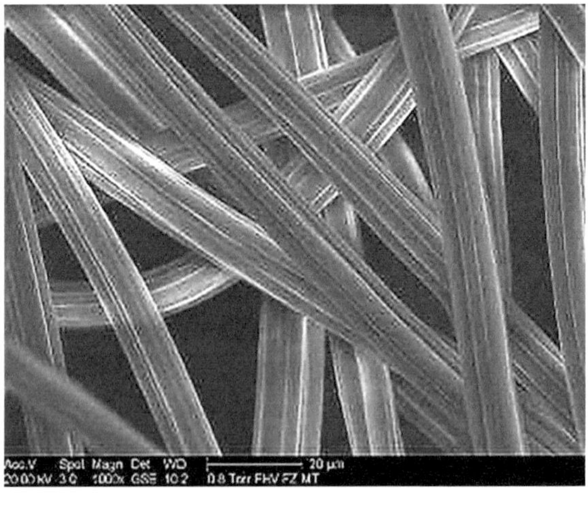

Fig. 11 SEM of viscose-rayon fibers (reused from Široky et al. [92] under CC-BY Permission)

Numerous efforts have been made to replace carbon disulfide (CS_2) in producing regenerated cellulose fibers. A successful solution emerged with the cellulose-dissolving system based on N-methylmorpholine-N-oxide (NMMO), which became the foundation for Lyocell fiber production. Lyocell fibers demonstrate high modulus and tenacity even when wet, with structural and property distinctions from viscose. Their crystallinity reaches about 42%, and their crystallites are longer (15–45 nm) and thinner compared to viscose (11–25 nm). These elongated crystallites (91% of the structure) are interconnected by small, well-oriented amorphous regions. The fiber cross-section is oval-shaped due to radial compression. As a new generation of regenerated cellulose fiber, Lyocell offers enhanced properties over viscose (see Table 3).

Viscose fibers are susceptible to mold damage; therefore, maintaining a dry environment is necessary [107]. Regenerated cellulose, also referred to as viscose or lyocell, is considered to be one of the most environmentally friendly chemical fibers. Compared to polyester, its production generates approximately 1.2 kg of CO_2 equivalent, compared to the 3.8 kg generated by producing polyester [112].

Nonwoven composites developed using viscose fibers have achieved thermal conductivity as low as 0.028 W/m K when embedded with nanoporous amorphous silica via sol-gel processing [44]. The incorporation of air in aerogel systems allows for even better thermal insulation, such as that documented by Li et al. using regenerated all-cellulose sponge-aerogel fibers, which demonstrated ultralow thermal conductivity while maintaining excellent toughness [58]. Multi-layered thermal applications incorporating viscose nonwoven fabrics as mid-layers have shown thermal conductivity values of approximately 0.0462 W/m K [54]. Similar to other thermal insulation materials investigated, the most common bonding method for the non-woven batts involving viscose was needle punching.

Table 3 Comparison of viscose and lyocell properties

Parameter	Viscose (standard)	Lyocell
Cross section	Multilobal	Circular, oval
Radial structure	Core/Sheet differences	Homogeneous
Crystallinity	Low 27%	High 42%
Crystallites length	Smaller 11–25 nm	Higher 15–45 nm
Crystallites width	Higher 5–11 nm	Smaller 3.5–5.5 nm
Orientation crystalline	High	High
Orientation amorphous	Low	High

Reused from Militký [62] under CC-BY Permission

4 Synthetic Fibers

Synthetic fibers do not require water and soil like their natural fiber counterparts. Still, their environmental impact is more associated with pollution in the form of nondegradable waste and microplastic generation, as studies have shown that a single garment can release between 1900 and 700,000 fibers per wash. [23, 40, 59, 75, 87]. Due to their relatively low cost and desired properties, synthetic fibers make up about 60% of global fiber consumption. Therefore, it is no surprise that they comprise about 60% of global primary microplastic waste release [21]. Microplastics are produced by the manufacturing, wearing, and washing of synthetic fiber-based products. Research has shown that thermal insulation products also produce microplastics in their lifetime through handling and manufacturing. Microplastic generation is also dependent on fiber choice as well as the form of the textile, since stronger and smoother fibers release fewer microplastics, and woven textiles do so more than knitted ones due to lower inter-fiber friction [4]. Synthetic fibers are commonly blended with natural or other synthetic fibers, such as spandex, to impart moisture management or elastic properties. This makes recycling very difficult and often requires multiple steps involving mechanical followed by chemical separation. Mechanical methods typically involve shredding and filtering. Chemical methods mostly involve solvation, such as heating polyester/spandex textiles to 210 °C in a glycol solution with ZnO catalyst to depolymerize polyester and spandex into their respective monomers. Nylon can be dissolved in 90% formic acid, allowing for separation from other fibers such as cotton [115].

4.1 Polyester (PET)

Polyester fibers (PES) are defined as a manufactured fiber in which the fiber-forming substance is any long-chain synthetic polymer composed of at least 85% by weight of an ester with a substituted aromatic carboxylic acid, including but not restricted to terephthalate units, and para-substituted hydroxybenzoate units. Polyethylene terephthalate (PET) is the predominant polyester used for fiber production.

PET fibers have low moisture absorption, good elasticity and dimensional stability, excellent wear resistance, and good blending ability with cotton and other natural or chemical fibers. They are relatively resistant to elevated temperatures, resistant to microorganisms and insects, physiologically inert, and have good resistance to weathering and light. PET is a thermoplastic with a softening range of 230–245 °C and a melting point of 252–262 °C. The setting of PET fibers can be varied due to drawing and heat treatment conditions. The strength of PET can vary widely depending on the manufacturing conditions, and it has good abrasion resistance. PET is the condensation product of terephthalic acid and ethylene glycol (see Fig. 12), forming a semi-crystalline polymer (crystallinity about 40%) [63].

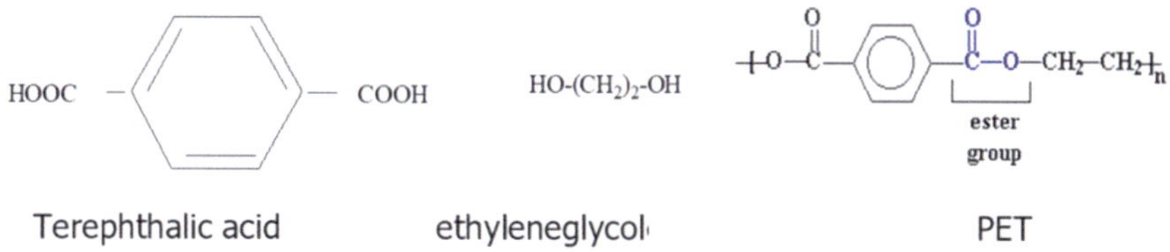

Fig. 12 Monomers and PET (reused from Militký [63] under CC-BY Permission)

For different PET fibers, the acidity caused by these end groups ranges from 2 × 10^{-2} to 4 × 10^{-2} mol. In addition, carboxyl groups are responsible for further degradation because they catalyze the hydrolysis of ester bonds. PET fibers contain a rigid benzene ring and flexible sequences of six aliphatic groups –CO–O–CH_2–CH_2–O–CO– in the main chain. The rigidity of benzene rings, in adjacent chains, allows for a coplanar arrangement. The cross-sectional area per PET chain is equal to 0.217 nm^2. The cohesion of PET chains is the result of hydrogen bonds and van der Waals interactions, caused by dipole interactions and dispersion forces between chains. The total magnitude of secondary forces in a PET unit is 1.37 kJ mol^{-1}. Of this, 1.02 kJ mol^{-1} is caused by dispersion forces induced by the benzene rings. The strength of PET fibers is therefore given primarily by the rigidity of the benzene rings (secondary van der Waals forces decrease exponentially by a magnitude of r^6, where r is the chain distance), which form an angle of only 12° with the plane of the ester bonds. The partial flexibility in the PET macromolecule is mainly caused by the ethylene group. The interactive forces create an inelastic, tight arrangement between the macromolecules, which leads to a high modulus, strength, and resistance of the fibers to moisture, dyes, and solvents [63].

The unusually high melting point of PET (compared to aliphatic polyesters) is attributed to the ester bonds. As a result, PET is difficult to crystallize. Rotation of the chains around the –CH_2–CH_2– bond leads to the formation of two types of conformations, the planar trans-conformation and the spatial gauche conformation. The trans-conformation corresponds to the arrangement with the longest elementary unit. It is a state in which the Van der Waals distances between the chains and individual groups of chains are maintained. Van der Waals distances are the distances at which the attractive and repulsive components of the dispersion force are in equilibrium. The PET unit is 1.075 nm long, which is slightly shorter than the length of a fully extended chain (1.09 nm) [29].

The crystalline unit of PET is triclinic with dimensions a = 0.456 nm, b = 0.594 nm, c = 1.075 nm, $\alpha = 98.5°$, $\beta = 118°$ and $\gamma = 112°$. The amorphous orientation in PET fibers is relatively low, with an orientation factor of 0.34–0.40, and the total fraction of trans conformation in PET fibers is almost 0.80. The trans-conformation fraction in the amorphous regions is approximately 0.64. The molar volume of PET V_{PET} = 144 cm^3/mol, and the density of the amorphous phase of PET is ρ_{aPET} = 1333 kg/m^3. The density of the crystalline phase of PET ρ_{cPET} = 1440 kg/m^3. PET has a glass transition temperature Tg of around 77–80 °C and a crystallization temperature of 180–190 °C. The cohesion energy of amorphous PET

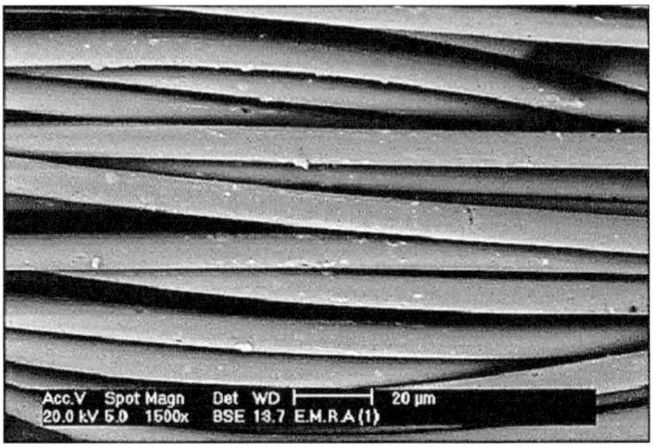

Fig. 13 SEM image of untreated polyester fibers (reused from Elnagar et al. [28] under CC-BY Permission)

and copolymers can be calculated from the molar attraction constant by the molar contribution method. For PET, the value was calculated to be 74.29 kJ mol^{-1}. The elastic modulus of the crystalline regions of PET in the direction parallel to the chain axis is 108 GPa [29].

PET fibers, after fixation, have a degree of crystallinity of about 0.4, the orientation factor of the crystalline phase is fc > 0.95, and the orientation factor of the amorphous phase is fa = 0.6. The basic structural units are relatively strong microfibrils with a diameter of 10–15 nm. These units are assembled into fibrils with a diameter of 30–45 nm. A very important role is played by the oriented non-crystalline phase (TTM)—"taut tie molecules" as it is responsible for the mechanical properties of PET fibers, with a proportion of TTM of around 0.1–0.05 [63]. A scanning electron microscope image of polyester is presented in Fig. 13.

The presence of rigid benzene rings in the terephthalate units and the ester bonds promotes partial crystallinity in PET fibers, which correlates with their inherently low thermal conductivity. Thus, supporting their effectiveness as insulating textile materials [56]. Polyester production generates around 3.8 kg of CO_2 equivalent due to fossil fuel use and polymerization [112].

Polyester's high hydrophobicity and molecular weight make it highly resistant to degradation by microorganisms, contributing to plastic waste accumulation [94]. Studies have shown that textiles made from polyester release the lowest amount of microplastics compared to other synthetic fibers; however, recycled polyester tends to release more microplastics than virgin polyester [4].

Studies have shown that needle-punched 100% polyester nonwoven batts had a thermal conductivity value as low as 0.033 W/m K at a fabric density of 142.4 kg/m^3. The authors noted a decrease in thermal insulation when batt density increased as the thermal conductivity increased to 0.05 W/m K at a density of fabric 125.5 kg/m^3 [2]. This is still well within the range of good thermal insulators. Polyester cutting waste has shown thermal conductivity values ranging between 0.0520 and 0.0603 W/m K. It was noted that the presence of Lycra in the waste cuttings reduced the material's insulating properties [98].

A similar observation regarding fiber density was made in another study by Valverde et al., where PET/polyurethane nonwovens of varying densities were measured. At first, thermal conductivity decreases as density increases due to the formation of more air gaps. This is until a critical insulation point is reached, since beyond this point, the voids get filled with fibers and the thermal conductivity increases. The authors obtained a minimum thermal conductivity of 0.041 W/m K at a fabric density of 396 kg/m^3 [101].

4.2 Polyamides

Polyamides are typically produced through the condensation reaction between diamines and a dicarboxylic acid, polymerizing amino acids or ring-opening polymerization using cyclic lactams. Depending on whether they have aromatic or aliphatic polymer chains, they typically come in the form of Nylon 6 (PA 6) or Nylon 6,6 (PA 6,6) with condensed formulas $(–NH–(CH_2)_5–CO–)_n$ and $(–NH–(CH_2)_6–NH–CO–(CH_2)_4–CO–)_n$, respectively.

Due to an odd number of CH2 groups in PA 6, a smaller number of hydrogen bonds occur and the arrangement of amide groups is head-to-head, i.e., CO–NH– … –CO–NH–. Fibers are structurally non-optimal with lower strength and a lower melting point of 220 °C. Due to an even number of CH_2 groups in PA 6.6, a higher number of hydrogen bonds occurs. The arrangement of amide groups is head-tail, i.e., CO–NH– … –NH–CO–. Fibers are structurally optimal with higher strength and a higher melting point of 256 °C. A more compact and crystalline structure in PA 6.6 causes differences in physical properties such as during boiling water shrinkage: PA 6: 10%, PA 6.6: 5.3%. Polyamides PA 6 and PA 6.6 have the same density, 1140 kg/m^3, and share many characteristics. However, they differ markedly in melting point by 40 °C due to differences in the alignment of molecular chains and crystallization behavior. The mean degree of polymerization *DP* for PA 6 is 100–200, and for PA 6.6 is 50–100 [65].

Polymeric chains in polyamide fibers are linked by weaker van der Waals—and strong intermolecular hydrogen bonds (between amide groups). Polymer chains have –COOH groups on one side and an $–NH_2$ group on the other side. In PA 6.6, the average content of terminal $–NH_2$ groups is 36–45 m eq kg^{-1}, and the content of terminal carboxylic groups –COOH is 75 m eq kg^{-1}. The isoelectric point is in the range of pH 3–3.5, i.e., significantly lower than at pH 4.8–4.9 for wool fiber. PA 6.6 fiber has melting point T_m = 250–260 °C, glass transition temperature Tg = 47–60 °C (depends on evaluation method), softening point 235 °C, dry heat-setting temperature 180–220 °C for 30–60 s. and saturated steam setting temperature 120–135 °C for 20–30 min. PA 6.6 is radially non-uniform as it has a core/shell structure [65].

In PA 6, the cohesive energy density CED = 774 J/cm^3, average content of terminal $–NH_2$ group is 54 m eq kg^{-1}, and content of terminal carboxylic groups –COOH is 100 m eq kg^{-1}. The isoelectric point is at pH 5.4. The PA 6 fiber has a melting point

Tm = 213–220 °C, softening point 170 °C, dry heat-setting temperature 165–192 °C for 30–60 s, saturated steam setting temperature 110–120 °C for 15–20 min, and dry glass transition temperature Tg = 29–50 °C. Properties of PA 6 are generally sensitive to moisture changes. Glass transition decreases rapidly due to increasing moisture content in the fiber. Heating the PA 6 fiber up to 90–100 °C causes a gradual loss of strength. Long-term exposure of PA 6 at a temperature of 150 °C leads to pale yellowing. A critical challenge in PA 6 processing is oligomer generation, particularly monomers. During fiber spinning, approximately 10% oligomers form, with 8% consisting of monomeric caprolactam. While most oligomers are eliminated through washing, residual water-soluble monomers persist and accelerate fiber degradation. To mitigate this, frequent washing at 60–80 °C is essential to dissolve and remove monomers effectively [65].

A scanning electron microscope image of PA 6 fibers is presented in Fig. 14. Like most synthetic fibers derived from fossil fuels, Nylon also has a large carbon footprint and a microplastic generation problem. Mechanical recycling is currently the simplest method of recycling Nylon, but research has shown that chemical recycling may deliver better results [31].

Neri et al. recycled facemasks from polyamide to form thermal insulation material. This type of waste was generated in significant amounts during the pandemic, and this method could prove a way to deal with the pandemic and uncontaminated medical waste of this type of material. The authors packed shredded polyamide waste into cardboard holders in fabric densities of 25, 30, 40, 50, and 60 kg/m^3. The lowest density sample had a thermal conductivity of 0.0354 W/m K, and all samples were lower than 0.04 W/m K, demonstrating the material's effectiveness as a thermal insulator [69].

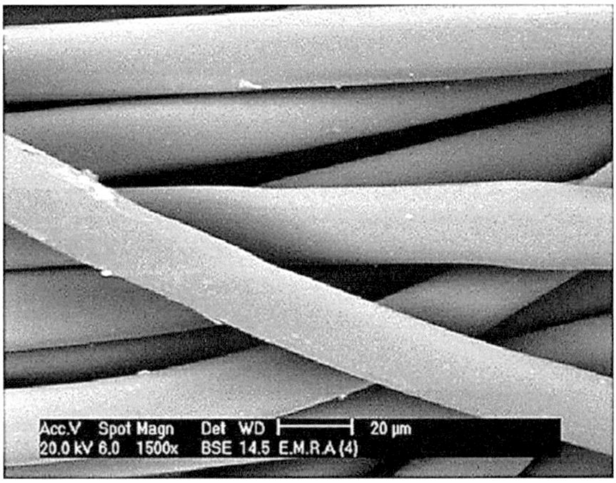

Fig. 14 SEM of untreated PA 6 fibers (reused from Elnagar et al. [28] under CC-BY Permission)

4.3 Acrylic Fibers

Acrylic or polyacrylonitrile-based copolymers, with the molecular formula of the main polymer $[-CH_2-CHCN-]_n$, are used as a synthetic alternative to wool fibers and have replaced it in many aspects due to its bulkiness and elasticity [43].

Acrylic fibers are defined as those containing more than 85% polyacrylonitrile, whereas modacrylic fibers contain less than 85% polyacrylonitrile. Up to 15% of acrylic fibers may consist of various modifier components—comonomers that are incorporated during the polymerization process to form the fibrous polymer. These comonomers are added to enhance specific properties of the resulting fibers. Since polyacrylonitrile decomposes before it melts, melt spinning is not feasible for acrylic fibers; instead, dry and wet solution spinning techniques are employed. The density of acrylic fibers typically ranges from 1130 to 1170 kg/m^3, making fabrics produced from them lighter than wool, which has a density of about 1320 kg/m^3.

In the wet spinning process for acrylic fibers, the undrawn fiber is solidified in a coagulation bath containing water and dimethylformamide (DMF) at temperatures between 20 and 30 °C. This process results in the formation of a porous fibrillar structure, with a pore volume of about 0.9×10^{-5} m^3/g. Subsequent simultaneous drawing in boiling water at a draw ratio of 2–3 increases the orientation and arrangement of fibrils. The coagulation process also creates microvoids, leading to a large internal relative surface area of 200–300 m^2/g. However, during final free-state drying, most of the porous structure collapses as the majority of pores disappear (see Fig. 15).

Dry spinning of acrylic fibers is performed by dissolving polyacrylonitrile (PAN) in DMF at 80–150 °C. The undrawn fiber is solidified in a cooling tube using air at 230 °C. After rinsing with water, the fibers undergo hot drawing at 140 °C (draw ratio 6–10), followed by heat setting in air at 140 °C or in steam at 115 °C. The internal relative surface area for dry-spun fibers is relatively low, typically below 1 m^2/g.

Acrylic fibers are held together primarily by weak van der Waals forces, making the polymeric system heat-sensitive. As a result, acrylic textile fabrics are among the most sensitive synthetic fibers to wrinkling and distortion at elevated temperatures.

Fig. 15 SEM polyacrylonitrile fiber **a** fiber bundle **b** individual fiber (reused from Ahn et al. [3] under CC-BY Permission)

Acrylic acts as an insulator since it has no free electrons to carry heat and has a weak phonon heat transfer system, giving it a low thermal conductivity. Acrylic has been shown to release the most microplastics compared to most synthetic fibers used in textiles [4].

Wanza et al. produced needle-punched nonwoven batts from recycled acrylic spinning waste and knitting waste with fabric densities of 14.571 and 11.333 kg/m^3, respectively. The authors obtained thermal conductivity values of 0.0431 and 0.0429 W/m K for both material types, respectively, which is almost 25% higher than that of the wool samples produced [108]. It was found that needle-punched recycled acrylic spinning waste at a fabric density and air permeability of 10.583 kg/m^3 and 1100 L/m^2/s had a thermal conductivity as low as 0.03827 W/m K [33, 43].

4.4 *Segmented Polyurethane Fibers*

A segmented polyurethane, commonly referred to as Elastane, Spandex, or Lycra, is an elastomeric fiber typically added in lower quantities to textiles containing other natural or synthetic fibers. For this reason, it is often overlooked for recycling since it typically makes up a smaller percentage of a garment's material [115]. Due to spandex typically being blended with other fibers, a comprehensive approach for recycling is needed. This typically involves a combination of mechanical and chemical methods. This requires first using physical techniques such as mechanical grinding to remove easily separable components, followed by chemical dissolution or degradation to further recover the remaining fiber [115].

From a molecular perspective, elastomeric behavior in a polymer refers to its ability to exist in two distinct states at ordinary temperatures: a relaxed state, where the polymer chains are coiled and packed together, and an extended state, where the chains are stretched out. For this elastomeric behavior to occur, the relaxed state must be stable, and the polymer must possess sufficient internal chain mobility to allow easy deformation and retraction—this requires that the glass transition temperature be below typical ambient temperatures. Polymers with highly flexible chains and weak interchain bonds are best suited for this purpose. For effective recovery after stretching, there must be relatively strong, long-range intermolecular attractions, while at the same time, there must be barriers to molecular flow in the extended state to prevent permanent deformation.

A number of segmented block copolymers fulfill these requirements to varying degrees. In such materials, soft, flexible segments alternate with rigid (often crystalline) segments. For fiber applications, segmented polyurethanes are especially suitable as elastomers, although technically they are copolymers in which polyurethane bridges connect the soft and hard segments. The soft segments typically consist of aliphatic polyethers or copolyesters and make up 65–90% of the fiber mass. During macroscopic stretching, these soft segments transform from a non-oriented, coiled configuration into an elongated form. As stretching proceeds, the individual chains come close enough to form ordered (crystalline) regions, which increases the stiffness and strength of the fiber while it is under tension. When the tension is released, the crystallites within the soft segments break apart (or "melt"), allowing the chains to return to their original, packed state. The driving force for this recovery is mainly entropic, similar to that in natural rubber, and the maximum elongation achievable is related to the length of the soft segments.

The hard segments are primarily composed of aromatic polyureas. These segments form relatively rigid clusters and regions held together by secondary bonds, such as hydrogen bonds or attractive forces between coplanar aromatic nuclei. These bonds must be strong enough to prevent molecular slippage during stretching, which would otherwise lead to plastic deformation. However, the rigid segments must not be so close together as to restrict the reversible deformation of the soft segments. Additionally, they must have a sufficiently high melting point to limit plastic flow at elevated temperatures. The connection between hard and soft segments is achieved through a urethane bond (–CO–O–NH–). In elastomeric fibers, the rigid segments typically measure 3–10 μm and are randomly oriented in the unstretched fiber. Partial crosslinking can be introduced to further inhibit molecular slippage at high elongations.

A notable drawback of polyurethane elastomers is that their recovery is time-dependent; oriented rigid segments in a stretched fiber return to a less oriented state more slowly. Spandex and Lycra fibers are produced from macroglycols and macrodiisocyanates, which form the building blocks of the soft and hard segments, respectively. These macro-radicals are connected by urethane bonds. The resulting polymer is soluble in solvents such as dimethylformamide or dimethylacetamide, allowing it to be spun into fibers using both wet and dry spinning methods. Reaction spinning is also possible, wherein a macrodiisocyanate prepolymer is spun while simultaneously undergoing polyaddition to form the final polymer. These fibers are capable of nearly complete recovery under cyclic loading within the range of 200–350% elongation.

Research has shown that compression molding technology can also be employed to directly recycle nylon/spandex-blended fabrics to produce thermal insulation materials using a mechanical process that this process requires neither large amounts of solvents nor complex pretreatment steps, thus reducing the environmental impact. The authors found that the best results were obtained with a blend of 60% Nylon/Spandex fabric shreds mixed with 40% Polyurethane shreds, obtaining a thermal conductivity of 0.0953 W/m K [25, 115].

5 Conclusion

Natural fibers are generally more resource-intensive due to the extensive agricultural inputs necessary to produce them, whereas synthetic fibers are more polluting due to the generation of microplastics and the fossil fuels required to produce them. Natural fibers tend to be more prone to biological attack than synthetic fibers, therefore requiring a dry and protected operating environment, making synthetic fibers more versatile. The sustainability and thermal properties of insulation materials are typically defined by the type of fiber used. However, thermal conductivity is mostly influenced by the amount of air in the material, either inside the fibers or between them. For synthetic materials, there is hardly any air inside them unless hollow fiber systems are used, and their thermal insulating effects are mostly dominated by the air between the fibers.

Acknowledgements This work was supported by the Czech Science Foundation (GACR)—project Advanced structures for thermal insulation in extreme conditions (Reg. No. 21-32510M) and the project Eco-clothing collection, reg. no. TQ01000450 is supported by the Technology Agency of the Czech Republic (Program for Support of Applied Research and Innovation SIGMA).

References

1. Abbate, S., Centobelli, P., Cerchione, R., Nadeem, S.P., Riccio, E.: Sustainability trends and gaps in the textile, apparel and fashion industries. Environ. Dev. Sustain. **26**(2), 1 (2023). https://doi.org/10.1007/S10668-022-02887-2
2. Abdel-Rehim, Z.S., Saad, M.M., El-Shakankery, M., Hanafy, I.: Textile fabrics as thermal insulators. Autex Res. J. **6**(3), 148–161 (2006). https://doi.org/10.1515/AUT-2006-060305
3. Ahn, H., Gwak, H.J., Kim, Y.M., Yu, W.R., Lee, W.J., Yeo, S.Y.: Microstructure analysis of drawing effect and mechanical properties of polyacrylonitrile precursor fiber according to molecular weight. Polymers **14**(13), 2625 (2022). https://doi.org/10.3390/POLYM14132625
4. Akyildiz, S.H., Fiore, S., Bruno, M., Sezgin, H., Yalcin-Enis, I., Yalcin, B., Bellopede, R.: Release of microplastic fibers from synthetic textiles during household washing. Environ. Pollut. **357**, 124455 (2024). https://doi.org/10.1016/J.ENVPOL.2024.124455
5. Alves, P.H.T.F., Clarke-Sather, A.R., Carlson, S., Martini, A.: Theoretical method for characterizing textile failure mechanics in mechanical recycling with carded drums, vol. 1, p. 1 (2023). https://doi.org/10.1115/MSEC2023-104361
6. America's Cotton Producers and Importers: Cotton morphology and cellulose chemistry—nonwovens technical guide (2025). https://www.cottoninc.com/quality-products/nonwovens/cotton-fiber-tech-guide/cotton-morphology-and-chemistry/
7. Angelotti, A., Alongi, A., Augello, A., Dama, A., De Antonellis, S., Ravidà, A., Zinzi, M., De Angelis, E.: Thermal conductivity assessment of cotton fibers from apparel recycling for building insulation. Energy Buildings **324**, 114866 (2024). https://doi.org/10.1016/J.ENBUILD.2024.114866
8. Asakura, T., Okonogi, M., Naito, A.: Toward understanding the silk fiber structure: 13C solid-state NMR studies of the packing structures of alanine oligomers before and after trifluoroacetic acid treatment. J. Phys. Chem. B **123**(31), 6716–6727 (2019). https://doi.org/10.1021/ACS.JPCB.9B04565
9. Averink, J.: Global water footprint of industrial hemp textile (2015)
10. Babu, K.M.: Silk fibres. In: Kozłowski, R.M. (ed.) Handbook of Natural Fibres, vol. 1, pp. 146–170. Woodhead Publishing (2012). https://doi.org/10.1533/9780857095503.1.146
11. Bhuiyan, M.A.R., Ali, A., Mohebbullah, M., Hossain, M.F., Khan, A.N., Wang, L.: Recycling of cotton apparel waste and its utilization as a thermal insulation layer in high performance clothing. Fashion Text. **10**(1), 1–12 (2023). https://doi.org/10.1186/S40691-023-00342-Y
12. Bilen, U.: The effect of linen and linen blends on the comfort properties of bedding fabrics. J. Nat. Fibers **18**(3), 430–441 (2021). https://doi.org/10.1080/15440478.2019.1624997
13. Brock, P.M., Graham, P., Madden, P., Alcock, D.J.: Greenhouse gas emissions profile for 1 kg of wool produced in the Yass Region, New South Wales: a life cycle assessment approach. Anim. Prod. Sci. **53**(6), 495–508 (2013). https://doi.org/10.1071/AN12208

14. Cao, H., Cobb, K., Yatvitskiy, M., Wolfe, M., Shen, H.: Textile and product development from end-of-use cotton apparel: a study to reclaim value from waste. Sustainability **14**(14) (2022). https://doi.org/10.3390/SU14148553
15. Chai, M.W., Bickerton, S., Bhattacharyya, D., Das, R.: Influence of natural fibre reinforcements on the flammability of bio-derived composite materials. Compos. B Eng. **43**(7), 2867–2874 (2012). https://doi.org/10.1016/J.COMPOSITESB.2012.04.051
16. Chen, F., Song, H., Liu, C., Ji, Y., Su, X., Sun, F.: Directly processing natural cotton into eco-friendly, highly thermal-insulating textiles via hierarchically self-locking structures. Text. Res. J **94**(21–22) (2024). https://doi.org/10.1177/00405175241247338
17. Chen, S., Chen, F., Zhu, L., Li, Q., Wang, X., Wang, L.: A spatial water footprint assessment of recycled cotton T-shirts: case of local impacts in selected China provinces. Sustainability **15**(1) (2023). https://doi.org/10.3390/SU15010817
18. Chen, Y., Qin, W., Shen, H., Ge, P., Guo, M., Huang, J., Yan, J., Zhang, H.: Development of waste silk-based aerogel: for continuous treatment of oily wastewater and thermal insulation. Int. J. Biol. Macromol. **288**, 138628 (2025). https://doi.org/10.1016/J.IJBIOMAC.2024.138628
19. Cheng, X.W., Guan, J.P., Chen, G., Yang, X.H., Tang, R.C.: Adsorption and flame retardant properties of bio-based phytic acid on wool fabric. Polymers **8**(4). https://doi.org/10.3390/POLYM8040122
20. Choudhury, K., Tsianou, M., Alexandridis, P.: Recycling of blended fabrics for a circular economy of textiles: separation of cotton, polyester, and elastane fibers. Sustainability **16**(14) (2024). https://doi.org/10.3390/SU16146206
21. De Falco, F., Di Pace, E., Cocca, M., Avella, M.: The contribution of washing processes of synthetic clothes to microplastic pollution. Sci. Rep. **9**(1) (2019). https://doi.org/10.1038/S41598-019-43023-X
22. Delhom, C.D., Kelly, B., Martin, V.: Physical properties of cotton fiber and their measurement. In: Cotton Fiber: Physics, Chemistry and Biology, pp. 41–73. Springer, Cham (2018). https://doi.org/10.1007/978-3-030-00871-0_3
23. Dijk, F., Song, S., van Eck, G.W.A., Wu, X., Bos, I.S.T., Boom, D.H.A., Kooter, I.M., Spierings, D.C.J., Wardenaar, R., Cole, M., Salvati, A., Gosens, R., Melgert, B.N.: Inhalable textile microplastic fibers impair lung repair (2021). https://doi.org/10.1101/2021.01.25.428144
24. Ding, L.: Research on the development and development of sustainable design in the textile and garment industry under the goal of. Mod. Econ. Manag. Forum **4**(6), 162 (2023). https://doi.org/10.32629/MEMF.V4I6.1499
25. Dissanayake, D.G.K., Weerasinghe, D.U., Wijesinghe, K.A.P., Kalpage, K.M.D.M.P.: Developing a compression moulded thermal insulation panel using postindustrial textile waste. Waste Manag. **79**, 356–361 (2018). https://doi.org/10.1016/J.WASMAN.2018.08.001
26. Dougherty, H.C., Oltjen, J.W., Mitloehner, F.M., Depeters, E.J., Pettey, L.A., Macon, D., Finzel, J., Rodrigues, K., Kebreab, E.: Carbon and blue water footprints of California sheep production. J. Anim. Sci. **97**(2), 945–961 (2019). https://doi.org/10.1093/JAS/SKY442
27. El Messiry, M., El-Tarfawy, S.Y., Ayman, Y., El Deeb, R.: Development of high-performance thermal insulation panels from flax fiber waste for building insulation. J. Ind. Text. **55** (2025). https://doi.org/10.1177/15280837251338160
28. Elnagar, K., Elmaaty, T.A., Raouf, S.: Dyeing of polyester and polyamide synthetic fabrics with natural dyes using ecofriendly technique. J. Text. **2014**(1), 363079 (2014). https://doi.org/10.1155/2014/363079
29. Fakirov, S., Seganov, I., Kurdowa, E.: Effect of chain composition of poly(ethylene terephthalate) structure and properties. Die Makromol. Chem. **182**(1), 185–197 (1981). https://doi.org/10.1002/MACP.1981.021820120
30. Fiedler, T., Pedersen, J.: Evaluating the thermal conductivity of hemp-based insulation. Materials **18**(8), 1723 (2025). https://doi.org/10.3390/MA18081723
31. Gama, N., Araújo, J., Godinho, B., Ferreira, A., Barros-Timmons, A.: Solvolysis of nylon: a pathway to sustainable recycling and circular economy. Sustainability **16**(22), 9725 (2024). https://doi.org/10.3390/SU16229725

32. Goswami, H., Das, D., Kumar, D.: Evaluation of efficacy of medicinal plant supplementation strategy against anemia. J. Reattach Therapy Dev Diversities (2023). https://doi.org/10.53555/JRTDD.V6I10S.2582
33. Gounni, A., El Wazna, M., El Alami, M., El Bouari, A., Cherkaoui, O., Mabouk, M.T., Kheiri, A.: Thermal performance evaluation of textile waste as an alternative solution for heat transfer reduction in buildings. J. Sol. Energy Eng. Trans. ASME **140**(2) (2018). https://doi.org/10.1115/1.4038786/367931
34. Guan, J., Lu, H., Chen, Y.: Apparel performance of flame retardant silk fabrics. J. Eng. Fibers Fabr. **8**(4), 30–35 (2013). https://doi.org/10.1177/155892501300800411
35. Guo, C., Li, C., Vu, H.V., Hanna, P., Lechtig, A., Qiu, Y., Mu, X., Ling, S., Nazarian, A., Lin, S.J., Kaplan, D.L.: Thermoplastic moulding of regenerated silk. Nat. Mater. **19**(1), 102–108 (2019). https://doi.org/10.1038/s41563-019-0560-8
36. Gutierrez-Moscardo, O., Canet, M., Gomez-Caturla, J., Lascano, D., Fages, E., Sanchez-Nacher, L.: Sustainable materials with high insulation capacity obtained from wastes from hemp industry processed by wet-laid. Text. Res. J. **92**(7–8), 1098–1112 (2021). https://doi.org/10.1177/00405175211046058
37. Haigler, C.H., Betancur, L., Stiff, M.R., Tuttle, J.R.: Cotton fiber: a powerful single-cell model for cell wall and cellulose research. Front. Plant Sci. **3**(May), 26402 (2012). https://doi.org/10.3389/FPLS.2012.00104
38. Harlin, A., Kajiwara, K., Suzuoka, A.: Challenges of polyester-cotton textiles co-fractionation. J. Text. Eng. Fashion Technol. **10**(1), 22–24 (2024). https://doi.org/10.15406/JTEFT.2024.10.00361
39. He, C.: Research on application of sustainable design thinking in clothing design. Highlights Art Des. **8**(1), 10–14 (2024). https://doi.org/10.54097/CBBJT683
40. Hernandez, E., Nowack, B., Mitrano, D.M.: Polyester textiles as a source of microplastics from households: a mechanistic study to understand microfiber release during washing. Environ. Sci. Technol. **51**(12), 7036–7046 (2017). https://doi.org/10.1021/ACS.EST.7B01750
41. Horne, M.R.L.: Bast fibres: hemp cultivation and production. In: Kozłowski, R.M., Mackiewicz-Talarczyk, M. (eds.) Handbook of Natural Fibres, vol. 1, 2nd edn, pp. 163–196. Woodhead Publishing (2020). https://doi.org/10.1016/B978-0-12-818398-4.00007-4
42. Huang, J., Yu, C.: Determination of cellulose, hemicellulose and lignin content using near-infrared spectroscopy in flax fiber. Text. Res. J. **89**(23–24), 4875–4883 (2019). https://doi.org/10.1177/0040517519843464
43. Islam, S., Bhat, G.: Environmentally-friendly thermal and acoustic insulation materials from recycled textiles. J. Environ. Manag. **251**, 109536 (2019). https://doi.org/10.1016/J.JENVMAN.2019.109536
44. Jinde, P., Naik, R., Rakshit, A.: Characterization and synthesis of polyester and viscose nonwovens fabrics embedded with nanoporous amorphous silica. J. Text. Inst. **110**(7), 972–979 (2018). https://doi.org/10.1080/00405000.2018.1534305
45. Jirsak, O., Sadikoglu, T.G., Ozipek, B., Pan, N.: Thermo-insulating properties of perpendicular-laid versus cross-laid lofty nonwoven fabrics. Text. Res. J. **70**(2), 121–128 (2000). https://doi.org/10.1177/004051750007000206
46. Johnson, S., Echeverria, D., Venditti, R., Jameel, H., Yao, Y.: Supply chain of waste cotton recycling and reuse: a review. AATCC J. Res. **7**(1_suppl), 19–31 (2020). https://doi.org/10.14504/AJR.7.S1.3
47. Jönsson, C., Wei, R., Biundo, A., Landberg, J., Schwarz Bour, L., Pezzotti, F., Toca, A., Jacques, L.M., Bornscheuer, U.T., Syrén, P.O.: Biocatalysis in the recycling landscape for synthetic polymers and plastics towards circular textiles. ChemSusChem **14**(19), 4028–4040 (2021). https://doi.org/10.1002/CSSC.202002666
48. Juanicó, L.E., González, A.D.: Thermal insulators with multiple air gaps: performance, cost and embodied impacts. J. Buil. Eng. **12**, 188–195 (2017). https://doi.org/10.1016/J.JOBE.2017.06.005
49. Kırbaş, H., Battal, Ö., Demir, S., Yatgın, U., Başlar, A., Uzun, M.: Industrial hemp: a comprehensive analysis from hemp seeds to a sustainable alternative raw material for textiles and

industrial applications. Tekstilna Industrija **73**(1), 4–25 (2025). https://doi.org/10.5937/TEKSTIND2501004K
50. Kosiński, P., Brzyski, P., Duliasz, B.: Moisture and wetting properties of thermal insulation materials based on hemp fiber, cellulose and mineral wool in a loose state. J. Nat. Fibers **17**(2), 199–213 (2020). https://doi.org/10.1080/15440478.2018.1477086
51. Kozasowski, R.M., Mackiewicz-Talarczyk, M., Allam, A.M.: Bast fibres: flax. In: Handbook of Natural Fibres (2012). https://doi.org/10.1533/9780857095503.1.56
52. Kuffner, H., Popescu, C.: Wool fibres. In: Handbook of Natural Fibres Types, Properties and Factors Affecting Breeding and Cultivation, vol. 1, pp. 171–195. Woodhead Publishing Series in Textiles (2012)
53. Kumar, N.M., Thilagavathi, G., Periasamy, S., Vinoth, V.: Development of needle punched nonwovens from natural fiber waste for thermal insulation application. J. Nat. Fibers **19**(14), 9580–9588 (2022). https://doi.org/10.1080/15440478.2021.1990175
54. Kursun Bahadir, S., Sahin, U.K., Acikgoz Tufan, H.: Design of a thermal waist-pad. IOP Conf. Ser. Mater. Sci. Eng. **254**(18) (2017). https://doi.org/10.1088/1757-899X/254/18/182006
55. Lacoste, C., El Hage, R., Bergeret, A., Corn, S., Lacroix, P.: Sodium alginate adhesives as binders in wood fibers/textile waste fibers biocomposites for building insulation. Carbohyd. Polym. **184**, 1–8 (2018). https://doi.org/10.1016/J.CARBPOL.2017.12.019
56. Lan, W.L., Kuo, C.F.J.: Development of a warming multi-functional fabric. Part I: The analytic hierarchy process combined with the technique for order preference by similarity to an ideal solution for the optimization of the multi-quality melt spinning parameters in far-infrared functional yarn. Text. Res. J. **89**(11), 2247–2259 (2019). https://doi.org/10.1177/0040517518790972
57. Leal Filho, W., Dinis, M.A.P., Liakh, O., Paço, A., Dennis, K., Shollo, F., Sidsaph, H.: Reducing the carbon footprint of the textile sector: an overview of impacts and solutions. Text. Res. J. **94**(15–16), 1798–1814 (2024). https://doi.org/10.1177/00405175241236971
58. Li, Q., Yuan, Z., Zhang, C., Hu, S., Chen, Z., Wu, Y., Chen, P., Qi, H., Ye, D.: Tough, highly oriented, super thermal insulating regenerated all-cellulose sponge-aerogel fibers integrating a graded aligned nanostructure. Nano Lett. **22**(9), 3516–3524 (2022). https://doi.org/10.1021/ACS.NANOLETT.1C03943
59. Lima, A.R.A., Ferreira, G.V.B., Barrows, A.P.W., Christiansen, K.S., Treinish, G., Toshack, M.C.: Global patterns for the spatial distribution of floating microfibers: Arctic Ocean as a potential accumulation zone. J. Hazard. Mater. **403** (2021). https://doi.org/10.1016/J.JHAZMAT.2020.123796
60. Liu, K., Takagi, H., Yang, Z.M.: Effect of lumen size on transverse thermal conductivity of unidirectional natural fiber-polymer composite via finite element method. Mater. Sci. Forum **675**(677), 431–434 (2011). https://doi.org/10.4028/www.SCIENTIFIC.NET/MSF.675-677.431
61. Manaia, J.P., Manaia, A.T., Rodriges, L.: Industrial hemp fibers: an overview. Fibers **7**(12), 106 (2019). https://doi.org/10.3390/FIB7120106
62. Militký, J.: In: Hes, L. (ed.) Textile Fibers-Classical and Special (Textilní vlákna-klasická a speciální). TUL Press (2012)
63. Militký, J.: Tensile failure of polyester fibers. In: Handbook of Properties of Textile and Technical Fibres (2018). https://doi.org/10.1016/B978-0-08-101272-7.00013-4
64. Militký, J., Venkataraman, M., Křemenaková, D.: Tepelná vodivost textilií obsahujících dutá vlákna. In: Křemenaková, D., Militký, J., Šesták, J. (eds.) Vlákenné struktury pro speciální aplikace, pp. 257–278. TUL Press (2013)
65. Militký, J., Venkataraman, M., Mishra, R.: The chemistry, manufacture, and tensile behavior of polyamide fibers. In: Handbook of Properties of Textile and Technical Fibres (2018). https://doi.org/10.1016/B978-0-08-101272-7.00012-2
66. Muniyasamy, S., Patnaik, A.: Biodegradable behavior of waste wool and their recycled polyester preforms in aqueous and soil conditions. J. Renew. Mater. **9**(10), 1661–1671 (2021). https://doi.org/10.32604/JRM.2021.014904

67. Nam, S., Condon, B.D., Liu, Y., He, Q.: Natural resistance of raw cotton fiber to heat evidenced by the suppressed depolymerization of cellulose. Polym. Degrad. Stab. **138**, 133–141 (2017). https://doi.org/10.1016/J.POLYMDEGRADSTAB.2017.03.005
68. Nam, S., Hillyer, M.B., Condon, B.D.: Method for identifying the triple transition (glass transition-dehydration-crystallization) of amorphous cellulose in cotton. Carbohyd. Polym. **228**, 115374 (2020). https://doi.org/10.1016/J.CARBPOL.2019.115374
69. Neri, M., Cuerva, E., Levi, E., Pujadas, P., Müller, E., Guardo, A.: Thermal, acoustic, and fire performance characterization of textile face mask waste for use as low-cost building insulation material. Dev. Built Environ. **14**, 100164 (2023). https://doi.org/10.1016/J.DIBE.2023.100164
70. Norovsuren, L., Ganbat, G., Myagmar, C.: An electron scan microscope analysis result of damage on woollen fiber by clothes moth (*Tineola bisselliella*). Mongolian J. Agric. Sci. **13**(2), 48–50 (2014). https://doi.org/10.5564/MJAS.V13I2.514
71. Orlando, J.R.: Cellulose Chemistry and Properties: Fibers, Nanocelluloses and Advanced Materials, vol. 271, p. 341. Springer (2016)
72. Parlato, M.C.M., Porto, S.M.C.: Organized framework of main possible applications of sheep wool fibers in building components. Sustainability **12**(3) (2020). https://doi.org/10.3390/SU12030761
73. Pervaiz, A., Azam, F., Ahmad, A., Ahmad, F., Ahmad, S., Ullah, T., Nawab, Y., Shaker, K.: An investigation of static and dynamic mechanical properties of eco-friendly textile PLA composites reinforced by flax woven fabrics. Int. J. Polym. Sci. **2024**(1), 2821777 (2024). https://doi.org/10.1155/IJPS/2821777
74. Pini, T., Sambucci, M., Valente, M.: Potential use of silk waste in sustainable thermoplastic composite material applications: a review. Fibers **13**(1), 6 (2025). https://doi.org/10.3390/FIB13010006
75. Rackov, J., Erceg, T., Živković, M., Teofilović, V.: Spectroscopic analysis of microplastic fibers released during laundry washing cycle. Adv. Eng. Lett. **2**(2), 71–79 (2023). https://doi.org/10.46793/ADELETTERS.2023.2.2.5
76. Reddy, N., Thillainayagam, V.A., Yang, Y.: Dyeing natural cellulose fibers from cornhusks: a comparative study with cotton fibers. Ind. Eng. Chem. Res. **50**(9), 5642–5650 (2011). https://doi.org/10.1021/IE200217W
77. Rengasamy, R.S., Kawabata, S.: Computation of thermal conductivity of fibre from thermal conductivity of twisted yarn. Indian J. Fibre Text. Res. **27**(4) (2002)
78. Romanovskiy, S.: Experience of application of heat-insulating plates based on plant fibers. Bull. Belgorod State Technol. Univ. Named After V G Shukhov 31–43 (2022). https://doi.org/10.34031/2071-7318-2021-7-1-31-43
79. Romanovskiy, S., Bakatovich, A.: Effect of modified liquid glass on absorption humidity and thermal conductivity of flax fiber slabs. IOP Conf. Ser. Mater. Sci. Eng. **660**(1) (2019). https://doi.org/10.1088/1757-899X/660/1/012072
80. Rubino, C., Bonet-Aracil, M., Liuzzi, S., Martellotta, F., Stefanizzi, P.: Thermal characterization of innovative sustainable building materials from wool textile fibers waste. Tec. Ital.-Ital. J. Eng. Sci. **63**(2–4), 277–283 (2019). https://doi.org/10.18280/TI-IJES.632-423
81. Sadrmanesh, V., Chen, Y.: Bast fibres: structure, processing, properties, and applications. Int. Mater. Rev. **64**(7), 381–406 (2019). https://doi.org/10.1080/09506608.2018.1501171
82. Sair, S., Oushabi, A., Kammouni, A., Tanane, O., Abboud, Y., El Bouari, A.: Mechanical and thermal conductivity properties of hemp fiber reinforced polyurethane composites. Case Stud. Constr. Mater. **8**, 203–212 (2018). https://doi.org/10.1016/J.CSCM.2018.02.001
83. Sair, S., Oushabi, A., Kammouni, A., Tanane, O., Abboud, Y., Oudrhiri Hassani, F., Laachachi, A., El Bouari, A.: Effect of surface modification on morphological, mechanical and thermal conductivity of hemp fiber: characterization of the interface of hemp—polyurethane composite. Case Stud. Therm. Eng. **10**, 550–559 (2017). https://doi.org/10.1016/J.CSITE.2017.10.012
84. Sakthivel, S., Melese, B., Edae, A., Abedom, F., Mekonnen, S., Solomon, E.: Garment waste recycled cotton/polyester thermal and acoustic properties of air-laid nonwovens. Adv. Mater. Sci. Eng. **2020**(1) (2020). https://doi.org/10.1155/2020/8304525

85. Sakthivel, S., Senthil Kumar, S., Mekonnen, S., Solomon, E.: Thermal and sound insulation properties of recycled cotton/polyester chemical bonded nonwovens. J. Eng. Fibers Fabr. **15** (2020). https://doi.org/10.1177/1558925020968819
86. Samardzioska, T., Jovanoska, M., Grujoska, V.: Sustainable thermal insulation derived from recycled textile waste. Adv. Environ. Eng. Res. **04**(01), 1–8 (2023). https://doi.org/10.21926/AEER.2301004
87. Sanchez-Vidal, A., Thompson, R.C., Canals, M., De Haan, W.P.: The imprint of microfibres in southern European deep seas. Plos One **13**(11) (2018). https://doi.org/10.1371/JOURNAL.PONE.0207033
88. Şardağ, S., Türk, M.: Thermo-physiological properties and burning behaviours of automotive upholstery fabrics woven with aramid and wool yarns. J. Ind. Text. **51**(3), 486–510 (2021). https://doi.org/10.1177/1528083720974042
89. Sashina, E.S., Yakovleva, O.I.: The current state and prospects of recycling silk industry waste into nonwoven materials. Fibers **11**(6) (2023). https://doi.org/10.3390/FIB11060056
90. Seagull, R.W., Oliveri, V., Murphy, K., Binder, A., Kothari, S.: Cotton fiber growth and development 2. Changes in cell diameter and wall birefringence. J. Cotton Sci. **4**(2) (2000)
91. Shaker, K., Nawab, Y.: Fibers for protective textiles. In: Topics in Mining, Metallurgy and Materials Engineering, pp. 65–91. Springer, Cham (2020). https://doi.org/10.1007/978-3-030-49224-3_4
92. Široky, J., Široka, B., Bechtold, T., Široky, J., Široka, B., Bechtold, T.: Alkali treatments of woven Lyocell fabrics. Woven Fabrics (2012). https://doi.org/10.5772/38697
93. Sonnier, R., Dumazert, L., Regazzi, A., Deborde, L., Lanos, C.: Flammability of thick but thermally thin materials including bio-based materials. Molecules **28**(13), 5175 (2023). https://doi.org/10.3390/MOLECULES28135175/S1
94. Sui, B., Wang, T., Fang, J., Hou, Z., Shu, T., Lu, Z., Liu, F., Zhu, Y.: Recent advances in the biodegradation of polyethylene terephthalate with cutinase-like enzymes. Front. Microbiol. **14**, 1265139 (2023). https://doi.org/10.3389/FMICB.2023.1265139
95. Sun, C.: True density of microcrystalline cellulose. J. Pharm. Sci. **94**(10) (2005). https://doi.org/10.1002/jps.20459
96. Téraube, O., Choupas, S., El Feggouri, L., Agopian, J.C., Charlet, K., Dubois, M.: Towards natural fibers resistant to mold and termites thanks to fluorine. Colloids Surf. A **702**, 134926 (2024). https://doi.org/10.1016/J.COLSURFA.2024.134926
97. Tong, T., Li, Y., Wu, C., Ma, C., Yang, J., Wei, Z.: Thermal conductivity of single silk fibroin fibers measured from the 3ω method. Int. J. Therm. Sci. **185**, 108057 (2023). https://doi.org/10.1016/J.IJTHERMALSCI.2022.108057
98. Trajković, D., Jordeva, S., Tomovska, E., Zafirova, K.: Polyester apparel cutting waste as insulation material. J. Text. Inst. **108**(7), 1238–1245 (2017). https://doi.org/10.1080/00405000.2016.1237335
99. Ude, A.U., Eshkoor, R.A., Zulkifili, R., Ariffin, A.K., Dzuraidah, A.W., Azhari, C.H.: *Bombyx mori* silk fibre and its composite: a review of contemporary developments. Mater. Des. **57**, 298–305 (2014). https://doi.org/10.1016/J.MATDES.2013.12.052
100. Valtere, M., Bezrucko, T., Blumberga, D.: Analysis of textile circularity potential. Environ. Climate Technol. **135** (2023). https://doi.org/10.7250/CONECT.2023.104
101. Valverde, I.C., Castilla, L.H., Nuñez, D.F., Rodriguez-Senín, E., De La Mano Ferreira, R.: Development of new insulation panels based on textile recycled fibers. Waste Biomass Valorization **4**(1), 139–146 (2013). https://doi.org/10.1007/S12649-012-9124-8/TABLES/3
102. Van Der Velden, N.M., Patel, M.K., Vogtländer, J.G.: LCA benchmarking study on textiles made of cotton, polyester, nylon, acryl, or elastane. Int. J. Life Cycle Assess. **19**(2), 331–356 (2014). https://doi.org/10.1007/S11367-013-0626-9
103. Vėjelis, S., Skulskis, V., Kremensas, A., Vaitkus, S., Kairytė, A.: Performance of thermal insulation material produced from Lithuanian sheep wool. J. Nat. Fibers **20**(2) (2023). https://doi.org/10.1080/15440478.2023.2237682
104. Vėjelis, S., Vaitkus, S., Skulskis, V., Kremensas, A., Kairytė, A.: Performance evaluation of thermal insulation materials from sheep's wool and hemp fibres. Materials **17**(13), 3339 (2024). https://doi.org/10.3390/MA17133339

105. Wang, D., Hu, S., Křemenáková, D., Militký, J., Zhu, G.: Effect of textile structure on heat transfer performance. In: Advanced Structured Materials, vol. 201 (2023). https://doi.org/10.1007/978-981-99-6002-6_8
106. Wang, L., Wang, X.: 6—Effect of structure–property relationships on fatigue failure in natural fibres. In: Miraftab, M. (ed.) Fatigue Failure of Textile Fibres, pp. 95–132. Woodhead Publishing (2009). https://doi.org/10.1533/9781845695729.2.95
107. Watanabe, T., Morita, K., Tsuwaki, H.: Morphological deterioration of cotton, wool, silk and rayon textiles by microorganisms (*Aspergillus* spp. and *Penicillium* spp.). Sen'i Gakkaishi **36**(9), T403–T408 (1980). https://doi.org/10.2115/FIBER.36.9_T403
108. Wazna, M.E.L., Gounni, A., Bouari, A.E.L., Alami, M.E.L., Cherkaoui, O.: Development, characterization and thermal performance of insulating nonwoven fabrics made from textile waste. J. Ind. Text. **48**(7), 1167–1183 (2019). https://doi.org/10.1177/1528083718757526
109. Wojciechowska, M., Kowaluk, G.: Challenges and opportunities in recycling upholstery textiles: enhancing high-density fiberboards with recycled fibers. Fibers **12**(12) (2024). https://doi.org/10.3390/FIB12120105
110. Xia, J., Kong, X., Li, L., Zhang, Z., Chen, Y., Li, M., Qin, Y., Cai, T., Dai, W., Fang, S., Yi, J., Lin, C.T., Nishimura, K., Jiang, N., Yu, J.: High thermal conductivity and radiative cooling designed boron nitride nanosheets/silk fibroin films for personal thermal management. ACS Appl. Mater. Interfaces **16**(6), 7732–7741 (2024). https://doi.org/10.1021/ACSAMI.3C16602
111. Yong, L., Jian, L., Xian, L., Bei, W.: Test and analysis of the porosity of cotton fiber assembly. J. Eng. Fibers Fabr. **16** (2021). https://doi.org/10.1177/15589250211024225
112. Zhang, S., Chen, C., Duan, C., Hu, H., Li, H., Li, J., Liu, Y., Ma, X., Stavik, J., Ni, Y.: Regenerated cellulose by the Lyocell process, a brief review of the process and properties (2018)
113. Zheng, C., Li, D., Ek, M.: Improving fire retardancy of cellulosic thermal insulating materials by coating with bio-based fire retardants. Nordic Pulp Pap. Res. J. **34**(1), 96–106 (2019). https://doi.org/10.1515/NPPRJ-2018-0031
114. Zhou, X., Zheng, F., Li, H., Lu, C.: An environment-friendly thermal insulation material from cotton stalk fibers. Energy Buildings **42**(7), 1070–1074 (2010). https://doi.org/10.1016/J.ENBUILD.2010.01.020
115. Zhu, M., Gao, C., Wang, S., Shi, S., Zhang, M., Su, Q.: Recycling of spandex: broadening the way for a complete cycle of textile waste. Sustainability **17**(8), 3319 (2025). https://doi.org/10.3390/SU17083319

Green and Sustainable Synthesis of Nanoparticles for Functionalization of Fibrous Structures

Muhammad Zaman Khan, Jiří Militký, Jakub Wiener, Azam Ali, and Blanka Tomková

1 Introduction

Nanotechnology involves synthesizing, characterizing, exploring, and applying materials at the nanoscale, specifically those sized between 1 and 100 nanometers (nm). This field focuses on materials whose structures exhibit significantly novel and enhanced physical, chemical, and biological properties due to their small size. In textiles, nanotechnology enables the development of multifunctional fabrics with special properties such as antibacterial effects, UV protection, photocatalytic activity, superhydrophobicity, self-cleaning capabilities, electromagnetic interference (EMI) shielding, flame resistance, and improved comfort. Nanotechnology involves manipulating structures at the nanoscale to create materials or devices with dimensions in this range. It accomplishes this by precisely controlling the arrangement of atoms and molecules, which enables the development of functional materials, devices, and systems with outstanding properties [1, 2]. Materials at nanoscale can exhibit different properties as compared to macro scale level. Properties of material change quickly when it reduces to nanoscale. Nanostructure materials have grain size less than 100 nm, which increases the surface area of the material and cause excellent

M. Z. Khan · J. Militký (✉) · J. Wiener · A. Ali · B. Tomková
Department of Material Engineering, Faculty of Textile Engineering, Technical University of Liberec, Liberec, Czech Republic
e-mail: jiri.militky@tul.cz

M. Z. Khan
e-mail: muhammad.zaman.khan@tul.cz

J. Wiener
e-mail: jakub.wiener@tul.cz

A. Ali
e-mail: azam.ali@tul.cz

B. Tomková
e-mail: blanka.tomkova@tul.cz

J. Militký and M. Venkataraman (eds.), *Fibrous Structures for Sustainable Future*, Engineering Materials, https://doi.org/10.1007/978-981-95-1382-6_8

change in physical and surface properties [3]. Nanostructures can be prepared from many materials such as proteins, polysaccharides and synthetic polymers. The selection of materials in mainly depended on factors like size of the nanoparticles required, inherent properties such as solubility and stability, surface characteristics, charge and permeability, degree of biodegradability, biocompatibility and toxicity etc. Atoms, molecules, clusters, fullerenes, supramolecular structures, their crystals, nanotubes, nanowires, nanorodes, their arrays and photonic crystals serve as nanotechnology objectives [4, 5]. Nanostructures can be polymer, metals and ceramics having different morphologies and sizes. Isotropic and anisotropic nanomaterials are categorized on the basis of dimensionality. A diverse spectrum of anisotropic nanomaterials, such as [6, 7] nanowires [8], nanotubes belongs to 1D nanostructures [9]. Triangles, plates, sheets and ribbons belong to 2D [10, 11]. Pyramids, stars, flowers, multipods, nanourchins, tadpole, nanocages, nanorice, nanocorns, nanoboxes, nanocubes, triangular nanoframes, nanodumbbells are belongs to 3D nanostructures [12, 13]. Various wet chemical techniques are mostly used for functionalization and deposition of nanostructures onto substrate. Nanoparticles manufacturing is an essential part of nanotechnology because many important properties are obtained at nanoparticle, nanocrystal, nanolayer level and assembling of nanoparticles. Nanotechnology is recognized as an emerging technology for the twenty-first century [14]. Nanoparticles (NPs) are typically produced using two primary methods: the bottom-up and the top-down approaches. The top-down approach involves breaking down bulk materials into nanoparticles, while the bottom-up approach focuses on assembling atoms to create nanoparticles. As shown in Fig. 1, the top-down strategy employs various preparation techniques, including lithography, ball milling, etching, and sputtering, to produce nanomaterials and nanoparticles. In contrast, the bottom-up approach utilizes several methods, such as chemical vapor deposition, sol-gel processes, spray pyrolysis, laser pyrolysis, and atomic or molecular condensation [15, 16].

The environmentally friendly synthesis of metal and metal oxide nanoparticles has recently attracted significant interest due to its growing applications across various fields, reduced reliance on harmful chemicals, and straightforward experimental procedures. This method aims to minimize the use of toxic substances by substituting them with biomolecules derived from plants, which include terpenoids, aldehydes, vitamins, alkaloids, and polysaccharides that function as reducing, capping, and stabilizing agents in the formation of the desired nanoparticles. The titanium dioxide nanomaterials are recognized for their wide applications, notably their photocatalytic properties and numerous uses in advanced devices like photovoltaic cells. Furthermore, they find various applications in the domains of biomedicine and dentistry. The challenge of creating environmentally friendly, sustainable nanostructures with controlled size and shape, along with their integration into fibrous materials, remains an issue in textiles and nanoscience. Nanostructures display unique or enhanced properties that depend on specific attributes such as size distribution, orientation, and morphology. The sustainable green synthesis methodology for producing ZnO, TiO_2, and silver nanostructures utilizes various approaches, including microwave synthesis, solvothermal, hydrothermal, and precipitation techniques, employing bioactive compounds and biomass byproducts [17, 18].

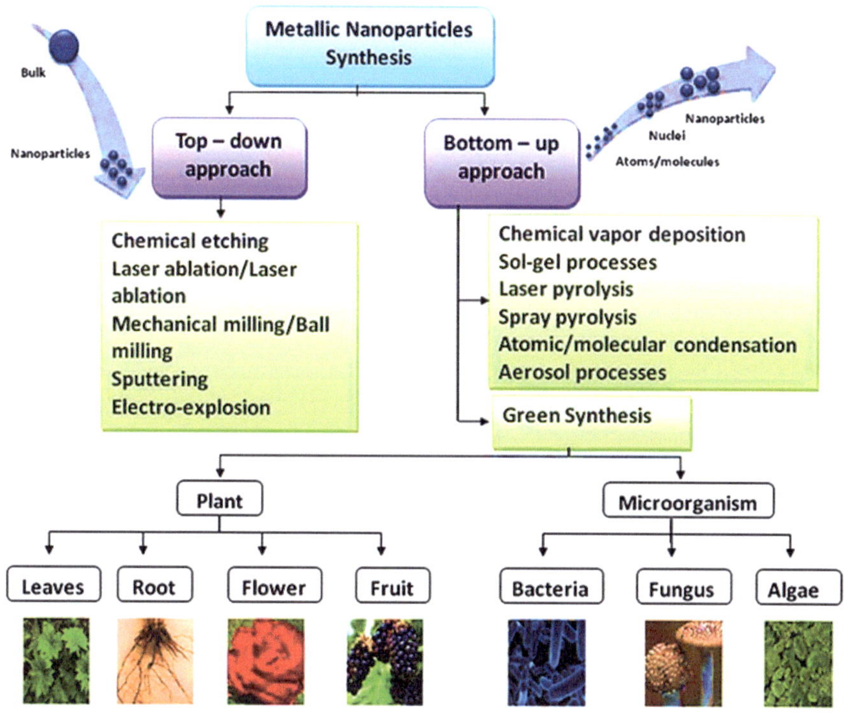

Fig. 1 Different methods for synthesizing metal nanoparticles (reused from Singh et al. [16] under CC-BY Permission)

2 Synthesis of Green Nanomaterials

Biosynthesized metal oxide nanoparticles reveal distinct physicochemical characteristics like thermal, optical, surface, and electrical properties. Natural environmental sources such as algae, fungi, bacteria, and plants contribute to the green-mediated synthesis of nanoparticles. Various plant parts, including stems, roots, leaves, flowers, seeds, and latex, produce nanoparticles with diverse shapes and sizes. Most research focuses on plant components such as leaves, seeds, and flowers, primarily due to their high efficacy in reducing metal ions. Additional advantages include the easy availability of these plants and their safe handling. Bioactive compounds sourced from plants encompass alkaloids, diterpenoids, lactones, glycosides, steroids, polysaccharides, terpenoids, flavonoids, phenolic acids, vitamins, enzymes, minerals, sugars, salicylic acids, saponins, amino acids, tannins, and sterols, all of which serve as reducing and stabilizing agents. Nanoparticles, including zinc, silver, titanium, tin, gold, aluminum, iron, magnesium, manganese, copper, and cobalt, have been successfully produced using green chemistry techniques. The practical application of ZnO and TiO_2 nanostructures, alongside the progress in controlled synthesis methods to create desired morphologies, is a vital aspect of material science and nanotechnology. These nanostructures can be formed using various physical and chemical processes; nevertheless, the green synthesis method stands out as a promising strategy to generate these nanostructures with specific properties while saving time and energy and/or minimizing the use of hazardous substances [19, 20].

The chemical approaches for creating metal oxide nanoparticles (NPs) can pose environmental risks. These synthesis methods often necessitate elevated temperatures and pressures, which restrict large-scale production of metal oxides. As a result, green nanotechnology has surfaced as a sustainable and eco-friendly alternative for creating metal oxide NPs, using reducing agents sourced from biological materials. This technique allows the same reducing agent to be utilized to produce various metallic compounds. Additionally, incorporating plants, their by-products, fruit extracts, and microorganisms in the synthesis process reduces dependency on harmful and expensive chemicals. This strategy also supports economical large-scale production of metal oxide NPs. As a result, green synthesis techniques are essential for producing stable, uniformly sized, and easily dispersed nanoparticles while enhancing energy efficiency [21].

The synthesis of nanoparticles (NPs) using plants can be achieved through three main methods: intracellular synthesis, where the process occurs within the plant cells; extracellular synthesis, which involves using plant extracts; and synthesis that employs specific phytochemicals. Many plants have the ability to accumulate metals and convert them into nanoparticles aided by various biomolecules, including amino acids, flavonoids, alkaloids, and proteins, among others. These biomolecules play a crucial role in reducing metals into nanoparticles [22]. Typically, the NP synthesis process involves mixing plant biomass or extracts with a metal salt solution under controlled temperature and pH conditions. A noticeable color change in the solution often indicates successful NP synthesis. The experimental method for synthesizing NPs using plant biomass is depicted in Fig. 2 [23].

3 Green Synthesis of Metal Oxide Nanoparticles

Various species of plants from different taxonomic groups have been used to synthesize titanium dioxide nanoparticles (TiO_2 NPs). The unique properties of these nanoparticles are primarily determined by the varying concentrations and compositions of organic reducing and stabilizing agents found in different plant extracts. Among the various plant parts, leaves are particularly effective for nanoparticle synthesis due to their high concentrations of metabolites. Specifically, the fruit, leaf, and peel of Citrus lemon have been utilized to synthesize TiO_2 NPs [24]. The use of pepper wastes as source of polyphenols in the synthesis of NPs could be a smart alternative to add them value, also promoting their circular economy and reducing environmental pollution. Thus, the lower carbon footprint and price generated by the green method is justified by changing the use of chemical reducers for natural reducers that are waste [25].

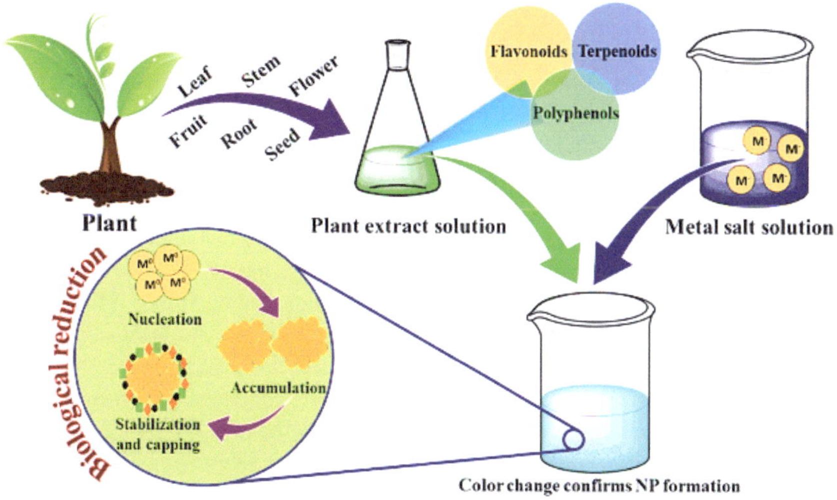

Fig. 2 Diagrammatic illustration of the biosynthesis of nanoparticles facilitated by plants (reused from Dikshit et al. [23] under CC-BY Permission)

Both plant and bacterial extracts have been employed for the biosynthesis of metal oxide nanoparticles (NPs). The metabolites found in bacterial biomass may aid in stabilizing and bio-reducing metal oxide NPs. Furthermore, fungal extracts have attracted significant attention over bacterial extracts due to advantages such as straightforward extraction, the potential for large-scale production, economic viability, and an enhanced surface area of the synthesized metal oxides. Various sizes and shapes of TiO_2 and ZnO NPs have been documented, with fungi possessing enzymes and metabolites that facilitate the reduction of bulk salts into ionic elements, demonstrating the benefits of green synthesis using microorganisms, including bacteria and fungi. Regarding bacterial utilization for TiO_2 synthesis, baker's yeast was employed as an economical approach to produce small-sized anatase TiO_2 NPs with high purity and stability. The impact of baker's yeast was noted for its effective antimicrobial properties. Beyond plants and microorganisms, various biological sources are also used for synthesizing TiO_2 NPs. Exploring biological sources for synthesis is still limited, but they have been recognized for their cost-effectiveness and environmental sustainability in NP production. Rice straw has been identified as a promising template for TiO_2 synthesis due to its green, simple, and soft template characteristics. Similarly, gelatin has been regarded as a biological material source for straightforward and efficient TiO_2 synthesis, offering increased surface area and enhanced hydrogen storage capacity attributed to gelatin. Likewise, peptides, arginine, and protamine have been reported for their role in the green synthesis of TiO_2 NPs, highlighting their cost-effectiveness and ease of use [21].

The chemical production of nanoparticles (NPs) often requires the use of expensive, toxic, and hazardous substances. The release of these substances into the environment poses significant ecological and toxicological risks. In contrast, green synthesis provides a straightforward, eco-friendly, and less toxic alternative for producing NPs, utilizing biodegradable materials such as plant extracts, microorganisms, and enzymes. Among these methods, synthesizing NPs with plant extracts is particularly effective as it reduces the risk of contamination, shortens reaction times, and preserves cellular integrity. Green synthesis is an emerging technique that is gaining increasing attention from scientists worldwide for its simplicity and environmental benefits in the production of nanoparticles [26]. Therefore, green synthesis meets the above requirements and is proposed as an alternative methodology for reducing the use of dangerous compounds and harsh reaction conditions in obtaining metal oxide powders and their corresponding nanoparticles (NPs). Green synthesis is one of the methods used in the present work. Additionally, despite nanotechnology developments, the adverse effects of nanomaterials are still relatively unknown. If the synthesis of the NPs is thus carried out using environmentally friendly and biocompatible processes, the toxicity of the material obtained and the environmental impact of the by-products might be reduced. Previously, the method was structured by adapting the controlled precipitation process using aqueous garlic and chili pepper extracts as solvents, bearing in mind the specific properties of these plant products. Garlic (*Allium sativum*) has been used since ancient times, and it has been considered an excellent homegrown antibacterial agent. The investigations show that allicin, an active biochemical component found in garlic, has antibacterial and antioxidant properties; the nature and structure of the bioactive compounds in garlic that could be "incorporated" into the ZnO particles would not be affected. Meanwhile, chili pepper (*Capsicum annuum*) was used because it is a plant product popularly known for its efficient antibacterial capacity [27].

Plants contain different components such as alkaloids, proteins, phenolic compounds, carbohydrates, and enzymes that play a key role as reducing or capping agents during the synthesis of nanoparticles. The choice of reducing agent, solvent medium, and non-toxic material influences the stabilization of nanoparticles. The most commonly used precursors for preparing Green synthesized TiO_2 nanoparticles are Titanium (IV) isopropoxide and Titanium oxysulfate. Green synthesized TiO_2 nanoparticles from fungi and bacteria. Fungi belong to biological sources known to be used in bioremediation, have the capability of mineralizing, and are known to be mediators in the formation of nanoparticles owing to their ability to produce vast amounts of biomass. Green synthesis has always yielded better results than existing synthesis due to its naturally embedded properties. Plant-mediated synthesis are also effective, low-cost, and environmentally friendly, moving us further towards a green environment without encouraging much use of chemicals [18].

Waste valorization is a term used to manage waste in the most sustainable way, and it has attracted significant attention in recent years. Various advanced technologies are available to employ agricultural wastes in novel functional nanomaterials. Different templates have been noted to produce metal oxide nanostructures. Additionally, surfactants, serving as secondary agents to templates, have been extensively utilized in fabricating various nanoparticles that exhibit favorable size distribution and dispersal. Adding diverse surfactants as capping agents into the reaction matrix can help synthesize monodispersed metal oxide nanoparticles. Many studies have investigated the effects of different biotemplates on the synthesis and properties of various metal oxide nanostructures. Gelatin, gum, starch, rice straw, eggshell membrane, bamboo membrane, DNA, yeast, dandelion pollen, and albumen were all investigated to synthesize metal oxide nanoparticles such as ZnO, TiO_2, and Ag nanoparticles [28].

The creation of nanomaterials through eco-friendly techniques is becoming increasingly popular due to its ease, rapidity, sustainability, and non-toxic nature. This method employs green solvents such as distilled water and ethanol, making it cost-effective. The synthesis procedure involves three main steps: selecting a suitable solvent, using environmentally safe reducing and capping agents, and choosing non-toxic materials to guarantee the stability of the nanoparticles [29]. Recently, plant-based synthesis of zinc oxide nanoparticles using the aqueous extracts of *M. fragrans* fruits has been reported. Green synthesis of ZnO-NPs has eco-friendly aspects and various biomedical applications. The metabolites found in the aqueous extract of *M. fragrans* act as an oxidizing, reducing, and capping agent for the synthesis of biogenic ZnO-NPs [30]. The eco-friendly synthesis of ZnO nanoparticles (NPs) is considerably safer than traditional chemical methods as it avoids generating harmful byproducts. Colloidal solutions containing ZnO NPs are employed in nanofertilizer applications. The potential biological uses of ZnO NPs have been investigated due to their antioxidant, antibacterial, antifungal, anticancer, and antidiabetic properties. Biological methods present promising, non-toxic alternatives to both physical and chemical synthesis. The production of ZnO NPs through biological synthesis involves using microorganisms such as bacteria, fungi, algae, bacteriophages, and plant extract DNA and proteins. The specific procedures for the biological synthesis of ZnO NPs are still not fully understood. Certain bacteria that utilize specific enzymes can generate ZnO NPs [31].

The study of nanostructures requires a better understanding of the crystal form with a suitable structure that can be achieved through a new design and simple synthesis methods. Biogenesis has grown rapidly in modern nanotechnology over the last few decades. Titanium dioxide nanoparticles can be synthesized using template, sol-gel, hydrothermal, thermal hydrolysis, electrochemical anodic oxidation, flame synthesis, and photochemical reduction methods. The most common precursors used to synthesize TiO_2 nanoparticles were titanium (IV) alkoxide, titanium tetra isopropoxide (TTIP), titanium butoxide, and titanium tetrachloride ($TiCl_4$). Titanium tetrachloride is the most toxic and corrosive of these precursors, while titanium (IV) alkoxide is both costly and soluble in organic solvents. The green route is the more suitable option for nanoparticle synthesis since it is more sustainable and eco-friendly, while physical and chemical methods for synthesis are also possible. Phytochemicals and bioactive compounds have been found in standard aqueous extracts from various medicinal plants that work as non-toxic capping and reducing agents and shape and size control agents for synthesized nanoparticles. *A. calamus* is a traditionally well-known medicinal plant used to treat many diseases. *A. calamus* was initially classified as a member of the Araceae family but is now designated as belonging to the Acoraceae family. The plant grows in northern temperate and subtropical regions of Asia, North America, Europe, and northern Jammu and Kashmir. Here, we present an innovative and creative sustainable technique for the fabrication of titania (TiO_2) using *Acorus calamus* (*A. calamus*) leaf extract as a new biogenic source, as well as a capping and reducing agent [32].

The extracts from their roots, leaves, stems, seeds, flowers, or fruits can be utilized when working with plants. Organic polymers serve as plants' fundamental components, including substances like starch, chitin, cellulose, hemicellulose, lignin, and various types of resins. When plant extracts are utilized, the organic polymers and biomolecules within them are responsible for their unique characteristics. Cellulose, the most prevalent organic polymer, is a polysaccharide found in plants' primary cell walls and many algae varieties. Lignocellulose biopolymers nourish the plant's cell walls and are composed of cellulose, hemicellulose, and lignin. Lignin makes up approximately 30% of lignocellulose biomass and contains numerous phenylpropanoids. It has the potential to substitute petroleum, and its depolymerization presents significant opportunities for generating high-quality chemicals [33]. Nanostructured semiconductor metal oxides are essential in developing most electronic devices, such as solar cells, transistors, diodes, and sensors. Among various metal oxides, ZnO and TiO_2 NPs are of much interest in the scientific community due to the fact of their unique electronic, chemical, and physical properties, their high surface-to-volume ratio, and availability of more surface atoms for an immediate chemical reaction [19]. Plants possess a variety of biomolecules, including phenolic compounds, terpenoids, acids, and alkaloids, which can serve as reducing and stabilizing agents in the synthesis of nanostructured metal oxides. Oat (*Avena sativa*), a cereal grain that belongs to the Poaceae family, has a diverse phytochemical profile. Additionally, the chemical components found in oat biomass are utilized in the biosynthesis of nanomaterials, specifically metal and metal oxide nanoparticles [34]. *Punica granatum* peel extract was successfully used to produce ZnO NPs by functioning as a reducing and stabilizing agent during the synthesis process, and the obtained ZnO NPs effectively inhibited the growth of bacteria. The effect of synthesis temperature on the size and shape of zinc oxide (ZnO) nanoparticles (NPs) has been investigated using pineapple peel waste and the antibacterial activity of ZnO NPs in starch films [35]. However, nanoparticle titanium dioxide synthesis using these bio-derivatives has not been much explored like their counterparts [36]. In a previous study, the significance of titanium dioxide nanoparticles (TiO_2 NPs) in promoting plant growth and enhancing plant resilience to abiotic stress was examined. The TiO_2 NPs were synthesized through a green method, utilizing an aqueous extract of *Aloe vera* leaves as a capping agent and titanium tetrachloride as a precursor [37].

Nearly all agro-industrial and food by-products contain phenolic compounds derived from fruits, vegetables, oilseeds, nuts, cereals, and beverages. These compounds have specific functional groups that improve their reduction and stabilization abilities. The phenolic substances in fruits and vegetables can serve as alternatives to synthetic preservatives, as they are effective in scavenging free radicals and inhibiting oxidation reactions in food. Moreover, these by-products are rich in bioactive components, such as carotenoids, vitamins, oils, and enzymes [38]. Biomass waste generated from fruit residues is rich in flavonoids, which can chelate and reduce metal ions into nanoparticles (NPs). Consequently, these materials are utilized in nanoparticle production. Metal oxide nanoparticles (MONPs) are typically synthesized through the hydrolysis of metal salts at ambient temperatures or below 100 °C. Due to their small size and high corner or edge surface site density, metal oxide nanoparticles exhibit exceptional physical and chemical properties. Beneficial bioactive substances present in fruit and vegetable waste, including alkaloids, amino acids, enzymes, phenolics, proteins, polysaccharides, tannins, saponins, vitamins, and terpenoids, act as reducing agents in the synthesis of metal nanoparticles (NPs). It has been reported that metal oxide NPs can be produced from the peels of various fruits, such as bananas, *Citrus sinensis* (sweet orange), jackfruit, lemons, mangoes, *Musa paradisiaca* (plantain), pomegranates, tangerines, *Punica granatum*, *Garcinia mangostana* (mangosteen), *Citrus aurantifolia* (lime), and *Nephelium lappaceum* (rambutan). Furthermore, agricultural waste can also be used in nanoparticle synthesis. Metal and metal oxide nanoparticles can be derived from weeds in agricultural fields, which act as bioreactors for nanoparticle synthesis. For instance, copper nanoparticles can be synthesized from *Lantana camara*, silver nanoparticles from *Ipomoea carnea*, and copper oxide nanoparticles from *Gloriosa superba* L. [39, 40]. Plants are the preferred choice for the green and simple synthesis of nanoparticles due to their ability to promote the large-scale production of stable nanoparticles in various shapes and sizes. Utilizing natural plant extracts is an eco-friendly, straightforward, and cost-effective approach. These extracts often contain phytochemical compounds such as phenols, flavonoids, alkaloids, terpenes, saponins, and tannins, serving as reducing, capping, or stabilization agents in nanoparticle synthesis [41].

4 Future Perspectives

The creation of eco-friendly, sustainable, and precisely shaped nanostructures for incorporation into fibrous materials continues to pose a challenge in the textile and nanoscience industries. Nanostructures exhibit novel or improved characteristics influenced by traits like size distribution, orientation, and morphology. Recent studies have concentrated on the environmentally friendly synthesis of metal oxide nanostructures. Biological sources like plants, flowers, and bacteria stabilize and reduce agents, enabling control over the nanomaterial's morphology. Given the toxic nature of chemical synthesis approaches and the challenges with large-scale production, there is a strong inclination toward more sustainable and eco-conscious methods. Green nanotechnology has attracted considerable attention for producing titanium dioxide (TiO_2) and zinc oxide (ZnO) nanoparticles using naturally available biological resources, including plants, fruit extracts, microorganisms, and their byproducts. This technique not only lowers costs but also promotes environmental sustainability. Organic polymers such as starch, chitin, cellulose, hemicellulose, lignin, and various resins are the main building blocks derived from plants. Plant extracts and biopolymers sourced from plants are preferred for fabricating metal and metal oxide nanoparticles due to their ability to reduce and stabilize these particles. Additionally, agricultural waste can be harnessed for nanoparticle synthesis, with unwanted plants in farmland functioning as bioreactors. Conventional synthesis methods frequently result in pollution, emphasizing the urgent need for environmentally responsible techniques. Thus, generating nanoparticles from biodegradable waste presents an effective strategy for waste reduction and environmental pollution alleviation. Natural biological systems are vital in the synthesis of green materials for the production of nanomaterials. Biowaste is rich in secondary metabolites, including phenolic acids, terpenoids, alkaloids, and flavonoids. Almost all agricultural and food byproducts are sources of phenolic compounds, which can be found in fruits, vegetables, oilseeds, nuts, grains, and drinks. These compounds, characterized by specific functional groups, enhance nanoparticles' reduction and stabilization processes. The phenolic compounds in fruits and vegetables can serve as alternatives to synthetic preservatives, as they help scavenge free radicals and inhibit oxidation reactions in food. Furthermore, they include a range of other bioactive components, such as carotenoids, vitamins, oils, and enzymes. Waste biomass from fruit residues is rich in various flavonoids that can chelate and reduce metal ions into nanoparticles, making them significant for nanoparticle synthesis [42, 43].

5 Conclusions

Biological methodologies present promising, non-toxic alternatives to conventional physical and chemical synthesis techniques. The green chemistry method may synthesize NPs such as zinc, silver, titanium, tin, gold, aluminum iron, magnesium, manganese, copper, and cobalt. Significant advantages such as cost-effectiveness, less toxicity, environment friendliness, and better applications are the reasons for adopting green synthesis over chemical methods. Biosynthesized metal oxide nanoparticles exhibit different physicochemical properties such as thermal, optical, surface, and electrical. Bioactive and phytochemical compounds obtained from plant sources include alkaloids, diterpenoids, lactones, glycosides, steroids, polysaccharides, terpenoids, flavonoids, phenolic acids, vitamins, enzymes, minerals, sugars, salicylic acids, saponins, amino acids, tannins and sterols which acts as reducing, capping and stabilizing agents for the formation of the desired nanostructures. The sustainable green synthesis approach to synthesize the ZnO, TiO_2, and silver nanostructures uses different routes such as microwave synthesis, solvothermal, hydrothermal, microwave synthesis, and precipitation techniques using bioactive compounds and biomass waste.

References

1. Nakata, K., Fujishima, A.: TiO_2 photocatalysis: design and applications. J. Photochem. Photobiol. C: Photochem. Rev. **13**, 169–189 (2012). https://doi.org/10.1016/j.jphotochemrev.2012.06.001

2. Skompska, M., Zar, K.: Electrodeposition of ZnO nanorod arrays on transparent conducting substrates—a review. Electrochim. Acta **127**, 467–488 (2014). https://doi.org/10.1016/j.electacta.2014.02.049
3. Khan, M.Z., Taghavian, H., Fijalkowski, M., Militký, J., Tomková, B., Venkataraman, M., Adach, K.: Effect of microwave power on bactericidal and UV protection properties of the ZnO nanorods grown cotton fabrics. Colloids Surf. A: Physicochem. Eng. Aspects **664**(Feb), 131135 (2023). https://doi.org/10.1016/j.colsurfa.2023.131135
4. Pokropivny, V., Hussainova, I., Vlassov, S.: Introduction to Nanomaterials and Nanotechnology, p. 192. Tartu University Press (2007)
5. Yu, R., Lin, Q., Leung, S., Fan, Z.: Nanomaterials and nanostructures for efficient light absorption and photovoltaics. Nano Energy **1**(1), 57–72 (2012). https://doi.org/10.1016/j.nanoen.2011.10.002
6. Gan, X., Gao, X., Qiu, J., He, P., Li, X., Xiao, X.: TiO_2 Nanorod-derived synthesis of upstanding hexagonal kassite nanosheet arrays an intermediate route to novel nanoporous TiO_2 nanosheet arrays. Cryst. Growth Des. **12**(1), 289–296 (2012)
7. Khan, M.Z., Militký, J., Baheti, V., Fijalkowski, M., Wiener, J., Adach, K.: Growth of ZnO nanorods on cotton fabrics via microwave hydrothermal method: effect of size and shape of nanorods on superhydrophobic and UV-blocking properties. Cellulose **27**, 10519–10539 (2020). https://doi.org/10.1007/s10570-020-03495-x
8. Feng, X., Shankar, K., Varghese, O.K., Paulose, M., Latempa, T.J., Grimes, C.: Vertically aligned single crystal TiO_2 nanowire arrays grown directly on transparent conducting oxide coated glass: synthesis details and applications. Nano Lett. **8**(11), 3781–3786 (2008). https://doi.org/10.1021/nl802096a
9. Tamilselvan, V., Yuvaraj, D., Rakesh Kumar, R., Narasimha Rao, K.: Growth of rutile TiO_2 nanorods on TiO_2 seed layer deposited by electron beam evaporation. Appl. Surf. Sci. **258**(10), 4283–4287 (2012). https://doi.org/10.1016/j.apsusc.2011.12.079
10. Ganguly, P., Byrne, C., Breen, A., Pillai, S.C.: Antimicrobial activity of photocatalysts: fundamentals, mechanisms, kinetics and recent advances. Appl. Catal. B Environ. **225**(Oct 2017), 51–75 (2018). https://doi.org/10.1016/j.apcatb.2017.11.018
11. Musial, J., Krakowiak, R., Mlynarczyk, D.T., Goslinski, T.: Titanium dioxide nanoparticles in food and personal care products—what do we know about their safety? Nanomaterials **10**, 1110 (2020)
12. Khan, M.Z., Baheti, V., Militký, J., Wiener, J., Ali, A.: Self-cleaning properties of polyester fabrics coated particles and trimethoxy (octadecyl) silane. J. Ind. Text. **50**(4), 543–565 (2020). https://doi.org/10.1177/1528083719836938
13. Wu, Z., Wu, Q., Du, L., Jiang, C., Piao, L.: Progress in the synthesis and applications of hierarchical flower-like TiO_2 nanostructures. Particuology **15**, 61–70 (2014). https://doi.org/10.1016/j.partic.2013.04.003
14. Roco, M.C.: Nanoparticles and nanotechnology research. J. Nanopart. Res. **1**, 1–6 (1999)
15. Abuzeid, H.M., Julien, C.M., Zhu, L., Hashem, A.M.: Green synthesis of nanoparticles and their energy storage, environmental, and biomedical applications. Crystals **13**, 1576 (2023)
16. Singh, J., Dutta, T., Kim, K.H., Rawat, M., Samddar, P., Kumar, P.: "Green" synthesis of metals and their oxide nanoparticles: applications for environmental remediation. J. Nanobiotechnol. **16**, 84 (2018). https://doi.org/10.1186/s12951-018-0408-4
17. Eid, A.M., Fouda, A., Niedbała, G., Hassan, S.E.: Endophytic *Streptomyces laurentii* mediated green synthesis of Ag-NPs with antibacterial and anticancer properties for developing functional textile fabric properties. Antibiotics **9**(10), 641 (2020)
18. Samhitha, S., Raghavendra, G., Quezada, C., Hima Bindu, P.: Green synthesized TiO_2 nanoparticles for anticancer applications: mini review. Mater. Today: Proc. **54**, 765–770 (2022). https://doi.org/10.1016/j.matpr.2021.11.073
19. Gonçalves, R.A., Toledo, R.P., Joshi, N., Berengue, O.M.: Green synthesis and applications of ZnO and TiO_2 nanostructures. Molecules **26**, 2236 (2021)
20. Miu, B.A., Dinischiotu, A.: New green approaches in nanoparticles synthesis: an overview. Molecules **27**, 6472 (2022)

21. Irshad, M.A., Nawaz, R., Zia ur Rehman, M., Adrees, M., Rizwan, M., Ali, S., Ahmad, S., Tasleem, S.: Synthesis, characterization and advanced sustainable applications of titanium dioxide nanoparticles: a review. Ecotoxicol. Environ. Saf. **212**, 111978 (2021). https://doi.org/10.1016/j.ecoenv.2021.111978
22. Nath, D., Banerjee, P.: Green nanotechnology—a new hope for medical biology. Environ. Toxicol. Pharmacol. **36**(3), 997–1014 (2013). https://doi.org/10.1016/j.etap.2013.09.002
23. Dikshit, P.K., Kumar, J., Das, A.K., Sadhu, S., Sharma, S., Singh, S., Gupta, P.K., Kim, B.S.: Green synthesis of metallic nanoparticles: applications and limitations. Catalysts **11**, 902 (2021)
24. Sunny, N.E., Mathew, S.S., Chandel, N., Saravanan, P., Rajeshkannan, R., Rajasimman, M., Vasseghian, Y.: Chemosphere green synthesis of titanium dioxide nanoparticles using plant biomass and their applications—a review. Chemosphere **300**(Mar), 134612 (2022). https://doi.org/10.1016/j.chemosphere.2022.134612
25. Jiménez-Rosado, M., Gomez-Zavaglia, A., Guerrero, A., Romero, A.: Green synthesis of ZnO nanoparticles using polyphenol extracts from pepper waste (*Capsicum annuum*). J. Cleaner Prod. **350**, 131541 (2022). https://doi.org/10.1016/j.jclepro.2022.131541
26. Goutam, S.P., Saxena, G., Singh, V., Yadav, A.K.: Green synthesis of TiO_2 nanoparticles using leaf extract of *Jatropha curcas* L. for photocatalytic degradation of tannery wastewater. Chem. Eng. J. **336**, 386–396 (2018). https://doi.org/10.1016/j.cej.2017.12.029
27. Benitez-Salazar, M.I., Niño-Castaño, V.E., Dueñas-Cuellar, R.A., Caldas-Arias, L., Fernández, I., Rodríguez-Páez, J.E.: Chemical synthesis versus green synthesis to obtain ZnO powders: evaluation of the antibacterial capacity of the nanoparticles obtained by the chemical method. J. Environ. Chem. Eng. **9**, 106544 (2021). https://doi.org/10.1016/j.jece.2021.106544
28. Ramimoghadam, D., Bagheri, S., Bee, S., Hamid, A.: Biotemplated synthesis of anatase titanium dioxide nanoparticles via lignocellulosic waste material. BioMed Res. Int. **2014**, 205636 (2014)
29. Bekele, E.T., Gonfa, B.A., Sabir, F.K.: Use of different natural products to control growth of titanium oxide nanoparticles in green solvent emulsion, characterization, and their photocatalytic application. Bioinorg. Chem. Appl. **2021**, 6626313 (2021)
30. Faisal, S., Jan, H., Shah, S.A., Shah, S., Khan, A., Akbar, M.T., Rizwan, M., Jan, F., Akhtar, N., Khattak, A., Syed, S.: Green synthesis of zinc oxide (ZnO) nanoparticles using aqueous fruit extracts of *Myristica fragrans*: their characterizations and biological and environmental applications. ACS Omega **6**, 9709−9722 (2021). https://doi.org/10.1021/acsomega.1c00310
31. Deka, B., Baruah, C., Babu, A., Kalita, P.: Biological and non-conventional synthesis of zinc oxide nanoparticles (ZnO-NPs): their potential applications. J. Nanotechnol. Nanomater. **3**(2), 79–89 (2022)
32. Ansari, A., Siddiqui, V.U., Rehman, W.U., Akram, K., Siddiqi, W.A., Alosaimi, A.M., Hussein, M.A., Rafatullah, M.: Green synthesis of TiO_2 nanoparticles using acorus calamus leaf extract and evaluating its photocatalytic and in vitro antimicrobial activity. Catalysts **12**, 181 (2022)
33. Bhullar, S., Goyala, N., Gupta, S.: Rapid green-synthesis of TiO_2 nanoparticles for therapeutic applications. RSC Adv. **11**, 30343–30352 (2021). https://doi.org/10.1039/d1ra05588g
34. Al-Mohaimeed, A.M., Al-Onazi, W.A., El-Tohamy, M.F.: Multifunctional eco-friendly synthesis of ZnO nanoparticles in biomedical applications. Molecules **27**, 579 (2022)
35. Basri, H.H., Talib, R.A., Sukor, R., Othman, S.H., Ariffin, H.: Effect of synthesis temperature on the size of ZnO nanoparticles derived from pineapple peel extract and antibacterial activity of ZnO—starch nanocomposite films. Nanomaterials **10**, 1061 (2020)
36. Arun, J., Goutham, S.N., Ram, R., Alagappan, P.: Synthesis and application of titanium dioxide photocatalysis for energy, decontamination and viral disinfection: a review. Environ. Chem. Lett. **21**(1), 339–362 (2023). https://doi.org/10.1007/s10311-022-01503-z
37. Abdalla, H., Adarosy, M.H., Hegazy, H.S., Abdelhameed, R.E.: Potential of green synthesized titanium dioxide nanoparticles for enhancing seedling emergence, vigor and tolerance indices and DPPH free radical scavenging in two varieties of soybean under salinity stress. BMC Plant Biol. **22**, 560 (2022). https://doi.org/10.1186/s12870-022-03945-7
38. Su, D., Zhang, R., Hou, F., Zhang, M., Guo, J., Huang, F., Deng, Y., Wei, Z.: Comparison of the free and bound phenolic profiles and cellular antioxidant activities of litchi pulp extracts

from different solvents. BMC Complement. Altern. Med. **14**(1), 9 (2014). https://doi.org/10.1186/1472-6882-14-9
39. Aswathi, V.P., Meera, S., Maria, C.G.A., Nidhin, M.: Green synthesis of nanoparticles from biodegradable waste extracts and their applications: a critical review. Nanotechnol. Environ. Eng. **8**, 377–397 (2022). https://doi.org/10.1007/s41204-022-00276-8
40. Suppiah, D.D., Julkapli, N.M., Sagadevan, S., Johan, M.R.: Eco-friendly green synthesis approach and evaluation of environmental and biological applications of iron oxide nanoparticles. Inorg. Chem. Commun. **152**, 110700 (2023). https://doi.org/10.1016/j.inoche.2023.110700
41. Naiel, B., Fawzy, M., Halmy, M.W.A., El Din Mahmoud, A.: Green synthesis of zinc oxide nanoparticles using Sea Lavender (*Limonium pruinosum* L . Chaz.) extract: characterization, evaluation of anti-skin cancer, antimicrobial and antioxidant potentials. Sci. Rep. **12**, 20370 (2022). https://doi.org/10.1038/s41598-022-24805-2
42. Rónavári, A., Igaz, N., Adamecz, D.I., Szerencs, B., Molnar, C.: Green silver and gold nanoparticles: biological synthesis approaches and potentials for biomedical applications. Molecules **26**, 844 (2021)
43. Kumar, H., Bhardwaj, K., Kuca, K., Kalia, A., Nepovimova, E., Verma, R., Kumar, D.: Flower-based green synthesis of metallic nanoparticles: applications beyond fragrance. Nanomaterials **10**, 766 (2020)

Biodegradability and Environmental Impact of Bacterial Cellulose

Sebnem Sozcu, Mohanapriya Venkataraman, Blanka Tomková, and Jiří Militký

1 Introduction

The textile industry poses significant environmental challenges due to its reliance on both synthetic fibers and natural fibers that require substantial resources. Petroleum-based synthetic fibers, such polyester and nylon, are a major source of greenhouse gas emissions during production and microplastic pollution during use and disposal. When synthetic clothing is laundered, for example, microplastics are released into waterways, where they build up in food chains and ecosystems [1, 2]. However, the cultivation and processing of natural fibers like cotton and wool demands enormous amounts of land, water, and chemical inputs. Up to 2700 gallons of freshwater are needed to produce one cotton t-shirt, underscoring the burden on the world's water supplies [2, 3]. Additionally, both fiber types contribute to industrial water pollution through dyeing and finishing processes that release harmful chemicals into the environment [4, 5]. The cumulative impact of these factors underscores the urgent need for sustainable approaches to textile production and consumption. Bacterial cellulose (BC), a highly versatile and sustainable biopolymer, is synthesized through fermentation by specific microbial species, primarily those belonging to the genera *Acetobacter* and *Gluconacetobacter*. Distinct from plant-derived cellulose obtained from

S. Sozcu · M. Venkataraman (✉) · B. Tomková · J. Militký
Department of Material Engineering, Faculty of Textile Engineering, Technical University of Liberec, Liberec, Czech Republic
e-mail: mohanapriya.venkataraman@tul.cz

S. Sozcu
e-mail: sebnem.sozcu@tul.cz

B. Tomková
e-mail: blanka.tomkova@tul.cz

J. Militký
e-mail: jiri.militky@tul.cz

J. Militký and M. Venkataraman (eds.), *Fibrous Structures for Sustainable Future*, Engineering Materials, https://doi.org/10.1007/978-981-95-1382-6_9

cotton or wood, BC exhibits unique properties such as high tensile strength (200–300 MPa) [6–9], excellent biodegradability, and superior water retention capacity (up to 100 cc/g) [10]. These characteristics make BC a highly attractive material for diverse applications across industries including food, biomedical, textile, and pharmaceutical sectors [11, 12]. Bacterial cellulose (BC) is synthesized through the microbial fermentation of glucose or alternative carbon sources, resulting in a biopolymer characterized by high biocompatibility and enhanced mechanical and structural properties relative to plant-derived cellulose [13].

As a biopolymer, BC represents a viable alternative to petroleum-based synthetic polymers, aligning with the increasing demand for environmentally sustainable and renewable materials. Moreover, its microbial biosynthesis enables precise control over production parameters, thereby improving process efficiency and further contributing to its overall sustainability [14]. Growing environmental concerns linked to textile manufacturing have intensified efforts to develop sustainable alternatives to petroleum-derived materials. As one of the largest industrial contributors to pollution, the textile sector is responsible for significant waste generation, water pollution, and greenhouse gas emissions [15].

The widespread use of synthetic fibers, predominantly derived from fossil fuels, poses significant environmental concerns, particularly regarding their contribution to microplastic pollution [16]. In contrast, bacterial cellulose (BC) has gained attention as a sustainable alternative, owing to its biodegradability, superior mechanical properties, and low environmental footprint [12]. Bacterial cellulose (BC) is a naturally derived polymer produced by specific bacterial strains, presenting a renewable and sustainable alternative for textile applications. Its biosynthesis demands fewer resources than conventional fiber production processes and can be tailored to achieve physicochemical properties comparable to those of synthetic textiles [17]. This chapter examines the potential of bacterial cellulose (BC) in addressing textile-related environmental challenges and decreasing reliance on petroleum-derived materials, highlighting its promise as a sustainable alternative in the development of environmentally responsible textile solutions.

- A comprehensive overview of the environmental impacts associated with conventional textiles, including synthetic fibers and resource-intensive natural fibers, is presented.
- Bacterial cellulose (BC) is introduced as a sustainable biopolymer synthesized by microbial processes.
- The role of BC in mitigating textile pollution and reducing dependency on fossil-based resources is emphasized, underscoring its significance in advancing sustainable textile technologies.

2 Bacterial Cellulose: Synthesis and Properties

Bacterial cellulose (BC) is an extracellular polymer produced largely by specific bacterial strains, particularly *Gluconacetobacter xylinus*. Unlike plant-derived cellulose, BC is manufactured in a highly pure form, free of lignin and hemicellulose, making it particularly appealing for high-value applications, including sustainable textiles [18].

2.1 *Synthesis of Bacterial Cellulose*

Bacterial cellulose (BC) is a pure, extracellular polymer distinct from plant cellulose, lacking lignin and hemicellulose. It has a high degree of polymerization, crystallinity, and a fine fiber (20–100 nm) network that increases its surface area (37 $m^2 g^{-1}$) [19–21]. BC production depends heavily on the nutrient composition of the medium, especially the carbon source. The Hestrin–Schramm (HS) medium is commonly used, though its high-cost limits commercial production. Therefore, using agro-industrial waste as a carbon source is a cost-effective alternative [22, 23]. Several bacteria, mainly Gram-negative strains like *Acetobacter*, *Pseudomonas*, and *Rhizobium*, and some Gram-positive species, can produce BC. Among them, *Acetobacter xylinum* is the most efficient producer, capable of converting nearly 50% of carbon sources into cellulose under optimal conditions [10].

BC production methods are crucial, with two main approaches: static and agitated conditions. In static culture, a continuous pellicle forms, while in agitated culture, faster growth results in spherical cellulose particles [24]. The synthesis of BC occurs through a fermentation process, where bacteria convert carbon sources—commonly glucose, fructose, or other simple sugars—into cellulose microfibrils. Under static culture conditions, the bacteria secrete cellulose into the surrounding medium, forming a pellicle at the air-liquid interface. In agitated conditions, BC can form irregularly shaped particles or fibers [25].

2.2 *Unique Characteristics of Bacterial Cellulose*

Bacterial cellulose has various features that set it apart from plant-derived cellulose and synthetic fibers, making it ideal for textile applications.

2.2.1 High Tensile Strength and Mechanical Stability

One of the most notable mechanical features of bacterial cellulose is its high tensile strength. BC nanofibers form a highly crystalline and entangled network, contributing

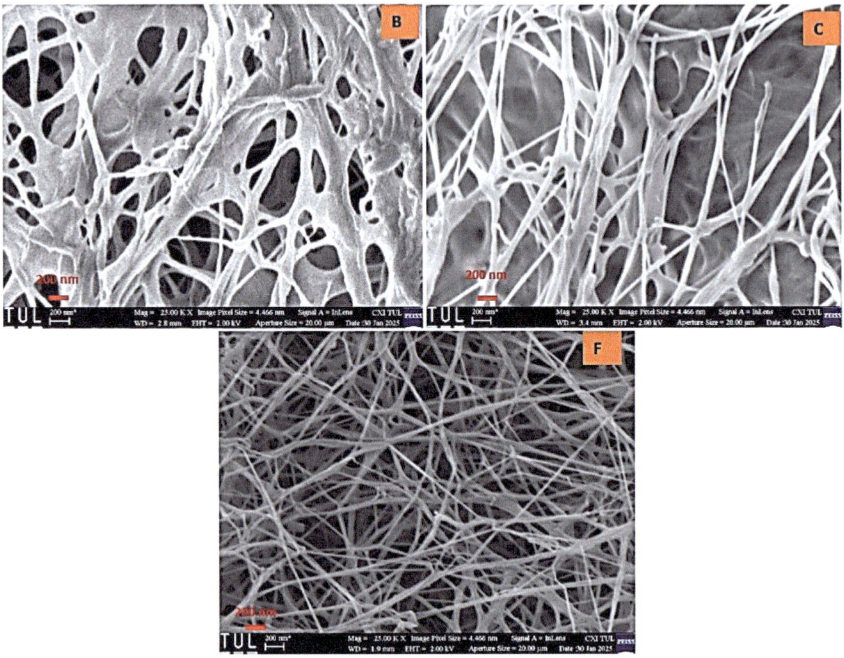

Fig. 1 Cross-sectional area of BCAs. (B, C) conventional (refrigerator) pre-freezing BCAs; (F) LN2 pre-freezing BCAs

to exceptional mechanical robustness due to an ultrafine nanofiber network with strong hydrogen bonding between fibrils [26]. The tensile strength of dried BC films ranges from 200 to 300 MPa, while the Young's modulus is reported to be around 15–30 GPa, depending on culture conditions and post-treatment processes [6, 7]. This strength surpasses many synthetic polymers and is significantly higher than that of conventional plant cellulose. In Fig. 1, the SEM images labeled B, C, and F display the typical nanofibrous network structure of bacterial cellulose (BC) aerogels at 25,000× magnification. These micrographs reveal a highly porous and interconnected 3D framework composed of entangled cellulose nanofibers, generally less than 100 nm in diameter. This fibrous architecture plays a key role in imparting BC aerogels with their distinctive properties, including high surface area, low density, and remarkable mechanical strength [27].

2.2.2 High Water Holding Capacity

Due to its highly porous network, BC can absorb and retain large amounts of water—up to 100 times its dry weight [6]. This property is advantageous for breathable and moisture-wicking textiles.

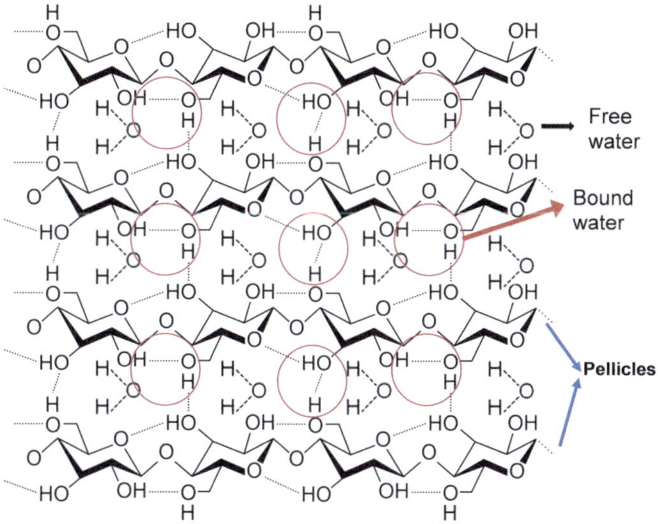

Fig. 2 Diagram illustrating the molecular structure of bacterial cellulose, including its bound and free water content (reused from Rebelo et al. [29] under CC-BY Permission)

Bacterial cellulose exhibits a remarkable capacity to retain water due to its three-dimensional nanofiber network and high surface area. The water holding capacity of BC can reach up to 100–200 times its dry weight [27, 28]. Its high hydrophilicity results from abundant hydroxyl groups on its surface, which form hydrogen bonds with water molecules. This property makes BC highly suitable for wound dressings, drug delivery systems, and hygiene products. Within the complex structure of BC, water molecules are held together by hydrogen bonds. Meanwhile, unbound or free water molecules can enter and leave the BC molecular network, as illustrated in Fig. 2 [29].

2.2.3 Biodegradability and Environmental Friendliness

Unlike petroleum-based synthetic fibers, BC is fully biodegradable under natural conditions without leaving harmful residues. This makes it an ideal material for minimizing textile waste at the end of a product's life cycle [30]. The production process is environmentally benign, requiring mild conditions and renewable substrates such as glucose or agricultural waste for microbial fermentation [31]. Unlike synthetic polymers, BC does not release microplastics or harmful degradation products, aligning with sustainable development goals and green chemistry principles [32].

2.2.4 High Purity

The absence of lignin, pectin, and hemicellulose reduces the need for extensive chemical processing, which is typically necessary in plant fiber refinement. This further enhances BC's environmental credentials [33].

2.2.5 Customizability

By altering the growth conditions or post-synthesis treatments, BC's physical characteristics, including thickness, porosity, and surface chemistry, can be customized during production [12]. This adaptability enables the material to be designed for particular textile purposes, including improved durability, stretchability, or breathability. These exceptional properties make BC a promising sustainable material for future textile innovations, which may result in the creation of high-performance apparel and accessories as well as a reduction in the material's adverse environmental effects.

Compared to plant-derived cellulose and synthetic fibers, bacterial cellulose has the following benefits that are given in Table 1.

BC outperforms plant cellulose in terms of purity and nanoscale structural control, which contributes to its mechanical strength and water absorption capacity. In contrast to synthetic fibers, BC is produced through a zero-waste, low-energy fermentation process, making it an ideal material for sustainable manufacturing. Its biodegradability and renewable feedstocks make it particularly attractive for green product design in line with circular economy principles [41].

3 Environmental Impact Reduction Across the Textile Lifecycle

The conventional textile sector has long been linked to detrimental effects on the environment, such as excessive carbon emissions, massive water use, and the production of non-biodegradable waste. Bacterial cellulose (BC), a sustainable substance that can address these environmental issues at every stage of the textile lifecycle, has gained popularity in recent years. When compared to traditional fibers like cotton, polyester, and nylon, BC exhibits significant environmental benefits based on an examination of its production methods, usage trends, and end-of-life behaviors.

Table 1 Comparison with bacterial cellulose, plant-based cellulose, and synthetic fiber, highlighting advantages in sustainability and performance

Property	Bacterial cellulose	Plant-based cellulose	Synthetic fibers (e.g., PET, nylon)	References
Purity	~ 99% cellulose	40–50% (with lignin, hemicellulose)	N/A—chemically synthesized	Iguchi et al. [6], Jonas and Farah [31], Nair et al. [34]
Fiber diameter	20–100 nm	5–50 μm	10–50 μm	Klemm et al. [35], Yamanaka et al. [36], Zhang et al. [37]
Tensile strength	200–300 MPa	50–170 MPa	20–100 MPa	Iguchi et al. [6], Klemm et al. [35], Yang et al. [38]
Young's modulus	15–30 GPa	5–13 GPa	1–10 GPa	Klemm et al. [35], Lee et al. [39]
Water holding capacity	100–200×	60–100×	< 5×	Begum et al. [40], Czaja et al. [18], Yamamoto et al. [28]
Biodegradability	Complete (weeks–months)	Partial/Slow	Poor (> 50 years)	Jonas and Farah [31], Shi et al. [41]
Environmental impact	Low (fermentation-based)	Medium (pulping, bleaching)	High (toxic chemical use)	Geyer et al. [42], Klemm et al. [35], Sozcu et al. [27]

3.1 *Reduced Carbon Emissions in Manufacturing*

One of the most important environmental advantages of bacterial cellulose is its significantly lower carbon footprint during manufacture. Conventional synthetic fibers, such as polyester, are produced by petrochemical methods that are both energy-intensive and carbon-emitting [42]. Cotton farming, while natural, is heavily mechanized, uses fertilizer, and requires irrigation, all of which contribute significantly to greenhouse gas emissions.

Microbial fermentation of bacterial cellulose (BC) occurs at ambient pressure and moderate temperatures, substantially reducing the energy consumption typically associated with conventional fiber production processes [43]. Furthermore, BC biosynthesis can utilize renewable and low-cost carbon sources such as molasses, fruit peels, and whey, thereby decreasing its overall carbon footprint [44]. Life Cycle Assessment (LCA) analyses have demonstrated that BC produced from waste substrates exhibits a significantly lower global warming potential compared to both synthetic and natural fibers [21, 22].

Bacterial cellulose (BC) production systems possess the advantage of being decentralized and operable at smaller scales, which can significantly reduce transportation-related emissions and foster localized circular economic models [32]. This inherent flexibility positions bacterial cellulose as a resilient and sustainable biomaterial, well-aligned with global carbon neutrality goals as outlined in international frameworks such as the Paris Climate Accord.

3.2 Lower Water Usage than Traditional Textiles

The worldwide textile industry is well known for its large water impact. For example, producing 1 kg of cotton fiber requires around 10,000 L of water [45]. Water is consumed not only for irrigation, but also for scouring, bleaching, dyeing, and finishing. While polyester manufacture requires less irrigation, it does use a lot of water during the polymerization and textile processing stages [46].

In contrast, bacterial cellulose fermentation takes place in water-based environments, enabling controlled and significantly reduced water usage. Recent studies demonstrate that optimized bacterial cellulose production systems can recycle water from the fermentation media, reducing overall water consumption by as much as 70% compared to conventional cotton processing [47]. Additionally, the lack of harmful agricultural chemicals and pesticides in this process helps prevent contamination of water sources, which is a major issue in cotton cultivation [48, 49].

BC offers key advantages in sustainable manufacturing. Its natural whiteness and purity mean less intensive chemical processing is needed, which cuts down on both water consumption and wastewater production [50]. Additionally, BC's straightforward harvesting and drying processes highlight its promise as a material that conserves water, promoting more environmentally friendly production methods.

3.3 Optimized Biodegradability at End-of-Life to Limit Environmental Pollution

At the conclusion of their service life, conventional textile materials represent a substantial source of environmental contamination. Synthetic polymeric fibers, including polyester and nylon, exhibit persistence in the environment, with degradation times ranging from approximately 50 to 200 years. This prolonged durability contributes to the accumulation of microplastic pollutants in both terrestrial and marine ecosystems [42, 51]. Furthermore, natural fibers such as cotton, when subjected to treatments involving synthetic dyes and chemical finishing agents, demonstrate reduced biodegradability and may release ecotoxic compounds during their degradation process.

BC, characterized by its ultra-pure nanofibrillar architecture and the absence of lignin and hemicellulose, exhibits high biodegradability under both aerobic and anaerobic conditions [52]. Empirical composting studies have demonstrated that BC-based materials undergo complete degradation within a matter of weeks, without generating toxic byproducts [53, 54]. In contrast to conventional textile fibers, which frequently necessitate controlled industrial composting environments, BC-derived textiles are capable of biodegrading effectively under typical home composting conditions, thereby enhancing their environmental sustainability and applicability [55].

Furthermore, advancements in BC functionalization, including the incorporation of natural dyes and environmentally benign additives, maintain the intrinsic biodegradability of BC while broadening its potential applications in both fashion and technical textile sectors [56]. This improved degradability mitigates environmental accumulation in landfills and facilitates the release of organic constituents that enhance soil quality (Table 2).

Table 2 Comparable environmental impacts of the natural and synthetic textiles

Material	Carbon footprint (kg CO_2/kg)	Water usage (L/kg)	Degradation time (years)	References
Cotton	0.9	2.56×10^{32}	1–5 months	Chapagain et al. [45], Li et al. [57] Muthu [58], Provin et al. [59], Yu and Yang [60]
Polyester	1.19	100–200	Slight degradation after 6 months	Li et al. [57]; Tekin et al. [61], Zhu et al. [62]
Bacterial cellulose	34–296 kg	36,100	30 days	Bueno et al. [63], Forte et al. [64], Provin et al. [59], Taokaew [65]

4 Applications of Bacterial Cellulose in the Textile Industry

4.1 Limitations of Commercialization of Bacterial Cellulose

We indicated in previous sections that BC is generally renowned for its exceptional qualities, ranging from high purity to biodegradability. Despite these benefits, the complete commercialization of BC-based products confronts major barriers, particularly in the textile and healthcare industries. These largely concern economic feasibility, scalability, and the necessity for advances in downstream processing technologies such as dyeing and finishing.

One of the primary challenges hindering the commercial-scale production of bacterial cellulose (BC) is its elevated production cost. This is attributed to several factors, including the requirement for costly culture media, limited growth rates of cellulose-producing microbial strains, and relatively low cellulose yields compared to plant-based sources. Although various agricultural and food industry by-products have been explored as alternative substrates to reduce production expenses, their heterogeneous composition often results in inconsistencies in both yield and product quality [66]. Moreover, the transition from laboratory-scale to industrial-scale bacterial cellulose (BC) production presents substantial engineering challenges. Static cultivation techniques, commonly employed at the laboratory scale, are impractical for large-scale applications due to spatial constraints and non-uniform formation of cellulose pellicles. Although stirred or agitated fermentation methods offer higher productivity, they often compromise the structural integrity of the cellulose and necessitate stringent process control to mitigate risks of contamination and excessive foam generation [67, 68]. The lack of established production infrastructure, coupled with the requirement for tailored bioreactor configurations, presents a significant barrier to the commercialization of bacterial cellulose (BC) production systems. Furthermore, downstream processing steps, including harvesting, purification, and drying—remain intensive in terms of both energy and labor, thereby substantially elevating the total production cost.

To establish BC as a viable alternative for sustainable textile applications, advancements in environmentally benign dyeing and finishing technologies are imperative. Traditional dyeing methods predominantly utilize synthetic dyes and require substantial energy and water inputs, thereby diminishing the ecological advantages inherent to BC. Additionally, BC's intrinsic properties, including its high-water absorption capacity and dense nanofibrillar architecture, present significant challenges in achieving uniform and durable dye fixation [69]. Environmentally sustainable dyeing techniques—such as the use of natural dyes, supercritical CO_2 dyeing, and plasma treatments—are under investigation to align with circular economy principles and reduce the environmental impact of bacterial cellulose (BC)-based textiles. However, these technologies remain in the early stages of development. Additionally, chemical modifications or surface functionalization may be required to enhance dye uptake, potentially increasing process complexity and raising concerns related to the biodegradability and toxicity of the resulting materials [70]. It is quite likely that the use of BC in fashion, medical textiles, or functional wear will continue to be restricted to niche markets or experimental design applications if these technological improvements are not made.

4.2 *Usage of Bacterial Cellulose in the Development of Sustainable Textile Materials*

Bacterial cellulose (BC), a very pure biopolymer produced through microbial fermentation, has gained popularity in the textile and fashion industries as a sustainable alternative to traditional fibers. BC's great mechanical strength, high crystalline, biocompatibility, and excellent water retention capacity make it a promising material for a wide range of applications in textile engineering and functional fabric creation.

The fashion industry significantly contributes to global pollution, primarily due to the widespread use of synthetic fibers such as polyester, nylon, and acrylic. Bacterial cellulose (BC) presents a renewable, biodegradable, and non-toxic alternative, which can be engineered into nanofibrous structures through wet-processing techniques or surface modifications. Numerous studies have demonstrated the potential for integrating BC into textiles for apparel. For instance, BC has been blended with natural fibers, such as cotton, to improve mechanical properties and moisture management [71]. Furthermore, BC-based composites with alginate or gelatin have demonstrated excellent elasticity and comfort appropriate for garment applications [72, 73]. One of the most commercially promising applications for bacterial cellulose (BC) is as a sustainable alternative to traditional animal- or petroleum-derived leather. BC's film-forming capabilities, smooth surface texture, and mechanical robustness allow it to replicate both the aesthetic and feel features of leather. Advances in tanning-free processing procedures and surface treatments, such as wax coatings or dyeing with natural pigments, have made it possible to create vegan, biodegradable leather materials. Modern Meadow has been a pioneer in commercial-scale manufacture of BC-based leather replacements [39, 74, 75]. Furthermore, research has revealed the potential of bacterial cellulose (BC) in applications like as footwear and purses, owing to its beneficial features, including increased breathability and innate antimicrobial resistance [76–78]. Bacterial cellulose (BC) extends beyond conventional fabric applications, offering significant potential in high-performance biocomposites, particularly in fields requiring specialized mechanical or acoustic properties. When integrated with nanoparticles or polymers, BC-based composites can be engineered for use in applications such as acoustic insulation, wearable sensors, and medical textiles, including wound dressings and surgical meshes. For instance, BC composites incorporating silver nanoparticles demonstrate enhanced antibacterial properties, making them suitable for use in hospital gowns and bandages [79–81]. BC-lignin composites have also been investigated for their potential acoustic damping properties, suggesting their applicability in soundproof textile panel fabrication [82, 83].

These versatile biocomposites present opportunities for smart textiles and biomedical engineering.

5 Challenges and Future Perspectives

Bacterial cellulose (BC) is a biopolymer primarily synthesized by strains of *Komagataeibacter xylinus*. Owing to its unique nanofibrillar architecture, exceptional purity, notable mechanical strength, and intrinsic biodegradability, BC is emerging as a promising candidate in the development of sustainable, bio-based alternatives to conventional petroleum-derived fibers. However, despite its advantageous physicochemical and environmental properties, the large-scale integration of BC into the textile industry remains limited by significant technological, economic, and social constraints as mentioned in the section 'Limitations of Commercialization of Bacterial Cellulose'. As well as its large-scale production limitation, cost-effectiveness and consumer acceptance are still influencing the future aspect of BC. The economic feasibility of BC-based textiles is currently hampered by high production costs when compared to synthetic fibers like polyester or natural cotton. These expenses are determined by pricey culture media, extensive cultivation timeframes, and labor-intensive processing stages [32]. Although alternative feedstocks such as molasses, fruit peels, and industrial effluents have demonstrated potential, challenges related to standardization and regulatory approval continue to hinder their industrial implementation [67].

Consumer acceptance represents a critical factor influencing the broader adoption of bacterial cellulose (BC) as a textile material. Prevailing skepticism regarding its performance, aesthetic appeal, and durability can be attributed to limited public awareness and its current confinement to niche market segments. Additionally, perceived challenges related to cost premiums and functional performance may further impede widespread integration, particularly within fast fashion sectors where affordability and design versatility are prioritized [84–86].

Current research focuses on enhancing both the intrinsic and extrinsic properties of BC to improve its suitability for textile applications. One approach involves the genetic engineering of BC-producing bacteria to increase production yield, modulate crystallinity, and alter polymer chain length [26]. Similarly, the incorporation of BC with biopolymers (e.g., chitosan, alginate) or synthetic polymers (e.g., polyvinyl alcohol) has been shown to enhance its mechanical properties, water resistance, and dye affinity, thereby improving its suitability for both functional and fashion-oriented textile applications [87, 88]. An additional promising approach involves the integration of BC with conductive materials or nanoparticles to engineer smart textiles, wherein the intrinsic breathability and biocompatibility of BC render it an ideal substrate for bio-integrated sensors and stimuli-responsive fabrics [89–91].

6 Conclusion

Bacterial cellulose (BC) has emerged as a highly promising sustainable biomaterial for the textile and fashion industries, offering an eco-friendly alternative to synthetic and animal-derived fibers. Its exceptional mechanical strength, biocompatibility, water retention, and film-forming abilities enable diverse applications—from apparel and leather alternatives to smart textiles and biomedical fabrics. Advances in BC composites with natural polymers and nanoparticles have expanded its functionality, making it suitable for moisture-regulating garments, antimicrobial medical textiles, and sound-absorbing panels. As a renewable and biodegradable material, BC holds significant potential to drive innovation in sustainable and functional textile engineering.

However, its commercial integration remains constrained by high production costs, scalability issues, and limited consumer awareness. Current efforts aim to overcome these barriers through genetic engineering, incorporation with complementary polymers, and functionalization with conductive materials, enhancing BC's performance and expanding its application potential in both traditional and smart textiles.

References

1. Leal Filho, W., Perry, P., Heim, H., Dinis, M.A.P., Moda, H., Ebhuoma, E., Paço, A.: An overview of the contribution of the textiles sector to climate change. Front. Environ. Sci. **10** (2022). https://doi.org/10.3389/fenvs.2022.973102
2. The impact of textile production and waste on the environment (infographics). Topics | European Parliament (2020, Dec 29). https://www.europarl.europa.eu/topics/en/article/20201208STO93327/the-impact-of-textile-production-and-waste-on-the-environment-infographics
3. The environmental crisis caused by textile waste (n.d.). Retrieved 1 Apr 2025, from https://www.roadrunnerwm.com/blog/textile-waste-environmental-crisis
4. Aldalbahi, A., El-Naggar, M.E., El-Newehy, M.H., Rahaman, M., Hatshan, M.R., Khattab, T.A.: Effects of technical textiles and synthetic nanofibers on environmental pollution. Polymers **13**(1), 155 (2021). https://doi.org/10.3390/polym13010155
5. The textile environmental impact of the sector—hedgehog (n.d.). Retrieved 1 Apr 2025, from https://www.hhc.earth/knowledge-base/articles/the-environmental-impact-of-the-textile-sector
6. Iguchi, M., Yamanaka, S., Budhiono, A.: Bacterial cellulose—a masterpiece of nature's arts. J. Mater. Sci. **35**(2), 261–270 (2000). https://doi.org/10.1023/A:1004775229149
7. Klemm, D., Schumann, D., Udhardt, U., Marsch, S.: Bacterial synthesized cellulose—artificial blood vessels for microsurgery. Prog. Polym. Sci. **26**(9), 1561–1603 (2001). https://doi.org/10.1016/S0079-6700(01)00021-1
8. Potivara, K., Phisalaphong, M.: Development and characterization of bacterial cellulose reinforced with natural rubber. Materials **12**(14), 2323 (2019). https://doi.org/10.3390/ma12142323
9. Subtaweesin, C., Woraharn, W., Taokaew, S., Chiaoprakobkij, N., Sereemaspun, A., Phisalaphong, M.: Characteristics of curcumin-loaded bacterial cellulose films and anticancer properties against malignant melanoma skin cancer cells. Appl. Sci. **8**(7), Article 7 (2018). https://doi.org/10.3390/app8071188

10. Shoda, M., Sugano, Y.: Recent advances in bacterial cellulose production. Biotechnol. Bioprocess. Eng. **10**(1), 1–8 (2005). https://doi.org/10.1007/BF02931175
11. Amr, A., Ibrahim, H., Amr, A., Ibrahim, H.: Bacterial cellulose: biosynthesis and applications. In: Next-Generation Textiles. IntechOpen (2022). https://doi.org/10.5772/intechopen.107021
12. Gorgieva, S., Trček, J.: Bacterial cellulose: production, modification and perspectives in biomedical applications. Nanomaterials **9**(10), 1352 (2019). https://doi.org/10.3390/nano9101352
13. Kongruang, S.: Bacterial cellulose production by *Acetobacter xylinum* strains from agricultural waste products. In: Adney, W.S., McMillan, J.D., Mielenz, J., Klasson, K.T. (eds.) Biotechnology for Fuels and Chemicals, pp. 763–774. Humana Press. https://doi.org/10.1007/978-1-60327-526-2_70
14. Gregory, D.A., Tripathi, L., Fricker, A.T.R., Asare, E., Orlando, I., Raghavendran, V., Roy, I.: Bacterial cellulose: a smart biomaterial with diverse applications. Mater. Sci. Eng. R: Rep. **145**, 100623 (2021). https://doi.org/10.1016/j.mser.2021.100623
15. Thomas, K., Durrani, H., Brady, J., Ludwig, K., Yatvitskiy, M., Clarke-Sather, A.R., Cao, H., Cobb, K.: Fundamental challenges and opportunities for textile circularity. Sustainability **16**(24), Article 24 (2024). https://doi.org/10.3390/su162411117
16. Shen, L., Worrell, E., Patel, M.K.: Environmental impact assessment of man-made cellulose fibres. Resour. Conserv. Recycl. **55**(2), 260–274 (2010). https://doi.org/10.1016/j.resconrec.2010.10.001
17. Keshk, S.M.: Bacterial cellulose production and its industrial applications. J. Bioprocess. Biotech. **04**(02) (2014). https://doi.org/10.4172/2155-9821.1000150
18. Czaja, W., Krystynowicz, A., Bielecki, S., Brown, R.M.: Microbial cellulose—the natural power to heal wounds. Biomaterials **27**(2), 145–151 (2006). https://doi.org/10.1016/j.biomaterials.2005.07.035
19. Kim, D.-Y., Nishiyama, Y., Kuga, S.: Surface acetylation of bacterial cellulose. Cellulose **9**(3–4), 361–367 (2002). https://doi.org/10.1023/A:1021140726936
20. Manoukian, O.S., Sardashti, N., Stedman, T., Gailiunas, K., Ojha, A., Penalosa, A., Mancuso, C., Hobert, M., Kumbar, S.G.: Biomaterials for tissue engineering and regenerative medicine. In: Narayan, R. (ed.) Encyclopedia of Biomedical Engineering, pp. 462–482. Elsevier (2019). https://doi.org/10.1016/B978-0-12-801238-3.64098-9
21. Urbina, L., Corcuera, M.Á., Gabilondo, N., Eceiza, A., Retegi, A.: A review of bacterial cellulose: sustainable production from agricultural waste and applications in various fields. Cellulose **28**(13), 8229–8253 (2021). https://doi.org/10.1007/s10570-021-04020-4
22. Fernandes, I. de A.A., Pedro, A.C., Ribeiro, V.R., Bortolini, D.G., Ozaki, M.S.C., Maciel, G.M., Haminiuk, C.W.I.: Bacterial cellulose: from production optimization to new applications. Int. J. Biol. Macromol. **164**, 2598–2611 (2020). https://doi.org/10.1016/j.ijbiomac.2020.07.255
23. Sozcu, S., Frajova, J., Wiener, J., Venkataraman, M., Tomková, B., Militký, J.: Effect of drying methods on the thermal and mechanical behavior of bacterial cellulose aerogel. Gels **10**(7), Article 7 (2024). https://doi.org/10.3390/gels10070474
24. Muthu, S.S., Rathinamoorthy, R.: Bacterial Cellulose: Sustainable Material for Textiles. Springer (2021). https://doi.org/10.1007/978-981-15-9581-3
25. Ross, P., Mayer, R., Benziman, M.: Cellulose biosynthesis and function in bacteria. Microbiol. Rev. **55**(1), 35–58 (1991). https://doi.org/10.1128/mr.55.1.35-58.1991
26. Lin, S.-P., Loira Calvar, I., Catchmark, J.M., Liu, J.-R., Demirci, A., Cheng, K.-C.: Biosynthesis, production and applications of bacterial cellulose. Cellulose **20**(5), 2191–2219 (2013). https://doi.org/10.1007/s10570-013-9994-3
27. Sozcu, S., Frajova, J., Wiener, J., Venkataraman, M., Tomková, B., Militký, J.: Synthesis of *Acetobacter xylinum* bacterial cellulose aerogels and their effect on the selected properties. Gels **11**(4), Article 4 (2025). https://doi.org/10.3390/gels11040272
28. Yamamoto, H., Horii, F., Hirai, A.: Structural studies of bacterial cellulose through the solid-phase nitration and acetylation by CP/MAS 13C NMR spectroscopy. Cellulose **13**(3), 327–342 (2006). https://doi.org/10.1007/s10570-005-9034-z

29. Rebelo, A.R., Archer, A.J., Chen, X., Liu, C., Yang, G., Liu, Y.: Dehydration of bacterial cellulose and the water content effects on its viscoelastic and electrochemical properties. Sci. Technol. Adv. Mater. **19**(1), 203–211 (2018). https://doi.org/10.1080/14686996.2018.1430981
30. Helenius, G., Bäckdahl, H., Bodin, A., Nannmark, U., Gatenholm, P., Risberg, B.: In vivo biocompatibility of bacterial cellulose. J. Biomed. Mater. Res., Part A **76**(2), 431–438 (2006). https://doi.org/10.1002/jbm.a.30570
31. Jonas, R., Farah, L.F.: Production and application of microbial cellulose. Polym. Degrad. Stab. **59**(1), 101–106 (1998). https://doi.org/10.1016/S0141-3910(97)00197-3
32. Portela, R., Leal, C.R., Almeida, P.L., Sobral, R.G.: Bacterial cellulose: a versatile biopolymer for wound dressing applications. Microb. Biotechnol. **12**(4), 586–610 (2019). https://doi.org/10.1111/1751-7915.13392
33. Bielecki, S., Krystynowicz, A., Turkiewicz, M., Kalinowska, H.: Bacterial cellulose. In: Biopolymers Online. Wiley (2005). https://doi.org/10.1002/3527600035.bpol5003
34. Nair, G.R., Singh, A., Zimniewska, M., Raghavan, V.: Comparative evaluation of physical and structural properties of water retted and non-retted flax fibers. Fibers **1**(3), Article 3 (2013). https://doi.org/10.3390/fib1030059
35. Klemm, D., Heublein, B., Fink, H.-P., Bohn, A.: Cellulose: fascinating biopolymer and sustainable raw material. Angew. Chem. Int. Ed. **44**(22), 3358–3393 (2005). https://doi.org/10.1002/anie.200460587
36. Yamanaka, S., Watanabe, K., Kitamura, N., Iguchi, M., Mitsuhashi, S., Nishi, Y., Uryu, M.: The structure and mechanical properties of sheets prepared from bacterial cellulose. J. Mater. Sci. **24**(9), 3141–3145 (1989). https://doi.org/10.1007/BF01139032
37. Zhang, X., Yu, Y., Jiang, Z., Wang, H.: The effect of freezing speed and hydrogel concentration on the microstructure and compressive performance of bamboo-based cellulose aerogel. J. Wood Sci. **61**(6), Article 6 (2015). https://doi.org/10.1007/s10086-015-1514-7
38. Yang, H.-S., Wolcott, M.P., Kim, H.-S., Kim, H.-J.: Thermal properties of lignocellulosic filler-thermoplastic polymer bio-composites. J. Therm. Anal. Calorim. **82**(1), 157–160 (2005). https://doi.org/10.1007/s10973-005-0857-5
39. Lee, K.-Y., Buldum, G., Mantalaris, A., Bismarck, A.: More than meets the eye in bacterial cellulose: biosynthesis, bioprocessing, and applications in advanced fiber composites. Macromol. Biosci. **14**(1), 10–32 (2014). https://doi.org/10.1002/mabi.201300298
40. Begum, H.A., Tanni, T.R., Shahid, M.A.: Analysis of water absorption of different natural fibers. J. Text. Sci. Technol. **7**(4), Article 4 (2021). https://doi.org/10.4236/jtst.2021.74013
41. Shi, Z., Zhang, Y., Phillips, G.O., Yang, G.: Utilization of bacterial cellulose in food. Food Hydrocolloids **35**, 539–545 (2014). https://doi.org/10.1016/j.foodhyd.2013.07.012
42. Geyer, R., Jambeck, J.R., Law, K.L.: Production, use, and fate of all plastics ever made. Sci. Adv. **3**(7), e1700782 (2017). https://doi.org/10.1126/sciadv.1700782
43. Wang, J., Tavakoli, J., Tang, Y.: Bacterial cellulose production, properties and applications with different culture methods—a review. Carbohyd. Polym. **219**, 63–76 (2019). https://doi.org/10.1016/j.carbpol.2019.05.008
44. Hussain, Z., Sajjad, W., Khan, T., Wahid, F.: Production of bacterial cellulose from industrial wastes: a review. Cellulose **26**(5), 2895–2911 (2019). https://doi.org/10.1007/s10570-019-02307-1
45. Chapagain, A.K., Hoekstra, A.Y., Savenije, H.H.G., Gautam, R.: The water footprint of cotton consumption: an assessment of the impact of worldwide consumption of cotton products on the water resources in the cotton producing countries. Ecol. Econ. **60**(1), 186–203 (2006). https://doi.org/10.1016/j.ecolecon.2005.11.027
46. Shirvanimoghaddam, K., Motamed, B., Ramakrishna, S., Naebe, M.: Death by waste: fashion and textile circular economy case. Sci. Total Environ. **718**, 137317 (2020). https://doi.org/10.1016/j.scitotenv.2020.137317
47. Cruz, M.A., Flor-Unda, O., Avila, A., Garcia, M.D., Cerda-Mejía, L.: Advances in bacterial cellulose production: a scoping review. Coatings **14**(11), Article 11 (2024). https://doi.org/10.3390/coatings14111401

48. Carolin, F.C., Kamalesh, T., Kumar, P.S., Hemavathy, R.V., Rangasamy, G.: A critical review on sustainable cellulose materials and its multifaceted applications. Ind. Crops Prod. **203**, 117221 (2023). https://doi.org/10.1016/j.indcrop.2023.117221
49. Rathinamoorthy, R., Kiruba, T.: Bacterial cellulose—a potential material for sustainable eco-friendly fashion products. J. Natu. Fibers **19**(9), 3275–3287 (2022). https://doi.org/10.1080/15440478.2020.1842841
50. Molina-Ramírez, C., Castro, M., Osorio, M., Torres-Taborda, M., Gómez, B., Zuluaga, R., Gómez, C., Gañán, P., Rojas, O.J., Castro, C.: Effect of different carbon sources on bacterial nanocellulose production and structure using the low pH resistant strain *Komagataeibacter medellinensis*. Materials **10**(6), Article 6 (2017). https://doi.org/10.3390/ma10060639
51. Lebreton, L., Andrady, A.: Future scenarios of global plastic waste generation and disposal. Palgrave Commun. **5**(1), 1–11 (2019). https://doi.org/10.1057/s41599-018-0212-7
52. Zhao, X., Zhang, L., Liu, D.: Biomass recalcitrance. Part I: The chemical compositions and physical structures affecting the enzymatic hydrolysis of lignocellulose. Biofuels Bioprod. Biorefin. **6**(4), 465–482 (2012). https://doi.org/10.1002/bbb.1331
53. Betlej, I., Zakaria, S., Krajewski, K., Boruszewski, P.: Bacterial cellulose—properties and its potential application. Sains Malaysiana **50**, 493–505 (2021). https://doi.org/10.17576/jsm-2021-5002-20
54. Chu, Y., Sun, Y., Wu, W., Xiao, H.: Dispersion properties of nanocellulose: a review. Carbohyd. Polym. **250**, 116892 (2020). https://doi.org/10.1016/j.carbpol.2020.116892
55. Marestoni, L.D., Barud, H. da S., Gomes, R.J., Catarino, R.P.F., Hata, N.N.Y., Ressutte, J.B., Spinosa, W.A.: Commercial and potential applications of bacterial cellulose in Brazil: ten years review. Polímeros **30**, e2020047 (2021). https://doi.org/10.1590/0104-1428.09420
56. Absharina, D., Padri, M., Veres, C., Vágvölgyi, C.: Bacterial cellulose: from biofabrication to applications in sustainable fashion and vegan leather. Fermentation **11**(1), Article 1 (2025). https://doi.org/10.3390/fermentation11010023
57. Li, L., Frey, M., Browning, K.J.: Biodegradability study on cotton and polyester fabrics. J. Eng. Fibers Fabr. **5**(4), 155892501000500400 (2010). https://doi.org/10.1177/155892501000500406
58. Muthu, S.S.: Water in Textiles and Fashion: Consumption, Footprint, and Life Cycle Assessment. Woodhead Publishing (2018)
59. Provin, A.P., Cubas, A.L.V., Dutra, A.R. de A., Schulte, N.K.: Textile industry and environment: can the use of bacterial cellulose in the manufacture of biotextiles contribute to the sector? Clean Technol. Environ. Policy **23**(10), 2813–2825 (2021). https://doi.org/10.1007/s10098-021-02191-z
60. Yu, Z., Yang, Y.: Carbon footprint of global cotton production. Resour. Environ. Sustain. **20**, 100214 (2025). https://doi.org/10.1016/j.resenv.2025.100214
61. Tekin, P., Alıcı, H., Demirdelen, T.: A life cycle analysis of a polyester–wool blended fabric and associated carbon emissions in the textile industry. Energies **17**(2), Article 2 (2024). https://doi.org/10.3390/en17020312
62. Zhu, L., Chen, B., Liu, J., Chen, S., Zhang, Y., Wang, X., Wang, L.: Assessing baseline water footprints of natural fiber textile products in China. J. Clean. Prod. **379**, 134747 (2022). https://doi.org/10.1016/j.jclepro.2022.134747
63. Bueno, F., Fultz, L., Husseneder, C., Keenan, M., Sathivel, S.: Biodegradability of bacterial cellulose polymer below the soil and its effects on soil bacteria diversity. Polym. Degrad. Stab. **217**, 110535 (2023). https://doi.org/10.1016/j.polymdegradstab.2023.110535
64. Forte, A., Dourado, F., Mota, A., Neto, B., Gama, M., Ferreira, E.C.: Life cycle assessment of bacterial cellulose production. Int. J. Life Cycle Assess. **26**(5), 864–878 (2021). https://doi.org/10.1007/s11367-021-01904-2
65. Taokaew, S.: Bacterial nanocellulose produced by cost-effective and sustainable methods and its applications: a review. Fermentation **10**(6), Article 6 (2024). https://doi.org/10.3390/fermentation10060316
66. Jozala, A.F., Pértile, R.A.N., dos Santos, C.A., de Carvalho Santos-Ebinuma, V., Seckler, M.M., Gama, F.M., Pessoa, A.: Bacterial cellulose production by *Gluconacetobacter xylinus*

by employing alternative culture media. Appl. Microbiol. Biotechnol. **99**(3), 1181–1190 (2015). https://doi.org/10.1007/s00253-014-6232-3
67. Çakar, F., Özer, I., Aytekin, A.Ö., Şahin, F.: Improvement production of bacterial cellulose by semi-continuous process in molasses medium. Carbohyd. Polym. **106**, 7–13 (2014). https://doi.org/10.1016/j.carbpol.2014.01.103
68. Ul-Islam, M., Khan, T., Park, J.K.: Water holding and release properties of bacterial cellulose obtained by in situ and ex situ modification. Carbohyd. Polym. **88**(2), 596–603 (2012). https://doi.org/10.1016/j.carbpol.2012.01.006
69. Costa, A.F. de S., de Amorim, J.D.P., Almeida, F.C.G., de Lima, I.D., de Paiva, S.C., Rocha, M.A.V., Vinhas, G.M., Sarubbo, L.A.: Dyeing of bacterial cellulose films using plant-based natural dyes. Int. J. Biol. Macromol. **121**, 580–587 (2019). https://doi.org/10.1016/j.ijbiomac.2018.10.066
70. Morais, D.S., Guedes, R.M., Lopes, M.A.: Antimicrobial approaches for textiles: from research to market. Materials **9**(6), Article 6 (2016). https://doi.org/10.3390/ma9060498
71. Satchanska, G., Davidova, S., Petrov, P.D.: Natural and synthetic polymers for biomedical and environmental applications. Polymers **16**(8), Article 8 (2024). https://doi.org/10.3390/polym16081159
72. Aarstad, O., Heggset, E.B., Pedersen, I.S., Bjørnøy, S.H., Syverud, K., Strand, B.L.: Mechanical properties of composite hydrogels of alginate and cellulose nanofibrils. Polymers **9**(8), 378 (2017). https://doi.org/10.3390/polym9080378
73. Mbituyimana, B., Liu, L., Ye, W., Ode Boni, B.O., Zhang, K., Chen, J., Thomas, S., Vasilievich, R.V., Shi, Z., Yang, G.: Bacterial cellulose-based composites for biomedical and cosmetic applications: research progress and existing products. Carbohyd. Polym. **273**, 118565 (2021). https://doi.org/10.1016/j.carbpol.2021.118565
74. Rathinamoorthy, R., Kiruba, T.: Bacterial cellulose—a sustainable alternative material for footwear and leather products. In: Muthu, S.S. (ed.) Leather and Footwear Sustainability: Manufacturing, Supply Chain, and Product Level Issues, pp. 91–121. Springer (2020). https://doi.org/10.1007/978-981-15-6296-9_5
75. Payne, A.: Will we soon be growing our own vegan leather at home? Daily Bulletin (2016, Nov 23). https://www.dailybulletin.com.au/the-conversation/24701-will-we-soon-be-growing-our-own-vegan-leather-at-home
76. García, C., Prieto, M.A.: Bacterial cellulose as a potential bioleather substitute for the footwear industry. Microb. Biotechnol. **12**(4), 582–585 (2019). https://doi.org/10.1111/1751-7915.13306
77. Joseph, A., Umamaheswari, S., Vassou, M.C.: Bacterial cellulose: a versatile biomaterial for biomedical application. Carbohyd. Res. **552**, 109350 (2025). https://doi.org/10.1016/j.carres.2024.109350
78. Rotor, A.V.: Bacterial cellulose (Nata de coco) made into shoes—first of its kind in the world. Living with Nature—School on Blog by Dr. Abercio V. Rotor (2023, Aug 14). https://avrotor.blogspot.com/2023/08/bacterial-cellulose-nata-de-coco-made_14.html
79. Kim, J., Cai, Z., Lee, H.S., Choi, G.S., Lee, D.H., Jo, C.: Preparation and characterization of a bacterial cellulose/chitosan composite for potential biomedical application. J. Polym. Res. **18**(4), 739–744 (2011). https://doi.org/10.1007/s10965-010-9470-9
80. Shah, N., Ul-Islam, M., Khattak, W.A., Park, J.K.: Overview of bacterial cellulose composites: a multipurpose advanced material. Carbohyd. Polym. **98**(2), 1585–1598 (2013). https://doi.org/10.1016/j.carbpol.2013.08.018
81. Sulaeva, I., Hettegger, H., Bergen, A., Rohrer, C., Kostic, M., Konnerth, J., Rosenau, T., Potthast, A.: Fabrication of bacterial cellulose-based wound dressings with improved performance by impregnation with alginate. Mater. Sci. Eng., C **110**, 110619 (2020). https://doi.org/10.1016/j.msec.2019.110619
82. Mehrotra, N., Bhuvana, T., Tiwari, A., Chandraprakash, C.: Aerogel-like biodegradable acoustic foams of bacterial cellulose. J. Appl. Polym. Sci. **141**(35), e55881 (2024). https://doi.org/10.1002/app.55881

83. Muchlisinalahuddin, Dahlan, H., Mahardika, M., Rusli, M.: Cellulose-based material for sound absorption and its application—a short review. BIO Web Conf. **77**, 01003 (2023). https://doi.org/10.1051/bioconf/20237701003
84. da Silva, C.J.G., de Medeiros, A.D.M., de Amorim, J.D.P., do Nascimento, H.A., Converti, A., Costa, A.F.S., Sarubbo, L.A.: Bacterial cellulose biotextiles for the future of sustainable fashion: a review. Environ. Chem. Lett. **19**(4), 2967–2980 (2021). https://doi.org/10.1007/s10311-021-01214-x
85. Harmon, J., Fairbourn, L., Thibault, N.: Exploring the potential of bacterial cellulose for use in apparel. J. Text. Sci. Fashion Technol. **5**(2), 1–9 (2020). https://doi.org/10.33552/JTSFT.2020.05.000606
86. Galdino Jr., C.J., Medeiros, A.D., Amorim, J.D., Nascimento, H.A., Henrique, M.A., Costa, A.F., Sarubbo, L.A.: The future of sustainable fashion: bacterial cellulose biotextile naturally dyed. Chem. Eng. Trans. **86**, 1333–1338 (2021). https://doi.org/10.3303/CET2186223
87. Choudhary, P., Jaiswal, A., Singh, S., Gupta, S.K.: Bacterial cellulose based composites: preparation and characterization. In: Materials Science Forum, vol. 978, pp. 183–190. International Conference on Processing and Characterization of Materials (2020). https://doi.org/10.4028/www.scientific.net/MSF.978.183
88. Salari, M., Sowti Khiabani, M., Rezaei Mokarram, R., Ghanbarzadeh, B., Samadi Kafil, H.: Preparation and characterization of cellulose nanocrystals from bacterial cellulose produced in sugar beet molasses and cheese whey media. Int. J. Biol. Macromol. **122**, 280–288 (2019). https://doi.org/10.1016/j.ijbiomac.2018.10.136
89. Fernandes, M., Gama, M., Dourado, F., Souto, A.P.: Development of novel bacterial cellulose composites for the textile and shoe industry. Microb. Biotechnol. **12**(4), 650–661 (2019). https://doi.org/10.1111/1751-7915.13387
90. Huang, Y., Zhu, C., Yang, J., Nie, Y., Chen, C., Sun, D.: Recent advances in bacterial cellulose. Cellulose **21**(1), 1–30 (2014). https://doi.org/10.1007/s10570-013-0088-z
91. Müller, D., Mandelli, J.S., Marins, J.A., Soares, B.G., Porto, L.M., Rambo, C.R., Barra, G.M.O.: Electrically conducting nanocomposites: preparation and properties of polyaniline (PAni)-coated bacterial cellulose nanofibers (BC). Cellulose **19**(5), 1645–1654 (2012). https://doi.org/10.1007/s10570-012-9754-9

Sustainable Bioactive Textile Structures with Improved End-Use Properties

Azam Ali, Blanka Tomková, Mahvish Fatimah, Muhammad Zaman Khan, Jiří Militký, and Jakub Wiener

1 Introduction

Sustainable textiles intend to lessen the impact of environment during their whole life processes including the source of raw material, its production, usage as well as disposal. Usually, such textiles are produced from environment friendly fibers which include organic cotton, hemp, bamboo, recycled materials, biodegradable as well as bio-based polymers. Modern processing approaches such as waterless dyeing, enzymatic treatments, and closed-loop recycling, normally enhance the sustainability by minimizing some parameters which include energy, water, and chemical consumption [1]. Following the ancient times, textiles are commonly utilized as a clothing material for the purpose of covering human skin. But nowadays due to the advances in science, it can also be modified to address other skin-related concerns. Normally,

A. Ali · B. Tomková · M. Z. Khan · J. Militký (✉) · J. Wiener
Department of Material Engineering, Faculty of Textile Engineering, Technical University of Liberec, Liberec, Czech Republic
e-mail: jiri.militky@tul.cz

A. Ali
e-mail: azam.ali@tul.cz

B. Tomková
e-mail: blanka.tomkova@tul.cz

M. Z. Khan
e-mail: muhammad.zaman.khan@tul.cz

J. Wiener
e-mail: jakub.wiener@tul.cz

M. Fatimah
Department of Physics, Qassim University, Buraidah, Saudi Arabia
e-mail: m.aziz@qu.edu.sa

J. Militký and M. Venkataraman (eds.), *Fibrous Structures for Sustainable Future*, Engineering Materials, https://doi.org/10.1007/978-981-95-1382-6_10

skin-related problems such as the human skin dermatological and chemical complexities, the humidity impacts, microorganisms, pH, temperature, and wind can be easily overcome by using functional textile materials as a cover [2].

A common method named, textile finishing deals with introducing functional properties to the fabrics including flame resistance, water resistance, wrinkle resistance, UV resistance as well as antimicrobial resistance simply by incorporating different chemicals along with mechanical methods of finishing [2]. During the past years, different compounds such as N-Halamine siloxanes, polyhexamethylene biguanide, triclosan, and biopolymers, have been efficiently utilized for the purpose of manufacturing antimicrobial-based textiles [3]. Also recently, several nanoparticles like titanium dioxide, silver, and zinc oxide, have been effectively utilized as an incorporated material to provide different kinds of functional textiles such as flame-retardant textiles, UV-protective textiles, antimicrobial and antibacterial textiles, water and oil repellent textiles, anti-odor textiles, wrinkle-resistant textiles as well as antistatic textiles [4].

Water, air and soil contamination is actually a consequence of incompatible binding of different chemical and additives which are incorporated during the process of finishing. Textile industries use different methods of wastewater treatment including adsorption, ion exchange, membrane separation, coagulation, advanced oxidation process, and biological degradation which aims to minimize the harmful textile-based contaminants in the environment. Nevertheless, in spite of using all these methods, conventional techniques have verified to be quite incompetent, challenging, costly as well as unsustainable. The incompetence of the traditional techniques in the management of textile wastewater results in the increasing demand for innovative sustainable finishing concepts. However, the textile finishing has a considerable effect on the environment at every step in the process which includes the production and applications of different finishes incorporated into textiles along with its utilization and disposal [5]. Such finishing materials can lead to many environmental problems as they are easily released into the environment throughout the laundry process of functional textiles. Besides the harmful effects along with the use of functional textiles, its disposal also impacts water, soil, and surroundings resulting in a negative effect on the ecosystem. Textiles have been efficiently biofinished for the purpose of imposing antibacterial properties as well as good and desirable appearance. Recently, an innovative method has been presented for the production of medical functional textiles in which the enzymatic pre-activation of textile materials occurs though the sonochemical nano-coating along with different nanoparticles [6].

Different research works also explained the fabrication process of functional textiles bearing different properties such as antibacterial and UV-protection by using various enzymatic pre-treatments which include acid pectinase, acid cellulase, laccase, and their mixtures, these are usually worked by coating them with different types of nanoparticles involving ZnO and TiO_2 NPs [7]. Different studies also explained the various promising applications of biopolymers of functional finishing of textiles [8]. Recently, the interest in the utilization of biopolymers as functional finishing agents has increased due to their environmentally friendly nature which reduces the use of harmful chemicals in the fabrication of functional textiles. Several biopolymers including chitosan, cyclodextrins, lignin, sericin, casein, pectin, and whey protein have been utilized for imposing various functional properties which includes antimicrobial, anti-UV as well as anti-flame to the textiles [9].

Different studies also explained the various promising applications of biopolymers in functional finishing of textiles [8]. The most efficient antibacterial properties have been accomplished by the deposition of different antibacterial agents while utilizing the aforementioned techniques. But these techniques also pose some unwanted effects including discoloration or staining on the surface of treated fabrics affecting the aesthetic activities of the textiles. A few researches work includes the dyeing of the treated textile fabrics to overcome different problems such as discoloration and staining however, it can also reduce their antibacterial effect [10].

Therefore, the formation of efficient antimicrobial textiles having enhanced aesthetics is quite difficult. Following all the major concerns, a sustainable method having better textile structures has presented an innovative technique to form antibacterial fabric along with enhanced aesthetic appeal. First of all, it was treated along with some natural anti-pathogenic agents that work according to the different applications of natural antibacterial dyes. Furthermore, the treatment of synthetic dye with bioactive groups is another appropriate method [11].

2 Sustainability in Textile

The latest type of textile is known a s a functional textile. It can best explain the textile materials like unique functionalities and maintain multiple factors like temperature and humidity. It also controls the release of certain chemicals, and even controls the release of specific chemicals stitched into the textiles based on some external stimuli for the textiles. It is noted a significant increase of 30% CAGR, or compound annual growth rate, in the functional textiles sector from 2015 to 2020 [12]. Functional textiles can be obtained through the utilization of different finishing techniques and the whole process is known as textile finishing [4]. Many serious issues arise from the binding of chemicals along with several other additives are used in the finishing process including the rise of pollutants in air, soil as well as water. In many textile industries, several wastewater treatment techniques are utilized to reduce the pollutants and contaminants which usually involve adsorption, ion exchange, membrane separation, coagulation, advanced oxidation process, as well as biological degradation for the purpose of reducing the release of harmful chemicals in the environment [13]. Despite the using mechanism, traditional methods have been known to be quite inefficient, uneconomical and unsustainable. Hence, the aforementioned problem eventually leads to the significant need of a novel sustainable method for textile wastewater which is produced during the process of textile finishing. The finishing of textile process causes many environmental damages at almost every stage of the mechanism including production, application along with both use and disposal of the textile finished wastes [5].

Different techniques to produce functional textiles including different applications of finishing chemicals incorporating the textiles including heating, then drying as well as all the washing stages of finished textiles, are quite harmful to environment as they are extremely unsustainable. All these stages involved in the textile finishing process utilize a lot of chemicals, water, energy as well as air emissions, and huge solid waste which eventually leads to higher risk of environmental pollution [14]. The major pollution is known to be caused through textile finishing which impacts the development of agriculture. The major components which are used in wastewater are the formed COD, BOD [15]. In textile finishing process, many oxides of nitrogen as well as carbon lead toward the release of greenhouse gases. Therefore, SO_x along with NO_x and dust particles also cause air pollution [16]. Nevertheless, due to the industrial production enhancement, the employment rate also increases which leads as an important economic growth element for the nation. However, regardless of its benefit to the economy of the nation, it is still quite harmful and toxic for the environment [17]. The considerable amount of waste is caused due to the improper procedures as well as the challenges textile industries are facing [18], this requires the formation of several policies to reduce the pollution without decreasing the yield [19]. The increase in the awareness of sustainability as well as identification of environmental effects, every textile industry must work on the different procedures of production to reduce pollution [20]. The environmental impact of functional textiles is quite worse at the point of usage as well as its disposal rather than its formation step. Keeping in mind the three pillars of sustainability, the available functional textiles are not sustainable in any stage of its formation like utilization and disposal. To reduce the chances of environmental issues as well as its effects on human health, it is important to examine the activities of textile industries while working on the three elements of sustainability.

2.1 Sustainable Techniques for the Development of Bioactive Functionalization of Textiles

The advancement of bioactive functional fabrics emphasizes incorporating sustainable methods to improve efficiency while minimizing environmental influence [21]. Eco-friendly methods like botanical and microbial production of bioactive compounds, enzymatic surface alteration, and plasma processing are becoming increasingly significant in textile operations [22]. At present, eco-conscious products are attracting global interest, with a focus on utilizing environmentally sustainable materials. Recent findings have highlighted the importance of sustainability, leading to initiatives aimed at preserving existing processes through eco-friendly methods and prioritizing eco-friendly research design. Sustainable finishing methods, including UV and ozone treatments, enzymatic coatings, and layer-by-layer application, improve textile performance while reducing chemical waste. These eco-friendly strategies not only enhance the bioactivity of fabrics but also aid in environmental conservation, promoting a transition toward more sustainable and responsible textile manufacturing [23]. The creation of bioactive eco-friendly fabrics incorporates innovative and sustainable methods that improve textile performance while reducing ecological impact.

2.1.1 Use of Natural Fibers Having Bioactive Properties

Natural fibers are made up of different kinds of fibers, occurring in nature and derived periodically from plants i.e. fibers of cellulose, animals (fibers of protein) and minerals (e.g. asbestos) [24]. Different parts of plants such as seed, fruit, leaf, wood, stalk and grass produce different kinds of fibers which promote diverse functionality of plant fibers. For example, the cotton plant and kapok fibers are obtained from their seed and non-vegetative tissues. A lot of plants such as palm, flax, jute, hemp, kenaf, ramie and pineapple give useful fibers from their stem/leave (vegetative parts) [25].

Due to global environmental perception, the dependency on eco-friendly products has increased with the passage of time [26]. The advancement of reusable and green products has emerged as a new and most significant research area. Natural fibers obtained from organic sources has started to take place of synthetic fibers of petroleum-based materials (non-renewable resource) [27].

Composites, a combination of a strong load-carrying substance (reinforcement) embedded with a "weaker" substance (matrix), have a lot of significant characteristics. These are readily available and cost-feasible fibers, making them the most promising reinforcement for a polymer matrix [28]. The natural fibers tend to be inconsistent in their properties, as they change for every batch or group. This instability arises from different factors such as plant growth conditions, method of harvesting and fiber extraction procedures. Furthermore, natural fibers lack mechanical resilience which is an important feature in many applications such as the formation of a polymer matrix [29]. A bio-composite is a material that incorporates, minimum, one natural source such as any plant-based fiber or biodegradable polymer derived from a renewable resource. The addition of these fibers promotes the value of bio-composite, providing a vast range of physical, mechanical and biological characteristics [30].

The development of natural fiber reinforced polymer composite has made a significant contribution to the relevant research field of polymer composite. This sudden growth highlights the importance of these cost-friendly synthetic fiber composites, decreasing environmental concerns. Owing to potential applications across several engineering sectors, a lot of efforts have been made to promote and enhance the efficiencies of these composites. In recent years, natural fibers have been functionally modified to enhance their competence. This review highlights the importance of NFRP in the R&D sector and its efficiencies from the year 2000–2019. It also emphasizes the NFRP production, its characteristics such as mechanical strength, tribological, resistant behavior of water and chemicals, thermal effects, biodegradability and machining properties along with their trends, and the challenges arising

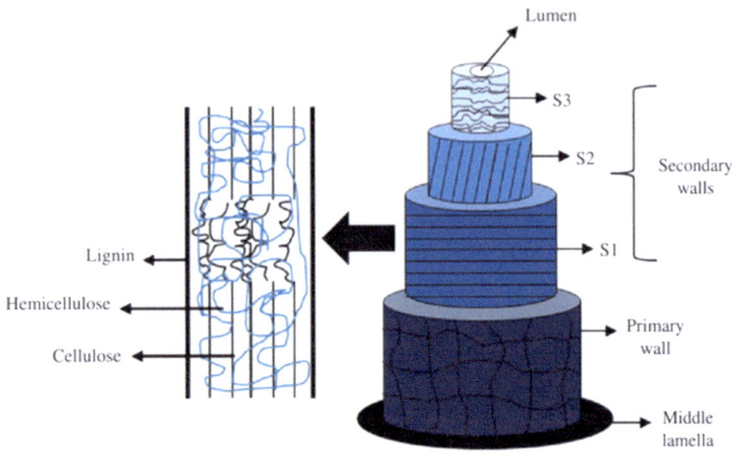

Fig. 1 Structure of natural fiber (reused from Pereira et al. [32] under CC-BY Permission)

with them and their future outcomes [31]. Figure 1 explains the structure of natural fiber [32].

2.1.2 Incorporation of Green Nanotechnology into Natural Fibers

Inorganic metallic nanoparticles have proven to be an attractive nanoscale platform for creating innovative and long-lasting functional coatings in textiles, offering features like self-cleaning, antibacterial effects, and UV protection while preserving essential qualities such as breathability, comfort, and mechanical strength [33]. Various eco-friendly manufacturing techniques for producing metal and metal oxide nanoparticles have been developed to align with environmental regulations and regulate the use of harmful substances, including capping agents, reducing agents, and organic solvents used in inorganic nanoparticle synthesis, while also minimizing excessive energy consumption [34]. Utilizing sustainable chemicals, physical techniques such as sonochemical-assisted processes, laser ablation, microwave-assisted methods, photochemical reactions, and bio-based strategies are among these innovative approaches. Bio-based methods for nanoparticle production are among the most beneficial techniques, offering features such as biocompatibility, non-toxicity, and affordability. Biopolymers, microorganisms, and plant extracts have been utilized as reducing agents in the biosynthesis of metal and metal oxide nanoparticles, capable of converting metal ions into zero-valent atoms. Due to the capability of biomolecules to act as capping agents throughout the biosynthetic process, there is no requirement to introduce additional stabilizers during nanoparticle synthesis. Moreover, biological pathways can functionalize biologically synthesized nanoparticles more efficiently than non-biological methods [35].

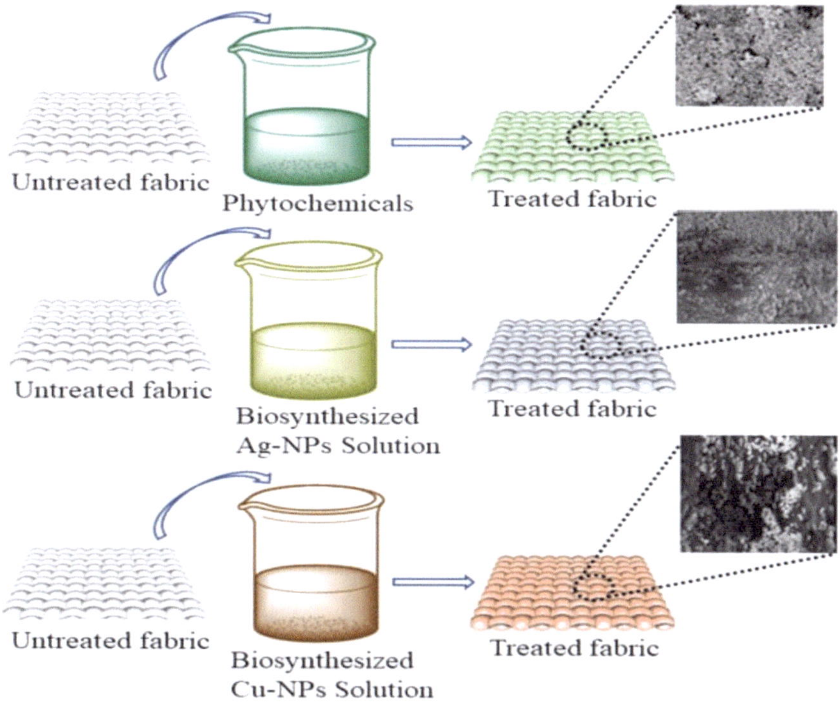

Fig. 2 Scheme representing the coating of different types of antimicrobial agents used for the fabrication of the textile (reused from Tahir et al. [36] under CC-BY Permission)

A recent study by Azam et al., implemented a sustainable approach for the green synthesis of silver nanoparticles, copper nanoparticles, and inspection of the phytochemical and biological testing of bark, leaves, and fruits of *Ehretia acuminata* (family Boraginaceae). Thereafter, the cotton fiber was loaded with prepared nanoparticles and extracted phytochemicals. These coated cotton fibers were examined for antibacterial, antiviral, and antifungal investigation. Different types of cotton fabric coated with an antimicrobial agent are illustrated in Fig. 2. Furthermore, Fig. 3 illustrates the application process of the prepared particles and phytochemicals onto the cotton fabric [36].

2.1.3 Enzymatic Functional Finishing

By primarily executing enzyme technology, the capabilities of natural as well as synthetic textile materials are improved and varied by means of the process of surface modification [37]. In order to improve the functionality of textiles, researchers have conducted a series of experiments to inspect the possibilities of modification of textile surfaces by utilizing enzymes [38]. Pre-activation of the surfaces of the textiles is

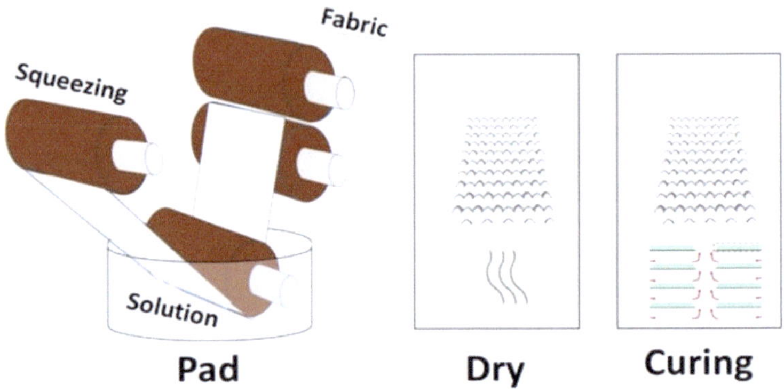

Fig. 3 Application of prepared particles and phytochemicals of cotton fabric (reused from Tahir et al. [36] under CC-BY Permission)

done by utilizing enzymes in order to improve the textile's affinity for finishing chemicals. For this objective, a number of enzymes like laccase, pectinase and cellulose have been utilized [38]. Hydrophilicity of textiles is improved by using these enzymes and introduction of various functional groups on fabric surface which in turn facilitate the action of functional finishing. In order to transmit antibacterial properties and an attractive, preferable aesthetic, textiles have been functionally biofinished. New research has been submitted by applying sonochemical Nano-coating using nanoparticles after the process of enzymatic pre-activation of textile materials in order to get functional medical textiles [39]. Previously, according to the literature, textiles have been fabricated in order to retain the durable functionalities like UV-protection and antibacterial properties by using pre-treatments with enzymes including laccase, acid pectinase and acid cellulose and also their mixtures. This fabrication process has been followed by coating with different Nano-particles like ZnO and TiO_2 NPs [7].

2.1.4 Irradiation Technology

Irradiation technology is proved to be modern and environment friendly technique for improving the sustainability and functionality of textiles. This method, actually works by imposing chemical modification without using harsh chemicals or excessive water and enhances the textiles' functionality by exposing to controlled radiation sources like, electron beams (EB), gamma rays or UV (Ultraviolet) rays [40]. Radiation-based technologies are gaining a lot of attention in textile industries as these are easy to use, less treatment speed, have very little impact on the environment and are energy efficient as these work on low temperature. Textile surface modifications obtained by the application of irradiation techniques involving gamma, ultrasonic and ultraviolet radiations as well as plasma, have undergone a versatile usage in textile functional finishing. These innovative methods which are being utilized do not affect the mechanical properties of textiles [41].

Plasma technology is utilized to enhance a number of properties like hydrophobicity/hydrophilicity, strength, adhesion and anti-static properties, strength, dyeing and printing performance of fabrics. A high number of reactive species are present in plasma and these all are capable of changing the chemical and physical properties of textile surfaces [42]. Plasma is said to be the "fourth state of matter" as it is different from gas, liquid and solid state. Neutral gas can generate plasma when it is exposed to heat or electromagnetic field. Plasma can be applied at low or high temperatures. Cold plasma (low-temperature plasma) has been utilized in textiles as most of the textile materials are heat sensitive. Plasma treatments can be executed with low-frequency (1–500 kHz), microwave (often 915 MHz or 2.45 GHz) energy, radio-frequency (RF) (usually 13.56 or 27.12 MHz), or by utilizing alternating current (AC) or direct current (DC) [43]. A Radiation-based technology, Ultrasound, is applied in a number of domains like medical field, environmental protection and textile applications for many years. Ultrasound waves which include sonic vibrations, have greater frequency of 17 kHz that is more than the human hearing range [44].

Same as electromagnetic waves, ultrasonic waves can be focused, refracted and reflected but they surely radiate through a medium having elastic properties. In literature, a number of favorable impacts of ultrasound technology are seen in textile wet-processing [45]. Additionally, a newly evolving combination of nanotechnology with ultrasonic-irradiation technology has amplified the significance of sonochemistry in the advancement of widespread multi-purpose textile materials [46]. Sonochemical irradiation is introduced to be convenient, low-cost, and pollution-free process for formation of nanomaterials and application of diverse nanoparticles on textile materials. According to recent studies, sonochemical applications of nanoparticles concluded to fabricate the durable and multifunctional textiles like generation of self-cleaning and UV-protective textiles. A versatile range of applications of ultrasound irradiation technology is found to be in the in-situ synthesis and coating of a number of nanoparticles like zinc oxide, titanium dioxide, silver and copper oxide etc. on the textiles for transmitting multiple functionalities [46]. Gamma rays, with energy greater than 100 keV and wavelengths below 10 pm, are termed as 'high-energy electromagnetic'. Gamma irradiation is deemed to be more promising for the applications of surface modification in textile materials. Along with a number of benefits in finishing and dyeing of textiles, these give improved absorbency and enhanced hydrophilicity. Accessible literature is found about gamma radiations-assisted fabrication of metal nanoparticles on textile surface for imparting antimicrobial properties to the textile [4].

2.1.5 Irradiation Technology Use as Functional Finish on Cotton Fabric

In textile industries, recent advances have been seen in the significance and quality of functional clothes as these have excellent reinforced performance and properties. For instance, performance is enhanced using some electromagnetic materials such as zinc oxide (ZnO) nanoparticles which act as electromagnetic interference-shielding material. Furthermore, cotton- coated ZnO fabrics could be utilized as a wearable cloth and are capable of protecting from high microwave absorptions when used as wearables by human beings. Cotton-coated ZnO fabrics have been utilized as flame retarders in previous studies [47]. On the other hand, still a detailed study of interactions between cotton and ZnO have not done. Exposure to gamma rays leads to the formation of radical on materials which is the root cause of chain scission or cross-linking. Thus, there is a need to figure out that either linkage is enhanced by means of gamma ray irradiation on ZnO and cotton structures or scission process will be dominated.

Textile substrates have mainly been employed in novel and innovative applications of ZnO films [48]. A number of constituents are present in cotton fiber. Among all of these constituents, cellulose is present in greater amounts being 88–96%. Riaz et al. presented a detailed overview about the applications of flexible textile-based nanomaterials in order to gain a number of functional properties [49], while Wang et al. discussed the applications of photocatalysts developed by the use of modified textiles [50]. Verbič et al. demonstrated a number of applications of ZnO on textile substrate with their practical implementations in thermal insulation, flame retardancy, hydrophobicity and so on [51]. Gupta et al. illustrated the applications of reduced graphene oxide/zinc oxide composites on cotton fabric in accordance with high microwave absorption [52]. This type of material is of greater concern nowadays as it provides a detailed understanding of the effects of gamma ray irradiation on cotton fibers.

In earlier researches, majority of groups focused on the effects of protons [53], electrons [54], gamma (γ) irradiations [55], and so forth on ZnO. In order to generate nanoscale metal at standard room temperature and also nanocomposites, gamma ray irradiation techniques have been proved to be very significant. Furthermore, this advanced technique is well-controlled, versatile, dynamic and capable of avoiding the participation of impurities into the matrix [56].

Reyhani et al. stated the impacts of gamma rays' exposure on ZnO nanowires with a number of diameters. Whereas the literature explains the gamma ray exposure to ZnO nanocomposites but there is still a gap between a number of aspects like variations in the mechanism of degradation and chemical and electronic structure modification, in particular the interaction between functional nanoparticles over gamma-irradiated, cotton-coated fabrics [56].

In this research work, our main objective is to study the ZnO (zinc oxide) based functional nanoparticles on cotton fabrics and the impact of irradiation of gamma rays with the strength of 9 kGv. We concluded from the experiment that gamma ray exposure on cotton fabrics bound with nanoparticles enhances the interactions between them. The reason for this enhanced impact is the oxygen deficiency or defects generated in the binding interface between cotton fabrics and ZnO after gamma irradiation and these oxygen deficiencies were detected by means of X-ray photoelectron spectroscopy (XPS) and Near-edge X-ray absorption fine structure. XPS results revealed that at the 1.5 kGv radiation dose, binding energy of Zn is shifted by 2 eV while at 9 kGv, it is shifted by 4 eV. This large shift is totally different from other researches because of the reactions between cotton fabrics and ZnO nanoparticles interface after exposure to the gamma rays. This research work determines that due to gamma ray exposure enhanced level of binding is proven to be significant for textile industries [57].

2.1.6 Biopolymers and Plant-Based Functional Materials

Biopolymers as sustainable alternatives to synthetic fibers and chemical treatments [8]. Biopolymers such as polylactic acid (PLA), chitosan, alginate, and silk fibroin are derived from renewable sources and offer excellent biodegradability, making them eco-friendly options for textile applications. These materials can be used to develop bioactive textiles with antimicrobial, moisture-wicking, and skin-friendly properties [9]. Chitosan is a kind of chitin which is an amino polysaccharide-based biopolymer produced from chitin after alkaline deacetylation. Due to its beneficial features such as non-toxicity, biocompatibility, and biodegradability, it has significant potential uses in a wide range of industries, including biomedical, culinary, agricultural, cosmetics, textiles, and pharmaceutical industries [58]. Chitosan and its derivatives have been used alone or in composites with organic or inorganic materials to confer antibacterial, flameproof, and multifunctional characteristics. Cyclodextrins (CDs) are classified as cyclic oligosaccharides that are formed when starch is degraded by enzymes. They are biodegradable, environmentally sustainable, and safe. CDs are widely used for functional finishing of textiles as UV-protective and antimicrobial finishes [59]. The application and fixation of these polymers were accomplished using a variety of methods. These techniques include pad-dry-curing with a crosslinking agent like citric acid or butanetetracarboxylic acid [60], a photoinitiator [61], chemical pre-treatments of textiles followed by the application of biopolymers [62].

Such biopolymers are believed to eventually replace the conventional chemically synthesized finishing agents soon since they are more cost-effective, perform better, and are less harmful to the environment. Plant-based bioactive chemicals include coloring pigments, phenolic compounds, quinones, flavonoids, tannins, polysaccharides, and essential oils. These materials are extracted from various parts of plants, such as the bark, leaves, roots, and flowers, have high antimicrobial activities [63]. The chemical compositions and structures of plant-based bioactive agents differ. These bioactive agents have been extensively studied as functional finishing agents for textile fabrics to give insect repellency, antibacterial activity, and UV protection. Plants are thought particularly to be important sources of natural antimicrobials [64].

2.1.7 Aesthetic Enhancements in Sustainable Bioactive Textiles

Sustainable bioactive textiles not only focus on environmental and functional benefits but also emphasize aesthetic appeal to meet consumer preferences. For this purpose, integrate aesthetic appeal with functionality through eco-friendly techniques such as natural dyeing, enzymatic finishing, and bio-inspired coatings could be applied [65]. These advancements ensure that bioactive textiles maintain their visual and tactile quality while offering antimicrobial, UV-protective, and self-cleaning properties, making them ideal for fashion and home textiles [66]. By combining sustainability with aesthetics, bioactive textiles offer a compelling solution for eco-conscious fashion, home decor, and functional applications, ensuring both style and performance without compromising environmental responsibility. A recent study was conducted to develop an environmentally friendly, low-price, easy and fast method for developing the antipathogenic (antibacterial, antifungal and antiviral) cuprous oxide coated multifunctional fabrics followed by modified reactive dye shown in Figs. 4 and 5. Fabrics were first sensitized with citric acid, and then Cu_2O particles were formed using the Fehling solution method. In second step, a reactive antibacterial dye was made (by reacting Reactive Blue 4 with triclosan). The molecular structure of modified dye was confirmed through FTIR. The resultant antibacterial dye was applied on copper-treated cotton fabrics through exhaust dyeing protocol. The dyed fabric was characterized through colorimetric data (L*, a*, b*, C, H, and K/S), levelness of dye, fastness properties as well as exhaustion and fixation rates. Cuprous oxide-coated fabrics were tested for antipathogenic activity using quantitative and qualitative measurement results shown in Fig. 6 [67].

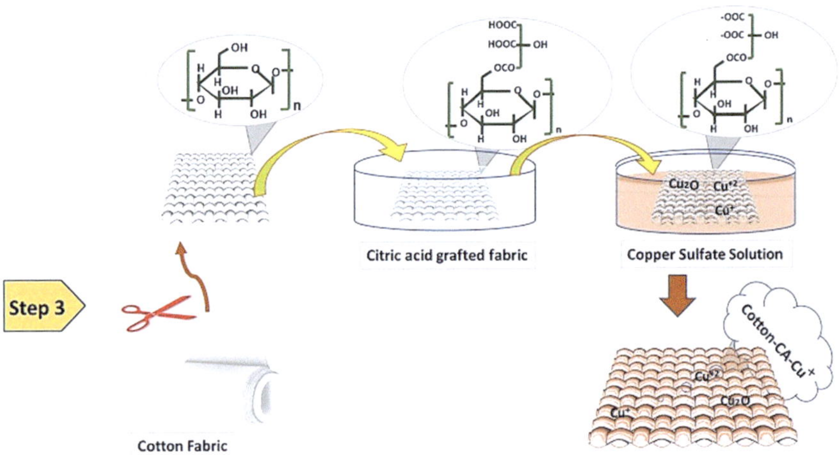

Fig. 4 Schematic illustration of the three-step method involved in the synthesis of cuprous oxide particles and their subsequent deposition on cotton (reused from Shahid et al. [67] under CC-BY Permission)

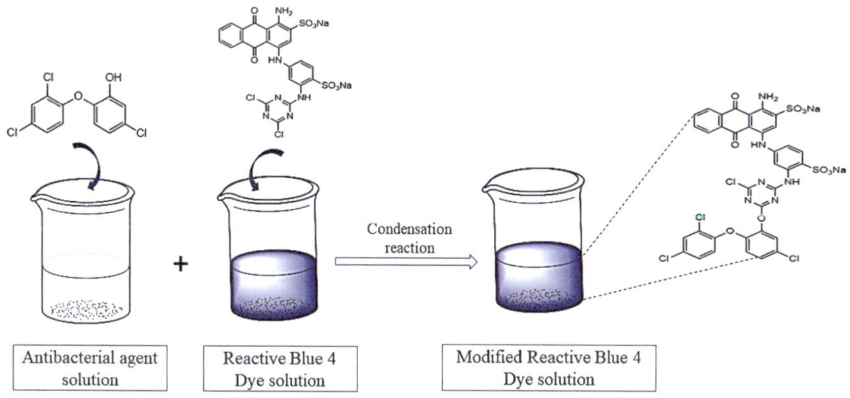

Fig. 5 Schematic illustration of the functionalization of dye (reused from Shahid et al. [67] under CC-BY Permission)

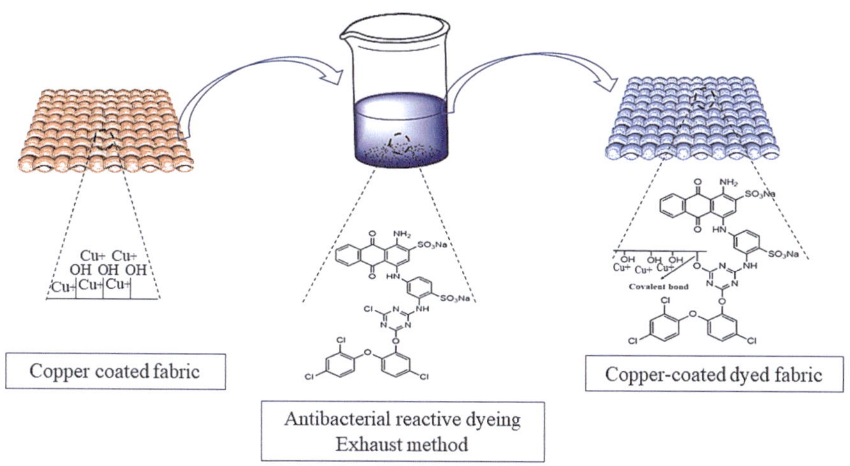

Fig. 6 Schematic illustration of the subsequent dyeing on Cu_2O coated fabrics (reused from Shahid et al. [67] under CC-BY Permission)

2.1.8 Bio-based Polymer Coatings on Natural Fibers to Produce Antibacterial Fabric

Because of their functional qualities, sustainability, and biodegradability, bio-based polymer coatings on natural fibers are becoming more and more popular for creating antibacterial textiles. These coatings improve the antibacterial qualities of natural fibers while keeping their eco-friendly attributes [68].

A significant and expanding industry with considerable growth in wound care products is new-generation medical textiles. The important features of fibers and bandages used in wound care include antiviral, bacteriostatic, and fungistatic characteristics, as well as being non-toxic, highly absorbent, non-allergenic, breathable, haemostatic, and biocompatible. Moreover, they can be tailored to incorporate medications while maintaining adequate mechanical characteristics. Compared to traditional materials, modified or blended products—often incorporating chitin/chitosan, alginate, carbon fibers and collagen, branched ferulate, offer significant advantages. Modern wound dressings come in a wide range of textile structures, including yarn, sliver, non-woven, woven, knitted, braided, crocheted, embroidered, and composite materials [69].

- **Biodegradable polyurethane embedded with natural antimicrobials**

Polyurethane (PU) coatings are extensively employed in several applications, including fabric coatings. Currently, landfill disposal remains the most common technique for handling PU waste. However, biodegradation offers an alternative method that helps decrease microplastic pollution linked with landfilling. This study focuses on the development of a biodegradable PU textile coating [70].

A recent study concentrated on developing a biodegradable PU fabric coating. The PU was examined using thermogravimetric analysis (TGA), Fourier-transform infrared (FT-IR) spectroscopy, and differential scanning calorimetry (DSC). It was recognized as a thermoplastic material with a melting point of nearly 33 °C. The coating's performance was assessed based on its water barrier characteristics and mechanical strength. The PU coating experienced complete disintegration, with biodegradation in soil reaching nearly 60%. Moreover, plant toxicity was evaluated by the tentative seedling emergence and growth [71].

2.1.9 Sonochemical Deposition (Ultrasound-Assisted Coating) Method to Deposit Antibacterial Agents to Produce the Antibacterial Fabrics

Annually, hospital-acquired infections impact numerous hospital-visiting patients, ranging from 1 to 10 [72]. Along with direct damage and uneasiness to the involved patients, it also affects overreached healthcare systems economically. Considering the range of the European Union, the cost of HAIs treatments has been predicted to be around 5 billion Euros annually [73]. Nevertheless, several infection control programs have been confirmed to reduce the HAIs incidence quite efficiently [74].

Better hygiene is the most essential feature of such infection control programs. Many basic procedures including consistently washing hands as well as cleaning of intensive care wards, are planned carefully to reduce transmission of such infections. Furthermore, the textile-based fabrics utilized in hospitals can easily act as a medium for such infections [75]. Therefore, the utilization of antibacterial fabrics in hospitals including curtains, uniforms of workers, dressings, bedding as well as pajamas, leads to minimizing the HAIs incidence.

According to an additional study, the hybrid antimicrobial coating was done on the surface of the cotton fabrics using a synchronized one-step sonochemical deposition of both chitosan as well as zinc-oxide nanoparticles. Furthermore, this process was optimized through controlling the concentrations of reagents as well as the time it takes to obtain the fabric with enhanced antimicrobial properties while considering the factor of their compatibility. Additionally, it was noted that the most excellent antibacterial properties of textile-based fabrics against both related species of bacteria were obtained when the sonochemical coating process of 2 mM zinc-oxide nanoparticles suspension was done for 30 min [76].

When both chitosan along with ZnO were deposited at the same time with the same ratio, the gained hybrid NPs coating product showed improved antibacterial effects such as 48% and 17% against both *S. aureus* as well as *E. coli*, respectively. Furthermore, the durability of the coating's antimicrobial impact was enhanced by about 21 and 40% for *S. aureus* as well as *E. coli*, respectively, simply through the incorporation of biopolymers which was then examined through various repeated cycles of washing at the hospital laundering regimes. Hence, the compatibility was enhanced up to 87% through the improvement in fibroblast viability, examined for the coating of hybrid ZnO/chitosan in comparison with the reduction in cell viability which was noted throughout a week along with fabrics that were only coated with ZnO [76].

2.1.10 Electroless Plating of Metals by Using Green Reductants to Produce Antibacterial Fabrics

A non-electrolytic process, named 'Electroless plating' is a method of deposition from solution (e.g. Cu Electroless deposition) by means of catalytic redox reaction between reducing agent (i.e. formaldehyde) dissolved in alkaline medium at elevate [77]. But, about formaldehyde, it was reported that during the plating process, there is a risk of production of hazardous gaseous-phase products [78]. Some of the research quoted the palladium metal utilized as catalyst to start the process of Electroless plating but palladium is found to be expensive [79]. The other disadvantage of traditional Electroless plating is that it is time taking procedure as surface activation and surface sensitization are necessary before the plating of actual metal. It is now necessary to explore new dimensions in order to fabricate Cu-coated fabrics under inexpensive, safe and reliable ways [80].

Furthermore, a recent research work was done on green materials and procedures in order to prepare antibacterial fabrics. In this study, Electroless plating of Cu was done on cotton fabrics skipping the process of sensitization of the surface of fabric. In order to activate the surface of fabric, Electroless plating of Cu was done by further depositing the Cu or Ag particles over it. Such processing is not only expected to make these materials more sustainable (palladium, stannous and formaldehyde free), but also the base of Cu or Ag nanoparticles is predicted to give dense and uniform metal plating because of higher rate of penetration of nanoparticles from the fabric structure. Moreover, redox potential of Cu and Ag nanoparticles is a way that can avoid the exchange of host elements over the course of sequential Electroless plating of Cu. This is the reason, after metal plating, thickness or variation in electrical resistivity is avoided due to which homogenous surface is provided by base of Cu or Ag nanostructures [81].

During electroless plating by conventional method we need to use hydrochloric acid with stannous chloride which in turn affects the strength of cotton fabric. Instead of it, we are performing the electroless plating in alkaline bath using sodium hydroxide. Therefore, using green reductants describes more favourable deposition of copper in alkaline pH [82].

Researchers have been using formaldehyde and hypophosphite as a reducing agent during electroplating, which has serious disadvantages. Formaldehyde has severed disadvantages when use for electroplating. It is carcinogenic, irritant to skin, give harmful vapor etc. While using hypophosphite is not proper for copper, because copper is not a good catalyst for the oxidation of hypophosphite. So, as result gives not proper plating over copper. So, to overcome these issues glyoxalic acid appeared as a best alternative, providing more stable, higher deposition with very less environmental pollution [83]. Electroless plating method by using glyoxylic acid is more environmentally friendly and does not give off harmful vapors [83].

2.1.11 Electroplating of Metals on Fabric Without the Use of Any Hazardous Chemicals to Produce Antibacterial Fabrics

Certain metal-based hard surface fabric electrodes and needle electrodes are usually used in electrotherapy. However, concerns arise about several issues related to electrode surfaces when applied to the skin or introduced into the body. Self-adhesive electrodes, in particular, have numerous disadvantages due to their viscous nature, including discomfort, skin redness, rashes, excessive sweating, difficulty in cleaning poor hygiene, limited durability, and inadequate compatibility with the body's structure [84]. In the past, silver nanoparticles were incorporated onto cotton-lycra structures to create textile-based flexible electrodes. Due to their unique chemical and physical properties, silver nanoparticles are currently in use in a wide range of products, such as consumer goods, pharmaceutical and food industries, cosmetics, antibacterial agents, coatings for medical devices, diagnostics, optical sensors, drug delivery, and anticancer agents. To evaluate these nanoparticles' compatibility with skin and wound-healing qualities, they were also used and applied in various kinds of textiles, including wound dressings [78].

Azam Ali et al. developed wearable, electrically conductive hygienic electrodes for Transcutaneous Electrical Nerve Stimulation (TENS) applications. The electrode production followed a two-stage procedure: first, dispersing conductive activated carbon particles into a flexible elastomer, followed by silver electroplating. The silver-plating procedure was optimized using the Orthogonal Array Testing (OATS) technique. Fourier Transform Infrared Spectroscopy (FTIR) analysis confirmed the presence of activated carbon by presenting improved intensity in stretching bands. To measure electrode performance under different human body movements, resistivity changes due to stretching and repeated extension were observed. The findings specified that electrical resistivity persisted nearly unchanged within a 0–60% stretch range but significantly improved beyond 70% extension. Even after 100 extension cycles, resistivity remained stable, and no significant differences were detected over time under a constant current. Various pressure levels were used to reduce skin-electrode resistivity during electrotherapy; at 6 N/cm^2, a minimum resistance of 1 Ω mm was recorded. Silver particle deposition was also used to evaluate the electrodes' hygienic qualities, such as their antiviral, antibacterial, and antifungal efficacy. Finally, resistance to rubbing and washing was validated by durability testing. These results demonstrate the designed electrodes' potential for use in electrotherapy and electrostimulation [85].

3 Conclusion

In conclusion, the current review highlights the sustainable techniques for the development of durable bioactive textiles having improved antifungal, antimicrobial, as well as antiviral properties while incorporating reactive bioactive compounds inside the dye as well as finish structures. The obtained treated cotton-based fabrics showed brilliant functional properties along with high stability while surviving various washing cycles, which will eventually be used in healthcare as well as hygiene, or the medical field. To overcome different challenges of sustainability, different eco-friendly approaches can be used in various finishing as well as dying processes. Researchers should work on the use as well as the methods utilized for their optimized applications of biodegradable antimicrobial agents for environmentally friendly production on a large scale. Furthermore, life cycle evaluation as well as smart functional properties can also help in the development of next-generation textiles, which are sustainable in nature.

Acknowledgements This work was also supported by the project "Advanced structures for thermal insulation in extreme conditions" (Reg. No. 21-32510 M) granted by the Czech Science Foundation (GACR).

References

1. Saxena, S., Raja, A.S.M., Arputharaj, A.: Challenges in sustainable wet processing of textiles, pp. 43–79 (2017). https://doi.org/10.1007/978-981-10-2185-5_2
2. Weerasinghe, D., Perera, S., Dissanayake, D.: Application of biomimicry for sustainable functionalization of textiles: review of current status and prospectus. Text. Res. J. **89**(19–20), 4282–4294 (2019). https://doi.org/10.1177/0040517518821911
3. Emam, H.E.: Antimicrobial cellulosic textiles based on organic compounds. 3 Biotech **9**(1), 29 (2019). https://doi.org/10.1007/s13205-018-1562-y
4. Bhandari, V., Jose, S., Badanayak, P., Sankaran, A., Anandan, V.: Antimicrobial finishing of metals, metal oxides, and metal composites on textiles: a systematic review. Ind. Eng. Chem. Res. **61**(1), 86–101 (2022). https://doi.org/10.1021/acs.iecr.1c04203

5. Roy Choudhury, A.K.: Environmental impacts of the textile industry and its assessment through life cycle assessment, pp. 1–39 (2014). https://doi.org/10.1007/978-981-287-110-7_1
6. Rahaman, Md.T., Hossain Khan, Md.S.: Biomaterials for manufacturing environmentally sustainable textiles and apparel: sources, applications, challenges, enablers and future directions. Int. J. Environ. Sci. Technol. **22**(10), 9655–9710 (2025). https://doi.org/10.1007/s13762-025-06481-7
7. Ivanova, A., Ivanova, K., Perelshtein, I., Gedanken, A., Todorova, K., Milcheva, R., Dimitrov, P., Popova, T., Tzanov, T.: Sonochemically engineered nano-enabled zinc oxide/amylase coatings prevent the occurrence of catheter-associated urinary tract infections. Mater. Sci. Eng. C **131**, 112518 (2021). https://doi.org/10.1016/j.msec.2021.112518
8. Shaarawy, S.: A review on the development of innovative capabilities in the textile finishing of natural fibers. Egypt. J. Chem. **0**(0), 0–0 (2020). https://doi.org/10.21608/ejchem.2019.19009.2169
9. Patti, A., Acierno, D.: Towards the sustainability of the plastic industry through biopolymers: properties and potential applications to the textiles world. Polymers **14**(4), 692 (2022). https://doi.org/10.3390/polym14040692
10. Shahid, M., Shahid-ul-Islam, Rather, L.J., Manzoor, N., Mohammad, F.: Simultaneous shade development, antibacterial, and antifungal functionalization of wool using *Punica granatum* L. peel extract as a source of textile dye. J. Nat. Fibers **16**(4), 555–566 (2019)
11. Cooper, D.M.: Optimizing wound healing. Nurs. Clin. North Am. **25**(1), 165–180 (1990). https://doi.org/10.1016/S0029-6465(22)00232-8
12. Kumar Singh, M.: Textiles functionalization—a review of materials, processes, and assessment. In: Textiles for Functional Applications. IntechOpen (2021). https://doi.org/10.5772/intechopen.96936
13. Mostafa, M.: Waste water treatment in textile industries—the concept and current removal technologies. Mostafa **501**(1), 501–525 (2015). http://www.innspub.net
14. Rukhaya, S., Yadav, S., Rose, N.M., Grover, A., Bisht, D.: Sustainable approach to counter the environmental impact of fast fashion. Pharma Innov. J. **8**, 517–523 (2021). http://www.thepharmajournal.com
15. Roy, C.K., Jahan, M.A.A., Rahman, S.S.: Characterization and treatment of textile wastewater by aquatic plants (Macrophytes) and algae. Eur. J. Sustain. Dev. Res. **2**(3) (2018). https://doi.org/10.20897/ejosdr/85933
16. Haseeb, M., Haouas, I., Nasih, M., Mihardjo, L.W.W., Jermsittiparsert, K.: Asymmetric impact of textile and clothing manufacturing on carbon-dioxide emissions: evidence from top Asian economies. Energy **196**, 117094 (2020). https://doi.org/10.1016/j.energy.2020.117094
17. Zhu, L., Chen, L., Wu, X., Ding, X.: Developing a greenhouse gas management evaluation system for Chinese textile enterprises. Ecol. Ind. **91**, 470–477 (2018). https://doi.org/10.1016/j.ecolind.2018.04.029
18. Madhav, S., Ahamad, A., Singh, P., Mishra, P.K.: A review of textile industry: wet processing, environmental impacts, and effluent treatment methods. Environ. Qual. Manag. **27**(3), 31–41 (2018). https://doi.org/10.1002/tqem.21538
19. Peters, G., Li, M., Lenzen, M.: The need to decelerate fast fashion in a hot climate—a global sustainability perspective on the garment industry. J. Clean. Prod. **295**, 126390 (2021). https://doi.org/10.1016/j.jclepro.2021.126390
20. Patti, A., Cicala, G., Acierno, D.: Eco-sustainability of the textile production: waste recovery and current recycling in the composites world. Polymers **13**(1), 134 (2020). https://doi.org/10.3390/polym13010134
21. Freitas, A.S., Oliveira, R., Ribeiro, A., Almeida-Aguiar, C.: Biofunctional textiles: antioxidant and antibacterial finishings of cotton with propolis and honey. Int. J. Mol. Sci. **25**(15), 8034 (2024). https://doi.org/10.3390/ijms25158034
22. Nosheen, A., Ashraf, M., Hussain, M.T., Aziz, H., Tahir, M.F., Akram, S.: Development of functional cotton fabric by simultaneous dyeing and finishing with a novel bioactive reactive dye. New J. Chem. **47**(4), 1872–1886 (2023). https://doi.org/10.1039/D2NJ05051J

23. Sarkar, S., Kumar, S., Singh, K., Mehra, D.: Documentation of recent advancements in finishing methods and chemicals in the textile industry (2024). https://doi.org/10.51201/JUSST/24/06241
24. Vanja, G., Bischof, S., Kovačević, Z.: Virginia mallow: the lost fiber of the future? Fibers **13**(5), 63 (2025). https://doi.org/10.3390/fib13050063
25. Fogorasi, M., Barbu, I.: The potential of natural fibres for automotive sector—review. IOP Conf. Ser. Mater. Sci. Eng. **252**, 012044 (2017). https://doi.org/10.1088/1757-899X/252/1/012044
26. Mohammed, L., Ansari, M.N.M., Pua, G., Jawaid, M., Islam, M.S.: A review on natural fiber reinforced polymer composite and its applications. Int. J. Polym. Sci. **2015**, 1–15 (2015). https://doi.org/10.1155/2015/243947
27. Ramamoorthy, S.K., Skrifvars, M., Persson, A.: A review of natural fibers used in biocomposites: plant, animal and regenerated cellulose fibers. Polym. Rev. **55**(1), 107–162 (2015). https://doi.org/10.1080/15583724.2014.971124
28. Thyavihalli Girijappa, Y.G., Mavinkere Rangappa, S., Parameswaranpillai, J., Siengchin, S.: Natural fibers as sustainable and renewable resource for development of eco-friendly composites: a comprehensive review. Front. Mater. **6** (2019). https://doi.org/10.3389/fmats.2019.00226
29. Chandramohan, D., Marimuthu, K.: Characterization of natural fibers and their application in bone grafting substitutes. In: Acta of Bioengineering and Biomechanics Original Paper, vol. 13, issue 1 (2011)
30. Ramakrishna, S., Huang, Z.-M.: Biocomposites. In: Reference Module in Materials Science and Materials Engineering. Elsevier (2016). https://doi.org/10.1016/B978-0-12-803581-8.00965-6
31. Vigneshwaran, S., Sundarakannan, R., John, K.M., Joel Johnson, R.D., Prasath, K.A., Ajith, S., Arumugaprabu, V., Uthayakumar, M.: Recent advancement in the natural fiber polymer composites: a comprehensive review. J. Clean. Prod. **277**, 124109 (2020). https://doi.org/10.1016/j.jclepro.2020.124109
32. Pereira, P.H.F., De Freitas Rosa, M., Cioffi, M.O.H., De Carvalho Benini, K.C.C., Milanese, A.C., Voorwald, H.J.C., Mulinari, D.R.: Vegetal fibers in polymeric composites: a review. Polímeros **25**(1), 9–22 (2015). https://doi.org/10.1590/0104-1428.1722
33. Rehan, M., Abdel-Wahed, N.A.M., Farouk, A., El-Zawahry, M.M.: Extraction of valuable compounds from orange peel waste for advanced functionalization of cellulosic surfaces. ACS Sustain. Chem. Eng. **6**(5), 5911–5928 (2018). https://doi.org/10.1021/acssuschemeng.7b04302
34. Shah, M.A., Pirzada, B.M., Price, G., Shibiru, A.L., Qurashi, A.: Applications of nanotechnology in smart textile industry: a critical review. J. Adv. Res. **38**, 55–75 (2022). https://doi.org/10.1016/j.jare.2022.01.008
35. Farooqi, Z.H., Akram, M.W., Begum, R., Wu, W., Irfan, A.: Inorganic nanoparticles for reduction of hexavalent chromium: physicochemical aspects. J. Hazard. Mater. **402**, 123535 (2021). https://doi.org/10.1016/j.jhazmat.2020.123535
36. Tahir, M.F., Khan, M.Z., Attacha, S., Asim, N., Tayyab, M., Ali, A., Militký, J., Tomková, B.: The comparative performance of phytochemicals, green synthesised silver nanoparticles, and green synthesised copper nanoparticles-loaded textiles to avoid nosocomial infections. Nanomaterials **12**(20), 3629 (2022). https://doi.org/10.3390/nano12203629
37. Rahman, M., Billah, M.M., Hack-Polay, D., Alam, A.: The use of biotechnologies in textile processing and environmental sustainability: an emerging market context. Technol. Forecast. Soc. Chang. **159**, 120204 (2020). https://doi.org/10.1016/j.techfore.2020.120204
38. Morshed, M.N., Behary, N., Bouazizi, N., Guan, J., Chen, G., Nierstrasz, V.: Surface modification of polyester fabric using plasma-dendrimer for robust immobilization of glucose oxidase enzyme. Sci. Rep. **9**(1), 15730 (2019). https://doi.org/10.1038/s41598-019-52087-8
39. Kumar, A., Verma, V., Dubey, V.K., Srivastava, A., Garg, S.K., Singh, V.P., Arora, P.K.: Industrial applications of fungal lipases: a review. Front. Microbiol. **14** (2023). https://doi.org/10.3389/fmicb.2023.1142536
40. Nogueira, L.F.B., Guidelli, É.J., Jafari, S.M., Ramos, A.P.: Green synthesis of metal nanoparticles by plant extracts and biopolymers. In: Handbook of Food Nanotechnology, pp. 257–278. Elsevier (2020). https://doi.org/10.1016/B978-0-12-815866-1.00007-8

41. Al-Sayed, W., Abdelrahman, S.H.: Sustainable chemistry in textile processes (pretreatment, coloration and chemical finishing). In: Green Chemistry for Sustainable Textiles, pp. 93–111. Elsevier (2021). https://doi.org/10.1016/B978-0-323-85204-3.00022-1
42. Rahman, N., Dafader, N.C.: Radiation technology for the modification of textiles. In: Fundamentals of Natural Fibres and Textiles, pp. 407–438. Elsevier (2021). https://doi.org/10.1016/B978-0-12-821483-1.00003-6
43. Tudoran, C., Roşu, M., Coroş, M.: A concise overview on plasma treatment for application on textile and leather materials. Plasma Processes Polym. **17**(8) (2020). https://doi.org/10.1002/ppap.202000046
44. Abou Elmaaty, T., Okubayashi, S., Elsisi, H., Abouelenin, S.: Electron beam irradiation treatment of textiles materials: a review. J. Polym. Res. **29**(4), 117 (2022). https://doi.org/10.1007/s10965-022-02952-4
45. Wixey, M., De Cristofaro, E., Johnson, S.D.: On the feasibility of acoustic attacks using commodity smart devices (n.d.). https://doi.org/10.1109/SPW50608.2020.00031
46. Golshan Abadi, B., Esmail, M.: Effects of ultrasonic waves on PTT dyeing with disperse red 60 dye. J. Chem. Chem. Eng. **40**(2) (2021)
47. Wang, Y.-W., Shen, R., Wang, Q., Vasquez, Y.: ZnO microstructures as flame-retardant coatings on cotton fabrics. ACS Omega **3**(6), 6330–6338 (2018). https://doi.org/10.1021/acsomega.8b00371
48. Kołodziejczak-Radzimska, A., Jesionowski, T.: Zinc oxide—from synthesis to application: a review. Materials **7**(4), 2833–2881 (2014). https://doi.org/10.3390/ma7042833
49. Riaz, S., Ashraf, M., Hussain, T., Hussain, M.T., Rehman, A., Javid, A., Iqbal, K., Basit, A., Aziz, H.: Functional finishing and coloration of textiles with nanomaterials. Color. Technol. **134**(5), 327–346 (2018). https://doi.org/10.1111/cote.12344
50. Wang, J., Zhao, J., Sun, L., Wang, X.: A review on the application of photocatalytic materials on textiles. Text. Res. J. **85**(10), 1104–1118 (2015). https://doi.org/10.1177/0040517514559583
51. Verbič, A., Gorjanc, M., Simončič, B.: Zinc oxide for functional textile coatings: recent advances. Coatings **9**(9), 550 (2019). https://doi.org/10.3390/coatings9090550
52. Gupta, S., Chang, C., Anbalagan, A.K., Lee, C.-H., Tai, N.-H.: Reduced graphene oxide/zinc oxide coated wearable electrically conductive cotton textile for high microwave absorption. Compos. Sci. Technol. **188**, 107994 (2020). https://doi.org/10.1016/j.compscitech.2020.107994
53. Moon, Y.-K., Moon, D.-Y., Lee, S., Park, J.-W.: Enhancement of ZnO thin film transistor performance by high-dose proton irradiation. Nucl. Instrum. Methods Phys. Res., Sect. B **268**(16), 2522–2526 (2010). https://doi.org/10.1016/j.nimb.2010.05.050
54. Kang, R., Noh, Y.-J., Yun, J.-M., Kim, H., Myoung, N., Lee, E.-H., Kim, T.-W., Na, S.-I., Oh, S.-H.: ZnO films using a precursor solution irradiated with an electron beam as the cathode interfacial layer in inverted polymer solar cells. RSC Adv. **7**(43), 26689–26696 (2017). https://doi.org/10.1039/C7RA02415K
55. Reyhani, A., Gholizadeh, A., Vahedi, V., Khanlary, M.R.: Effect of gamma radiation on the optical and structural properties of ZnO nanowires with various diameters. Opt. Mater. **75**, 236–242 (2018). https://doi.org/10.1016/j.optmat.2017.10.027
56. Baydogan, N., Ozdemir, O., Cimenoglu, H.: The improvement in the electrical properties of nanospherical ZnO: Al thin film exposed to irradiation using a Co-60 radioisotope. Radiat. Phys. Chem. **89**, 20–27 (2013). https://doi.org/10.1016/j.radphyschem.2013.02.042
57. Anbalagan, A.K., Gupta, S., Kumar, A., Haw, S.-C., Kulkarni, S.S., Tai, N.-H., Tseng, F.-G., Hwang, K.C., Lee, C.-H.: Gamma ray irradiation enhances the linkage of cotton fabrics coated with ZnO nanoparticles. ACS Omega **5**(25), 15129–15135 (2020). https://doi.org/10.1021/acsomega.0c01050
58. Dai, Y., Row, K.H.: Isolation and determination of beta-carotene in carrots by magnetic chitosan beta-cyclodextrin extraction and high-performance liquid chromatography (HPLC). Anal. Lett. **52**(11), 1828–1843 (2019). https://doi.org/10.1080/00032719.2019.1570245
59. Rouhani Shirvan, A., Shakeri, M., Bashari, A.: Recent advances in application of chitosan and its derivatives in functional finishing of textiles. In: The Impact and Prospects of Green

Chemistry for Textile Technology, pp. 107–133. Elsevier (2019). https://doi.org/10.1016/B978-0-08-102491-1.00005-8
60. Rather, L.J., Zhou, Q., Ganie, S.A., Li, Q.: Environmental Profile of Nano-finished Textile Materials: Implications on Public Health, Risk Assessment, and Public Perception, pp. 57–83 (2020). https://doi.org/10.1007/978-981-15-3669-4_3
61. Bajpai, M., Gupta, P., Bajpai, S.K.: Silver(I) ions loaded cyclodextrin-grafted-cotton fabric with excellent antimicrobial property. Fibers Polym. **11**(1), 8–13 (2010). https://doi.org/10.1007/s12221-010-0008-x
62. Cabrales, L., Abidi, N., Hammond, A., Hamood, A.: Cotton fabric functionalization with cyclodextrins. J. Mater. Environ. Sci. **3**(3), 561–574 (2012)
63. Ibrahim, N.A., E-Zairy, W.R., Eid, B.M.: Novel approach for improving disperse dyeing and UV-protective function of cotton-containing fabrics using MCT-β-CD. Carbohydr. Polym. **79**(4), 839–846. https://doi.org/10.1016/j.carbpol.2009.10.008
64. Aslam, M., Abdullah, A.Z., Rafatullah, M.: Recent development in the green synthesis of titanium dioxide nanoparticles using plant-based biomolecules for environmental and antimicrobial applications. J. Ind. Eng. Chem. **98**, 1–16 (2021). https://doi.org/10.1016/j.jiec.2021.04.010
65. Repon, Md.R., Islam, T., Islam, T., Ghorab, A.E., Rahman, M.M.: Cleaner pathway for developing bioactive textile materials using natural dyes: a review. Environ. Sci. Pollut. Res. **30**(17), 48793–48823 (2023). https://doi.org/10.1007/s11356-023-26131-0
66. Militký, J., Venkataraman, M., Periyasamy, A.P.: Introduction to textile materials used in health care. In: Textiles and Their Use in Microbial Protection, pp. 3–52. CRC Press (2021). https://doi.org/10.1201/9781003140436-2
67. Shahid, M., Ali, A., Zahid, N., Anjam, M.S., Militký, J., Wiener, J., Palanisamy, S., Tomková, B.: Copper-treated environmentally friendly antipathogenic cotton fabric with modified reactive blue 4 dye to improve its antibacterial and aesthetic properties. Coatings **13**(1), 133 (2023). https://doi.org/10.3390/coatings13010133
68. Szadkowski, B., Piotrowska, M., Rybiński, P., Marzec, A.: Natural bioactive formulations for biodegradable cotton eco-fabrics with antimicrobial and fire-shielding properties. Int. J. Biol. Macromol. **237**, 124143 (2023). https://doi.org/10.1016/j.ijbiomac.2023.124143
69. Petrulyte, S.: Advanced textile materials and biopolymers in wound management. Danish Medical Bulletin, 2008 europepmc.org (n.d.). Retrieved 5 June 2025, from https://europepmc.org/article/med/18321446
70. Wang, Y., Song, H., Ge, H., Wang, J., Wang, Y., Jia, S., Deng, T., Hou, X.: Controllable degradation of polyurethane elastomer via selective cleavage of C–O and C–N bonds. J. Clean. Prod. **176**, 873–879 (2018). https://doi.org/10.1016/j.jclepro.2017.12.046
71. De Smet, D., Verjans, J., Bader, M., Mondschein, A., Vanneste, M.: Design of biodegradable PU textile coating. Polymers **16**(16), 2236 (2024). https://doi.org/10.3390/polym16162236
72. Plowman, R., Graves, N., Griffin, M.A.S., Roberts, J.A., Swan, A.V., Cookson, B., Taylor, L.: The rate and cost of hospital-acquired infections occurring in patients admitted to selected specialties of a district general hospital in England and the national burden imposed. J. Hosp. Infect. **47**(3), 198–209 (2001). https://doi.org/10.1053/jhin.2000.0881
73. Beddow, J., Singh, G., Blanes, M., Molla, K., Perelshtein, I., Gedanken, A., Joyce, E., Mason, T.: Sonochemical coating of textile fabrics with antibacterial nanoparticles, pp. 400–403 (2012). https://doi.org/10.1063/1.3703213
74. Weinstein, R.A.: Controlling antimicrobial resistance in hospitals: infection control and use of antibiotics. Emerg. Infect. Dis. **7**(2), 188–192 (2001). https://doi.org/10.3201/eid0702.010206
75. Neely, A.N., Maley, M.P.: Survival of enterococci and staphylococci on hospital fabrics and plastic. J. Clin. Microbiol. **38**(2), 724–726 (2000). https://doi.org/10.1128/JCM.38.2.724-726.2000
76. Petkova, P., Francesko, A., Fernandes, M.M., Mendoza, E., Perelshtein, I., Gedanken, A., Tzanov, T.: Sonochemical coating of textiles with hybrid ZnO/chitosan antimicrobial nanoparticles. ACS Appl. Mater. Interfaces **6**(2), 1164–1172 (2014). https://doi.org/10.1021/am404852d

77. Ali, A., Baheti, V., Militký, J., Khan, Z., Gilani, S.Q.Z.: Comparative performance of copper and silver coated stretchable fabrics. Fibers Polym. **19**(3), 607–619 (2018). https://doi.org/10.1007/s12221-018-7917-5
78. Ali, A., Baheti, V., Militký, J., Khan, Z.: Utility of silver-coated fabrics as electrodes in electrotherapy applications. J. Appl. Polym. Sci. **135**(23) (2018). https://doi.org/10.1002/app.46357
79. Naeem, S., Baheti, V., Tunakova, V., Militký, J., Karthik, D., Tomková, B.: Development of porous and electrically conductive activated carbon web for effective EMI shielding applications. Carbon **111**, 439–447 (2017). https://doi.org/10.1016/j.carbon.2016.10.026
80. Irene, G., Georgios, P., Ioannis, C., Anastasios, T., Diamantis, P., Marianthi, C., Philippe, W., Maria, S.: Copper-coated textiles: armor against MDR nosocomial pathogens. Diagn. Microbiol. Infect. Dis. **85**(2), 205–209 (2016). https://doi.org/10.1016/j.diagmicrobio.2016.02.015
81. Ali, A., Baheti, V., Vik, M., Militký, J.: Copper electroless plating of cotton fabrics after surface activation with deposition of silver and copper nanoparticles. J. Phys. Chem. Solids **137**, 109181 (2020). https://doi.org/10.1016/j.jpcs.2019.109181
82. Barker, D.: Electroless deposition of metals. Trans. Inst. Met. Finish. **71**(Pt 3), 121–124 (1993). https://doi.org/10.1080/00202967.1993.11871003
83. Yu, L., Guo, L., Preisser, R., Akolkar, R.: Autocatalysis during electroless copper deposition using glyoxylic acid as reducing agent. J. Electrochem. Soc. **160**(12), D3004–D3008 (2013). https://doi.org/10.1149/2.002312jes
84. Keller, T., Kuhn, A.: Electrodes for transcutaneous (surface) electrical stimulation. J. Autom. Control **18**(2), 35–45 (2008). https://doi.org/10.2298/JAC0802035K
85. Ali, A., Tomková, B., Islam, S., Farooq, A., Howari, H., Subrova, T., Wiener, J., Militký, J.: Silver-plated stretchable elastomeric electrodes for electrotherapy applications. J. Ind. Text. **54** (2024). https://doi.org/10.1177/15280837241287561

MIX
Papier aus verantwortungsvollen Quellen
Paper from responsible sources
FSC® C105338

If you have any concerns about our products,
you can contact us on
ProductSafety@springernature.com

In case Publisher is established outside the EU,
the EU authorized representative is:
Springer Nature Customer Service Center GmbH
Europaplatz 3, 69115 Heidelberg, Germany

Printed by Libri Plureos GmbH
in Hamburg, Germany